The Application of Electrochemical Methods in Water Treatment

The Application of Electrochemical Methods in Water Treatment

Guest Editors

Minhua Cui

Guoshuai Liu

MDPI

Basel • Beijing • Wuhan • Barcelona • Belgrade • Novi Sad • Cluj • Manchester

Guest Editors

Minhua Cui
School of Environment & Ecology
Jiangnan University
Wuxi
China

Guoshuai Liu
School of Environment & Ecology
Jiangnan University
Wuxi
China

Editorial Office
MDPI AG
Grosspeteranlage 5
4052 Basel, Switzerland

This is a reprint of the Special Issue, published open access by the journal *Water* (ISSN 2073-4441), freely accessible at: https://www.mdpi.com/journal/water/special_issues/electrochemistry_water_wastewater.

For citation purposes, cite each article independently as indicated on the article page online and as indicated below:

Lastname, A.A.; Lastname, B.B. Article Title. *Journal Name* **Year**, *Volume Number*, Page Range.

ISBN 978-3-7258-5323-6 (Hbk)
ISBN 978-3-7258-5324-3 (PDF)
https://doi.org/10.3390/books978-3-7258-5324-3

Contents

About the Editors vii

Min-Hua Cui and Guo-Shuai Liu
The Application of Electrochemical Methods in Water Treatment
Reprinted from: *Water* **2025**, *17*, 2551, https://doi.org/10.3390/w17172551 1

Zhipeng Zhang, Chenghan Tang, Hongbin Wang, Ming Zhong, Pengchao Ge, Wenlai Xu and Yiyang Chen
Synthesis and Characterization of Potassium Bicarbonate and Urea-Modified Biochar from Rape Straw: Application in the Removal of Tetracycline from Aqueous Solution
Reprinted from: *Water* **2024**, *16*, 2522, https://doi.org/10.3390/w16172522 5

Liangang Hou, Feng Huang, Zhengwei Pan, Wei Chen and Xiujie Wang
Characteristics of Nitrogen Removal and Functional Gene Transcription of Heterotrophic Nitrification-Aerobic Denitrification Strain, *Acinetobacter* sp. JQ1004
Reprinted from: *Water* **2024**, *16*, 1799, https://doi.org/10.3390/w16131799 32

Hui Hong, Shiwei Xie, Aoxue Qiu, Yuming Yao, Wenzhe Jiang, Jilei Li, et al.
Efficient H_2O_2 Production and Activation by Air Diffusion Cathode Combined with Ultraviolet for Lake Water Treatment: A Long-Term Evaluation
Reprinted from: *Water* **2024**, *16*, 1658, https://doi.org/10.3390/w16121658 46

Ze-Chong Guo, Lu Zhang, Yong Chen, Cheng Huang and Zhi-Min Liao
Effect of UV-LED Wavelength on Reactive Species Photogeneration from Dissolved Organic Matter
Reprinted from: *Water* **2024**, *16*, 635, https://doi.org/10.3390/w16050635 61

Cheng Huang, Lixian Wang, Lingyi Fan and Yong Chen
Co-Pyrolysis of Fenton Sludge and Pomelo Peel for Heavy Metal Stabilization: Speciation Mechanism and Risk Evaluation
Reprinted from: *Water* **2023**, *15*, 3733, https://doi.org/10.3390/w15213733 72

Feng Ouyang, Yujie Liu, Jiao Chen, Chenghan Tang, Aojie Wang, Yixin Lu and Yiping Yuan
Study on Preparation of Rabbit Manure Biochar and Activation of Peroxymonosulfate for Rhodamine B Degradation
Reprinted from: *Water* **2023**, *15*, 2015, https://doi.org/10.3390/w15112015 85

Ling Wang, Chang Liu, Xing Fan, Chunxue Yang, Xiaolin Zhou and Zechong Guo
Methane Promotion of Waste Sludge Anaerobic Digestion: Effect of Typical Metal Meshes on Community Evolution and Electron Transfer
Reprinted from: *Water* **2022**, *14*, 3129, https://doi.org/10.3390/w14193129 107

Shuangshuang Yao, Lei Liu, Shiyang Zhang and Xinhua Tang
Nitrate Removal from Groundwater by Heterotrophic and Electro-Autotrophic Denitrification
Reprinted from: *Water* **2022**, *14*, 1759, https://doi.org/10.3390/w14111759 120

Zechong Guo, Lu Zhang, Min-Hua Cui and Aijie Wang
Electrode Microbial Communities Associated with Electron Donor Source Types in a Bioelectrochemical System Treating Azo-Dye Wastewater
Reprinted from: *Water* **2022**, *14*, 1505, https://doi.org/10.3390/w14091505 137

Yixin Lu, Chenghan Tang, Yujie Liu and Jiao Chen
Mechanism and Kinetic Analysis of the Degradation of Atrazine by O_3/H_2O_2
Reprinted from: *Water* **2022**, *14*, 1412, https://doi.org/10.3390/w14091412 149

About the Editors

Minhua Cui

Minhua Cui received his Ph.D. in Environmental Science and Engineering from Harbin Institute of Technology in 2017. He is currently an Associate Researcher at the School of Environment & Ecology, Jiangnan University. His research interests focus on the application of biocatalytic, electrocatalytic and photocatalytic technologies for the degradation of emerging contaminants and resource recovery, with a particular emphasis on fundamental theoretical studies and process optimization for practical applications. He currently serves as a member of the Youth Committee of the China Biogas Association, a young editorial board member of *Fermentation*, and an editorial board member of *Methane*.

Guoshuai Liu

Guoshuai Liu received his Ph.D. in Environmental Science and Technology from the Harbin Institute of Technology (2020). He currently works as an associate professor at School of Environment & Ecology, Jiangnan University. His current interests are the low-carbon electrochemical wastewater purification and resource recovery. In these areas, he has published more than 50 peer-reviewed articles (first author or corresponding author) in high-quality journals such as *Advanced Energy Materials, Advanced Functional Materials, Environmental Science & Technology, ACS ES&T Engineering*, and *Applied Catalysis B: Environment and Energy*.

water

MDPI

Editorial

The Application of Electrochemical Methods in Water Treatment

Min-Hua Cui * and Guo-Shuai Liu

Jiangsu Key Laboratory of Anaerobic Biotechnology, School of Environment & Ecology, Jiangnan University, Wuxi 214122, China; gsliu@jiangnan.edu.cn
* Correspondence: cuiminhua@jiangnan.edu.cn

Water resources form the cornerstone of sustainable development in human society. Over the past century, the accelerating pace of industrialization and urbanization has resulted in the discharge of significant pollutant loads stemming from anthropogenic activities. Simultaneously, the diversification of societal needs has prompted the widespread development and release of numerous synthetic compounds into aquatic environments, thereby increasing the complexity of water pollution and presenting formidable challenges for effective treatment. Against this backdrop, the advancement of wastewater treatment technologies becomes imperative for reducing contaminant concentrations to levels below ecologically acceptable thresholds, an essential prerequisite for safeguarding ecosystem health and ensuring the protection of public health.

For conventional pollutants commonly found in domestic wastewater, such as suspended solids, organic matter, nitrogen, and phosphorus, mature biochemical treatment technologies have been widely adopted in engineering practice, demonstrating robust performance and operational reliability [1]. However, the growing prevalence of emerging contaminants (ECs) in recent years has introduced substantial new challenges to the field of water treatment. ECs comprise a diverse array of substances, including persistent organic pollutants (POPs), endocrine-disrupting compounds (EDCs), antimicrobials, antibiotics, microplastics, and nanoparticles. Of particular concern are pharmaceuticals and personal care products (PPCPs), which represent a class of recalcitrant organic pollutants characterized by high chemical stability and low biodegradability. Even at trace concentrations, these substances can pose significant risks to both ecosystems and human health. Studies have shown that conventional wastewater treatment processes are often inadequate for the effective removal of many ECs, with residual contaminants frequently detected in secondary effluents or wasted sludge, thereby entering natural water bodies and exacerbating ecological risks [2]. As a result, the development of efficient, sustainable, and scalable treatment technologies specifically targeting these persistent and hazardous compounds has become a critical and urgent priority.

In response to these challenges, researchers have explored a broad range of advanced pollutant removal strategies, including membrane separation, adsorption, chemical oxidation, and electrochemical techniques. Among these, electrochemical technologies have garnered increasing attention due to their high removal efficiency, operational simplicity, and precise process control. Particularly noteworthy is the integration of electrochemical and biological processes within bioelectrochemical systems (BESs), which offers a promising pathway to overcome limitations commonly associated with conventional electrochemical methods, such as high energy consumption and the generation of undesirable byproducts. This synergistic approach not only mitigates inherent drawbacks but also broadens the applicability of electrochemical treatment to complex and variable wastewater matrices [3].

 https://doi.org/10.3390/w17172551

Against this backdrop, the Special Issue "Applications of Electrochemical Methods in Water Treatment (AEMWT)" was launched in the journal Water, with the objective of showcasing recent advances in electrochemical technologies and electron transfer mechanisms relevant to water treatment. This collection features cutting-edge research that bridges fundamental theoretical understanding with practical applications, addressing the urgent global need for effective wastewater treatment and the sustainable management of related solid waste.

This Special Issue presents a selection of exemplary studies that collectively underscore the versatility and potential of various technologies, especially the electrochemical technologies in tackling both conventional and emerging contaminants. Hong et al. developed an advanced oxidation system that integrates an air-diffusion cathode with ultraviolet (UV) irradiation. In this system, electrochemically generated hydrogen peroxide is activated by UV light, enabling the efficient mineralization of organic pollutants in lake water. The system exhibited excellent operational stability over extended periods and significantly enhanced the microbial safety of the treated water, offering a practical and scalable solution for water purification in regions experiencing water scarcity [contribution 8]. Guo et al. investigated the impact of different electron donor types on the microbial community structure and functional dynamics of electrode biofilms in BESs. It demonstrated that simple organic substrates (e.g., sodium acetate and glucose) rendered microbial communities more sensitive to electrode polarity, whereas complex substrates, such as actual domestic wastewater, promoted higher microbial diversity and community stability, with stronger spatial responses to electrode positioning. These findings provide important insights for optimizing the design and operational parameters of BESs [contribution 2]. Yao et al. proposed a hybrid denitrification system that couples solid-phase heterotrophic denitrification with electrochemical hydrogen autotrophic denitrification. Using polycaprolactone as a solid carbon source, the system achieved effective removal of nitrate from groundwater. Electrical stimulation was found to significantly enrich dominant denitrifying and electroactive microorganisms, with the reduction step from nitric oxide to nitrous oxide identified as the rate-limiting stage. This integrated approach offers an innovative strategy for mitigating nitrate pollution in groundwater environments [contribution 3]. Besides, other emerging studies have also provided unique insights into wastewater treatment and resource recovery applications. Lu et al. investigated the ozonation (O_3/H_2O_2) degradation of atrazine in phosphate buffer (pH 7), achieving a 92.59% removal rate under optimal conditions (25 °C, 20 mol/L O_3, 20 mol/L H_2O_2). Mechanism analysis revealed a synergistic 1:1 oxidation by HO· and O_3 under acidic conditions, with HO· playing the dominant role. Kinetics followed a pseudo-second-order model across various temperatures, pH, and H_2O_2 concentrations [contribution 1]. Wang et al. revealed that the addition of biocompatible high-mesh metal materials (Ni, Cu, stainless steel) significantly enhanced methane production from waste activated sludge by up to 61% (77.52 mL $gVSS^{-1}$) via promoting microbial cooperation and enriching syntrophic bacteria and methanogens, with electron transfer on metal surfaces further improving efficiency [contribution 4]. Ouyang et al. prepared rabbit manure biochar (RBC) at different pyrolysis temperatures (400–600 °C) and found that RBC600 exhibited the highest catalytic activity for activating peroxymonosulfate (PMS), degrading 93.38% of rhodamine B within 60 min via both free-radical ($SO_4^{-}\cdot$, ·OH) and non-radical (dominant $^1O_2$) pathways, while maintaining preferable reusability over five cycles, enabling waste reuse and efficient organic wastewater treatment [contribution 5]. Huang et al. developed a co-pyrolysis method for Fenton sludge and pomelo peel that significantly enhanced biochar surface area and aromaticity, effectively immobilizing heavy metals by converting them into stable fractions and reducing leachability, thereby lowering ecological risk from considerable pollution to moderate or clean levels [contribution 6]. Guo et al. demonstrated that UV365 was the most effective wavelength in generating reactive species ($^3DOM^*$, $^1O_2$, and ·OH) from dissolved organic matter (HA, FA, EfOM), achieving the highest formation rates, steady-state concentrations, and quantum yields, thereby enhancing understanding of DOM photochemistry in aquatic environments [contribution 7]. Hou et al. demonstrated that the heterotrophic nitrification–aerobic denitrification strain JQ1004 efficiently removed nitrogen primarily through assimilation (54.61%) and heterotrophic nitrification–aerobic denitrification, with maximum degradation rates of 7.93 (ammonia) and 4.08 mg/(L·h) (nitrate), and revealed a potential direct ammonia oxidation pathway (dominated by highly transcribed amoA) alongside high oxygen tolerance and pH sensitivity of key functional genes [contribution 9]. Zhang et al. prepared N and $KHCO_3$ co-modified biochar from rapeseed straw, which exhibited a high tetracycline adsorption capacity of 604.71 mg/g through multiple mechanisms including pore filling, π–π interaction, and hydrogen bonding, demonstrating great potential for antibiotic removal from water [contribution 10].

The contributions to the AEMWT Special Issue collectively reflect both significant advancements and enduring challenges in the application of electrochemical technologies for water treatment. These studies not only highlight important technological breakthroughs but also contribute to the establishment of a solid theoretical foundation and practical framework to guide future research and engineering implementation in this rapidly evolving field.

We extend our sincere gratitude to all the authors whose contributions to this Special Issue exemplify intellectual rigor and a deep commitment to advancing the field. Their works have significantly expanded the frontiers of knowledge in electrochemical water treatment. We are equally grateful to the expert reviewers for their meticulous evaluations and scholarly insights, which have been instrumental in ensuring the academic integrity and high quality of this publication. Special appreciation is also due to the Water editorial team for their steadfast support and seamless coordination, which enabled the successful and timely release of this issue. It is through this collective effort that we have assembled a forward-looking and practically impactful body of work.

In conclusion, the AEMWT Special Issue marks a significant milestone in the application of electrochemical technologies to address contemporary water pollution challenges. As researchers dedicated to this field, we are both encouraged and inspired by the achievements presented herein. We warmly invite colleagues around the world to join in fostering innovation and collaboration aimed at advancing electrochemical water treatment toward broader implementation and technological maturity.

Author Contributions: Writing—original draft preparation, M.-H.C.; writing—review and editing, M.-H.C. & G.-S.L. All authors have read and agreed to the published version of the manuscript.

Conflicts of Interest: The authors declare no conflicts of interest.

List of Contributions:

1. Lu, Y.; Tang, C.; Liu, Y.; Chen, J. Mechanism and Kinetic Analysis of the Degradation of Atrazine by O_3/H_2O_2. *Water* **2022**, *14*, 1412.
2. Guo, Z.; Zhang, L.; Cui, M.-H.; Wang, A. Electrode Microbial Communities Associated with Electron Donor Source Types in a Bioelectrochemical System Treating Azo-Dye Wastewater. *Water* **2022**, *14*, 1505. https://doi.org/10.3390/w14091505.
3. Yao, S.; Liu, L.; Zhang, S.; Tang, X. Nitrate Removal from Groundwater by Heterotrophic and Electro-Autotrophic Denitrification. *Water* **2022**, *14*, 1759. https://doi.org/10.3390/w14111759.
4. Wang, L.; Liu, C.; Fan, X.; Yang, C.; Zhou, X.; Guo, Z. Methane Promotion of Waste Sludge Anaerobic Digestion: Effect of Typical Metal Meshes on Community Evolution and Electron Transfer. *Water* **2022**, *14*, 3129.

5. Ouyang, F.; Liu, Y.; Chen, J.; Tang, C.; Wang, A.; Lu, Y.; Yuan, Y. Study on Preparation of Rabbit Manure Biochar and Activation of Peroxymonosulfate for Rhodamine B Degradation. *Water* **2023**, *15*, 2015.
6. Huang, C.; Wang, L.; Fan, L.; Chen, Y. Co-Pyrolysis of Fenton Sludge and Pomelo Peel for Heavy Metal Stabilization: Speciation Mechanism and Risk Evaluation. *Water* **2023**, *15*, 3733.
7. Guo, Z.-C.; Zhang, L.; Chen, Y.; Huang, C.; Liao, Z.-M. Effect of UV-LED Wavelength on Reactive Species Photogeneration from Dissolved Organic Matter. *Water* **2024**, *16*, 635.
8. Hong, H.; Xie, S.; Qiu, A.; Yao, Y.; Jiang, W.; Li, J.; Wan, Z.; Xiang, S.; Xi, C.; Xiao, J.; et al. Efficient H_2O_2 Production and Activation by Air Diffusion Cathode Combined with Ultraviolet for Lake Water Treatment: A Long-Term Evaluation. *Water* **2024**, *16*, 1658. https://doi.org/10.3390/w16121658.
9. Hou, L.; Huang, F.; Pan, Z.; Chen, W.; Wang, X. Characteristics of Nitrogen Removal and Functional Gene Transcription of Heterotrophic Nitrification-Aerobic Denitrification Strain, *Acinetobacter* sp. JQ1004. *Water* **2024**, *16*, 1799.
10. Zhang, Z.; Tang, C.; Wang, H.; Zhong, M.; Ge, P.; Xu, W.; Chen, Y. Synthesis and Characterization of Potassium Bicarbonate and Urea-Modified Biochar from Rape Straw: Application in the Removal of Tetracycline from Aqueous Solution. *Water* **2024**, *16*, 2522.

References

1. Wei, Z.; Qin, Y.; Li, X.; Gao, P. Resource recovery of high value-added products from wastewater: Current status and prospects. *Bioresour. Technol.* **2024**, *398*, 130521. [CrossRef] [PubMed]
2. Qian, Y.; Guan, L.; Ke, Y.; Wang, L.; Wang, X.; Yu, N.; Yu, Q.; Wei, S.; Geng, J. Unveiling intricate transformation pathways of emerging contaminants during wastewater treatment processes through simplified network analysis. *Water Res.* **2024**, *253*, 121299. [CrossRef] [PubMed]
3. Zhao, J.; Rao, M.; Zhang, H.; Wang, Q.; Shen, Y.; Ye, J.; Feng, K.; Zhang, S. Evolution of interspecific interactions underlying the nonlinear relationship between active biomass and pollutant degradation capacity in bioelectrochemical systems. *Water Res.* **2025**, *274*, 123071. [CrossRef] [PubMed]

water

MDPI

Article

Synthesis and Characterization of Potassium Bicarbonate and Urea-Modified Biochar from Rape Straw: Application in the Removal of Tetracycline from Aqueous Solution

Zhipeng Zhang [1,2], Chenghan Tang [3], Hongbin Wang [4], Ming Zhong [5], Pengchao Ge [6], Wenlai Xu [7] and Yiyang Chen [8,*]

1 Sichuan Geological Environment Survey and Research Center, Chengdu 610081, China; zzpwss123@126.com
2 Engineering & Technology Center of Groundwater Pollution Control for Environmental Protection in Sichuan, Chengdu 610081, China
3 College of Environmental Science and Engineering, Southwest Jiaotong University, Chengdu 611756, China; 15201822120@163.com
4 Sichuan Provincial Engineering Research Center of City Solid Waste Energy and Building Materials Conversion and Utilization Technology, Chengdu 610199, China; wanghongbin@cdu.edu.cn
5 National Postdoctoral Research Station, Haitian Water Group, Chengdu 610213, China; txgsfy@163.com
6 School of Materials and Environmental Engineering, Chengdu Technological University, Chengdu 611730, China; 15582326457@163.com
7 College of Ecology and Environment, Chengdu University of Technology, Chengdu 610059, China; xuwenlai2012@cdut.cn
8 Sichuan Ecology and Environment Industry Group Co., Ltd., Chengdu 610095, China
* Correspondence: chenyiyang@sdholding.com; Tel.: +86-13880902258

Abstract: Using rapeseed straw as a raw material and potassium bicarbonate ($KHCO_3$) and urea ($CO(NH_2)_2$) as modification reagents, the pyrolysis raw materials were mixed in a certain proportion, and the unmodified biochar GBC800, $KHCO_3$-modified biochar KGBC800, and ($KHCO_3$)/($CO(NH_2)_2$) co-modified biochar N-KGBC800 were, respectively, prepared using the one-pot method at 800 °C. The physicochemical properties, such as surface morphology, pore characteristics, functional group distribution, and elemental composition of the three biochars, were characterized, and the adsorption performance and mechanism of the typical antibiotic tetracycline (TC) in water were studied. The results showed that the surface of GBC800 was smooth and dense, with no obvious pore structure, and both the specific surface area and total pore volume were small; the surface of KGBC800 showed an obvious coral-like three-dimensional carbon skeleton, the number of micropores and the specific surface area were significantly improved, and the degree of carbonization and aromatization was enhanced; N-KGBC800 had a coral-like three-dimensional carbon skeleton similar to KGBC800, and there were also many clustered carbon groups. The carbon layer changed significantly with interlayer gaps, presenting a multi-level porous structure. After N doping, the content of N increased, and new nitrogen-containing functional groups were formed. When the initial TC concentration was 100 mg/L, pH $\approx$ 3.4, the temperature was 25 °C, and the dosage of the three biochars was 0.15 g/L, the adsorption equilibrium was reached before 720 min. The adsorption capacities of GBC800, KGBC800, and N-KGBC800 for TC were 16.97 mg/g, 294.86 mg/g, and 604.71 mg/g, respectively. Fitting the kinetic model to the experimental data, the adsorption of TC by the three biochars was more in line with the pseudo-second-order adsorption kinetic model, and the adsorption isotherm was more in line with the Langmuir model. This adsorption process was a spontaneous endothermic reaction, mainly chemical adsorption, specifically involving multiple adsorption mechanisms such as pore filling, electrostatic attraction, hydrogen bonds, $n-\pi$ interaction, Lewis acid–base interaction, $\pi-\pi$ stacking, or cation $-\pi$ interaction between the aromatic ring structure of the carbon itself and TC. A biochar-adsorption column was built to investigate the dynamic adsorption process of tetracycline using the three biochars against the background of laboratory pure water and salt water. The adsorption results show that the Thomas model and the Yoon–Nelson model both provide better predictions for dynamic adsorption processes. The modified biochars KGBC800 and N-KGBC800 can be used as preferred materials for the efficient adsorption of TC in water.

Keywords: biochar; potassium bicarbonate modification; nitrogen doping; tetracycline; adsorption mechanism

1. Introduction

China has a huge population and a strong demand for agricultural products. In 2021, China's grain output was 682.85 million tons, and the output of rice, wheat, corn, rapeseed, etc. accounted for about 20% of the global output [1]. Rapeseed is a major crop in China's oil economy [2]. In 2021, the output of rapeseed straw in China was approximately 29.147 million tons, the total amount of straw collected reached 23.684 million tons, and the comprehensive utilization amount was 20.591 million tons [3]. According to the statistics from the National Grain and Oils Information Center, by 2025, the planting area of rapeseed in China will reach about 80 million hectares. At that time, along with the planting and processing of rapeseed, its straw output will increase rapidly. The "14th Five-Year Plan for the Development of Circular Economy" clearly states that it is necessary to strengthen the resource utilization of wastes in the agricultural production process and promote the efficient utilization of agricultural and forestry wastes, such as crop straws, livestock and poultry manure, forestry wastes, and by-products of agricultural product processing [4]. In recent years, the resource utilization technologies of agricultural wastes at home and abroad and related research have significantly developed. In addition to the conventional "five utilization methods", namely, fertilization, feed utilization, combustion, substrate utilization, and raw material utilization. There are also deep processing and transformation technologies, such as solidification utilization technology, gasification utilization, and liquefaction utilization technology. The former has the disadvantage of a low utilization degree, while the latter has high requirements for equipment and technology and is difficult to popularize [5]. Therefore, exploring the efficient utilization of rapeseed straw has become the only way to support the sustainable development of the rapeseed industry.

Tetracycline (TC), an antibiotic with a strong antibacterial effect, low cost, and low allergenicity, is widely used in human health, animal disease control, and feed additives. China is the country with the largest production and usage of tetracycline worldwide [6]. Meanwhile, due to China's aquaculture area and aquatic product output accounting for more than 60% of the global total, according to the "2022 National Fishery Economic Statistics Bulletin", the national aquaculture area reached 7,107,500 hm^2, including 2,074,420 hm^2 of marine aquaculture area and 5,033,080 hm^2 of freshwater aquaculture area [7,8]. A large amount of tetracycline is used in the breeding process. Tetracycline is difficult for humans and animals to absorb. About 70–90% is discharged into water bodies or soil through urine and feces. Tetracycline is difficult to degrade in the natural environment, has poor biodegradability, and can also induce the production of resistant microorganisms and resistance genes, leading to serious biological environmental pollution and endangering public health [9]. Currently, the main methods for removing tetracycline from water include advanced oxidation methods [10], biological treatment methods [11], and adsorption methods [12]. Among them, the adsorption method has the advantages of low cost, simplicity, efficiency, and no harmful by-products and is considered one of the most effective wastewater treatment methods [13]. The original biochar has problems such as poor anti-interference ability and unsatisfactory adsorption capacity. Activation and modification can effectively improve the adsorption performance of biochar. Du et al. [14] prepared ginkgo biochar through ball mill modification, and its adsorption capacity for target pollutant increased from 16.47 mg/g to 182.63 mg/g; Zhang et al. [15] used corn cob waste as the experimental raw material and KOH as the modified activator. The adsorption capacity of CBC750 prepared for TC was three times that of ordinary biochar CBC. Sun et al. [16] prepared modified biochar with KOH and $FeCl_3$ to adsorb sulfonamide antibiotics in water. The maximum adsorption capacities for sulfadiazine, sulfamethazine, and sulfamethoxazole were 294.12 mg/g, 400.00 mg/g, and 454.55 mg/g, respectively, which were 5–7 times that of ordinary biochar. The nitrogen-doped biochar prepared by using bagasse as the carbon source, melamine as the nitrogen source, and sodium bicarbonate as the pore-forming agent was proved by Che et al. [17] to be a reliable wastewater treatment technology for adsorbing ciprofloxacin (CIP). In recent years, heteroatom doping has attracted much attention in improving the adsorption properties of carbon-based materials. For example, Lu et al. [18] used cellulose and ammonium oxalate to synthesize nitrogen-doped biochar with a layered porous structure, which enhanced the adsorption of toluene. Li et al. [19] prepared nitrogen-doped biochar with corn and urea for CO_2 adsorption, and studied that the introduction of surface N contributes to the generation of ultrafine pores. At the same time, since N has lone pair electrons and high electronegativity, the flow of π electrons in sp^2C is enhanced through conjugation, resulting in more abundant functional groups and surface defects [20,21]. Among the many current adsorbents, there are few studies on modifying rapeseed straw to remove TC in water, and its adsorption behavior characteristics and mechanism are still unclear.

Given this, this study aims to use rapeseed straw as the biomass raw material, potassium bicarbonate ($KHCO_3$) as the activator, and urea ($CO(NH_2)_2$) as the nitrogen source to prepare modified rapeseed straw biochar via one-pot oxygen-limited pyrolysis at 800 °C. The possibility and influencing factors of its application in the removal of TC from water bodies are discussed, and its adsorption mechanism is analyzed by combining kinetics, adsorption isotherms, thermodynamic models, and characterization methods. This provides a new and effective method for using rapeseed straw and efficiently removing TC from water.

2. Materials and Experimental Methods

2.1. Reagents and Apparatus Used in the Experiment

2.1.1. Test Reagents

Tetracycline hydrochloride, $KHCO_3$, urea, ethanol, HCl, NaOH, NaCl, $MgCl_2$, Na_2S, and $CaCl_2$ are analytically pure and were purchased from Cologne (Chengdu) Chemical Co.

2.1.2. Laboratory Instruments

The main instruments and equipment used in the experiment include the following: an ultraviolet–visible spectrophotometer (WFZ UV-4802, Shanghai, China), constant temperature oscillation incubator (BS-2E, Changzhou, Jiangsu, China), tube furnace (BEQ-1200C, Hefei, Anhui, China), precision bench top pH meter (pH S-3CU, Shanghai, China), analytical balance (FA224, Shanghai, China), electric blast drying oven (DHG-9240A, Shanghai, China), high-speed multifunctional crusher (SS-1022, Changzhou, Jiangsu, China), UPU ultrapure water system (UPHW-IV-90T, Chengdu, Sichuan, China), peristaltic pump (BT-100F, Baoding, HeBei, China), numerical control ultrasonic cleaner (CR-080S, Shenzhen, China), automatic sampler (CBS-A, Shanghai, China), scanning electron microscope (ZEISS Gemini SEM 300, Baden-Württemberg, Germany), BET-specific surface area analyzer (Quantachrome/ASiQwin, Boynton Beach, CA, USA), Fourier transform infrared spectrometer (Thermo Scientific Nicolet iS5, Waltham, MA, USA), and X-ray photoelectron spectrometer (Thermo Scientific K-Alpha, Waltham, MA, USA).

2.2. Preparation of Biochar

Experimental materials: rape straw from a farm in Pidu District, Chengdu City, Sichuan Province.

The process of biochar production using rapeseed straw is shown in Figure 1.

Using rape straw as the raw material for pyrolysis to produce biochar, after removing impurities such as stones and leaves, the straw was crushed using a crusher to pass through a 100-mesh sieve for sieving. An appropriate amount of the material was weighed and mixed at a ratio of biomass of $KHCO_3$:urea = 1:1.5:1 and then sent into a tube furnace purged with nitrogen for pyrolysis. The heating rate was 8 °C/min, the nitrogen flow rate was 50 mL/min, and the isothermal pyrolysis lasted 120 min. After the pyrolysis, the heating was stopped until it cooled to below 100 °C before removing it. The pyrolysis product was mixed with 1 mol/L HCl at a solid–liquid ratio of 1:10 (g:mL), and ultrasonic cleaning was performed for 15 min. Then, ultrapure water was added for cleaning, and this process was repeated multiple times until the pH of the supernatant was approximately 7 and remained stable. The resulting solid was collected, dried at 75 °C to a constant weight, and then ground using a ceramic mortar and passed through a 100-mesh sieve to ensure that the particle size of the material remained basically constant. After sieving, the solid was collected to obtain the modified biochar. That created without adding potassium bicarbonate and urea was named GBC800, the biochar with only potassium bicarbonate added was named KGBC800, and the one with both potassium bicarbonate and urea added was named N-KGBC800 biochar.

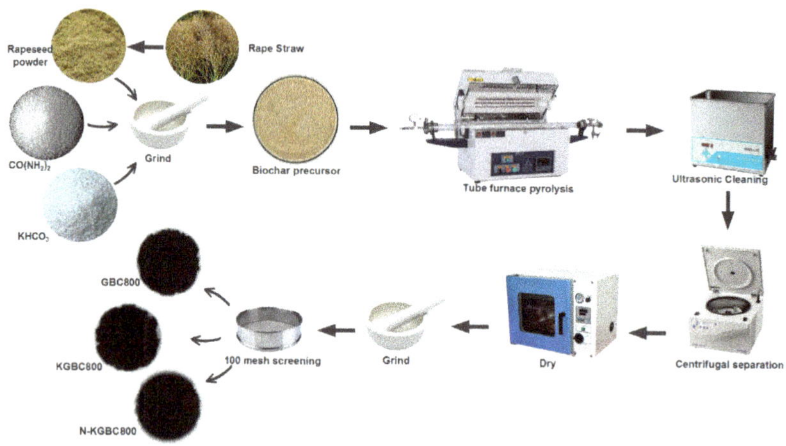

Figure 1. A flow chart of modified biochar preparation.

2.3. Adsorption Experiment

2.3.1. Batch Adsorption Experimental Method

First, a 1 g/L TC stock solution was prepared, which can be diluted to the required concentration of TC reaction solution as needed. Then, 200 mL of TC solution with a concentration of ρ_0 mg/L was taken and added to a 250 mL conical flask. Then, m g of GBC800, KGBC800, and N-KGBC800 was weighed and added. The original pH of the reaction solution was 3.4 (when exploring the effect of different pH on the adsorption performance of biochar, 1 mol/L of NaOH and HCl were used to change the pH of the solution). The solution was shaken in the dark at T °C and 150 r/min for t min. After the reaction, it was taken out and left to stand. A 5 mL syringe was used to draw 2 mL of the supernatant, and the biochar particles are filtered out through a 0.45 filter membrane. The filtrate was placed in a colorimetric tube containing 2 mL of pure water and stored at a low temperature for later use to determine its absorbance with an ultraviolet–visible spectrophotometer. When investigating the effect of the initial pH of the solution, ρ_0, m, T, and t were fixed at 100 mg/L, 0.15 g/L, 25 °C, and 720 min, respectively, and the pH = 3~11. When investigating the effect of the biochar dosage, ρ_0, pH, T, and t were fixed at 100 mg/L, 3.4, 25 °C, and 720 min, respectively, and the variation range of m was 0.05 to 0.4 g/L. When investigating the effect of the reaction time, ρ_0, m, pH, and T were fixed at 100 mg/L, 0.15 g/L, 3.4, and 25 °C, respectively, and the t values were 0, 5, 10, 20, 40, 70, 120, 180, 240, 360, 480, 600, and 720 min. Kinetic analysis was conducted based on the adsorption experimental data under different t values. When investigating the effect of the initial TC concentration of the solution and the reaction temperature, ρ_0, m, and pH were fixed at 100 mg/L, 0.15 g/L, and 3.4, respectively. The temperature varied at 15, 25, and 35 °C, and the initial concentrations of TC were 10, 20, 30, 40, 50, 60, 70, 80, 90, and 100 mg/L.

Adsorption isotherm models and adsorption thermodynamic analysis were conducted based on the experimental results.

2.3.2. Biochar Fixed-Bed Adsorption Experiment

The laboratory apparatus is illustrated in Figure 2. The chromatography column had an inner diameter of 10 mm and a length of 200 mm, and the filling sequence from bottom to top was 2.0 g of 30-mesh quartz sand, 0.1 g of biochar, and 8 g of 30-mesh quartz sand. A certain amount of TC sample was weighed and added to the brine (according to the mass ratio of NaCl (2.5%), MgCl (1.1%), $NaSO_4$ (0.4%), and $CaCl_2$ (0.16%)) to obtain simulated saline wastewater contaminated with TC (100 mg/L) as the reaction solution to be reacted. The conductor entered the tank through the worm pump, adsorption column, and automatic collector connection. The flow rate was adjusted to 2.0 mL/min of the reagent from the bottom up through the absorption column. The sampling time interval was 10 min, and the response time was 12 h. At the end of the reaction, the absorption rate was determined using an ultraviolet spectrometer to calculate the outflow concentration of TC.

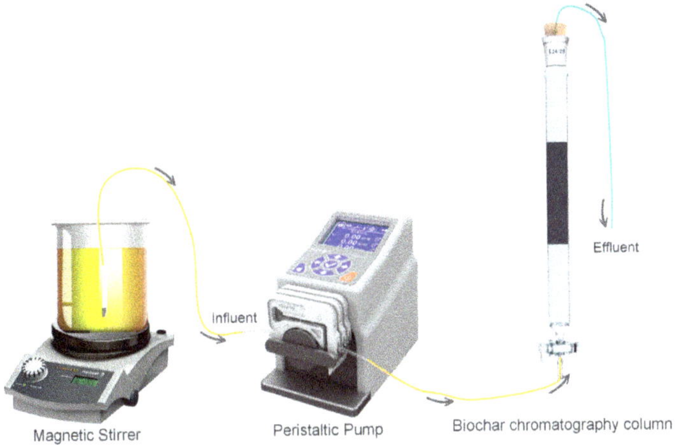

Figure 2. The biochar fixed-bed adsorption experimental device.

2.4. UV-Vis Spectrophotometric Analysis of Tetracycline

The detection method for TC: The absorbance of TC was 358 nm and was tested using a UV–visible spectrophotometer. The formula for calculating the adsorption capacity is shown below:

$$q_t = \frac{(C_0 - C_t) \times V}{m} \quad (1)$$

q_t is the adsorption capacity at time t, mg/g; C_0 is the initial concentration of the pollutant, mg/L; C_t is the concentration of the pollutant in the solution after a specific time of adsorption, mg/L; m is the amount of adsorbent dosing, g; and V is the volume of adsorbent solution, L.

3. Results and Discussion

3.1. Material Characterization Results

3.1.1. Analysis of Physical Properties of Materials

GBC800, KGBC800, and N-KGBC800 were placed under a scanning electron microscope to observe their structural features and microscopic patterns; the findings are displayed in Figure 3.

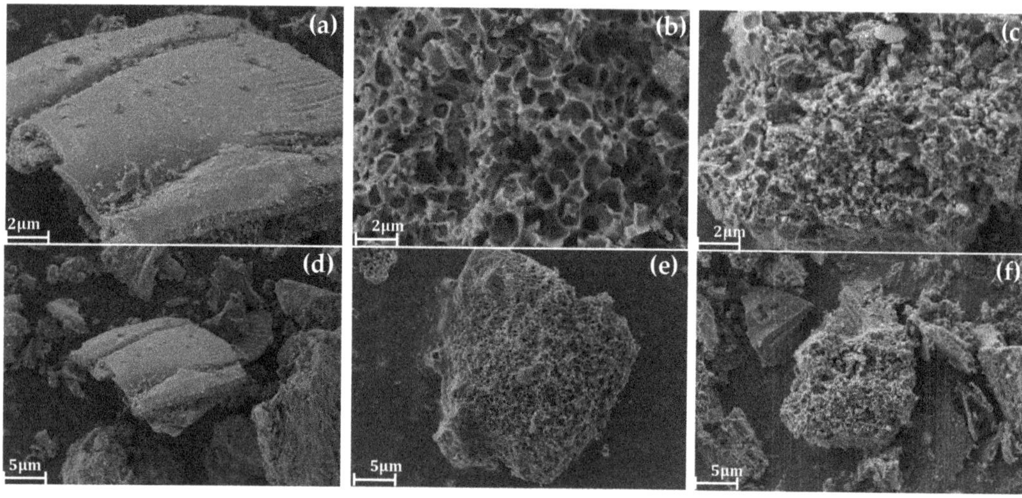

Figure 3. Scanning electron microscope images of biochar. (**a**,**d**) show the surface morphology of GBC800, (**b**,**e**) show the surface morphology of KGBC800, and (**c**,**f**) show the surface morphology of N-KGBC800.

In Figure 3a,d demonstrate the surface morphology of pristine rapeseed straw biochar GBC800 without activation and heteroatom doping, which has a smooth and compact surface without obvious pores and exhibits a dense carbon skeleton laminar structure with no obvious pores present at the edge fracture.

The morphology of KGBC800 prepared by $KHCO_3$ activation pyrolysis is shown in Figure 3b,e, which exhibits an obvious pore structure on the carbon surface, with a randomly generated coral-like carbon skeleton similar to a sphere and more extremely small pores on the surface. This is because the gas generated by the thermal decomposition of $KHCO_3$ will continuously impact the surface of the biochar to make it more uneven. At the same time, the thermal decomposition of K_2CO_3 generated by the heat into the cellulose weakens the intermolecular hydrogen bonding within the straw, making the biomass structure more easily damaged and forming a richer microporous structure. Around 700~800 °C, elemental K reacts with C to produce metal K, thereby continuing to etch the pore structure [22]. The vapor precipitation of potassium metal squeezes into the carbon interlayer to promote the activation reaction, forming a richer multilevel pore structure and a larger specific surface area, presenting a continuous sheet of a spherical, coral-like, three-dimensional pore structure [23].

Figure 3c,f demonstrate the surface morphology of N-KGBC800 prepared in the case of $KHCO_3$ activation and $CO(NH_2)_2$ doping, which exhibits a more complete stacked structure with the same spherical coral-like pore structure as that of KGBC800, with the pore sizes exhibiting a more irregular and inhomogeneous shape and the presence of a certain thickness of gaps between its carbon layers, generating more fine particles and pore channels. A two-dimensional network is generated between urea and H_2O due to strong hydrogen bonding, and $KHCO_3$ and $CO(NH_2)_2$ molecules can enter the pores and thermally decompose inside to form more micropores [24]. In addition, there are many fine massive carbon layers with rough outer surfaces and extremely tiny cluster structures on the surface, which may be formed due to chemical vapor deposition during the activation process. At around 320 °C, urea can decompose into HCNO and NH_3 [25,26]. NH_3 reacts chemically with the hydroxyl and carboxyl groups on the surface of the carbon layer to produce nitrogen-containing functional groups such as $-CO-NH_2$, $-C-NH_2$, etc. [27]. When the pyrolysis temperature exceeds 500 °C, NH_3 begins to decompose continuously, producing all kinds of chemical groups, which attack carbon-active sites on the surface of the biochar, leading to the formation of a larger surface area and a slight increase in the volume of micropores of the modified biochar [24,28].

N_2 adsorption/desorption was performed on all three biochars at a temperature of 77.3 K. The results are presented in Figure 4, and the parameters are presented in Table 1. With the addition of $KHCO_3$ and urea, the specific surface area and total pore volume of the biochar were significantly enhanced and clearly differentiated, with the specific surface area of N-KGBC800 being 1.4 and nearly 180 times higher than that of KGBC800 and GBC800, respectively, and the total pore volume being 1.3 and 26 times higher than that of KGBC800 and GBC800, respectively. The specific surface area and the total pore volume of GBC800 were small, and the overall pores were dominated by mesopores. However, it is noteworthy that the volume of micropores of KGBC800 was larger than that of N-KGBC800, and in general, the introduction of urea, to a certain extent, should open up the micropores inside the biochar even further. However, the microporous volume of N-KGBC800 decreases instead. This is mainly because the carbon layer, under the impact of CO_2 and NH_3 and the etching of metallic potassium monomers, undergoes a vaporization reaction with tiny potassium-containing compound crystals as the core, which cracks the surrounding carbon-containing compounds and swells the micropores into medium- or large-sized pores or leads to the loosening of its layers and fragility, which triggers the pore collapses.

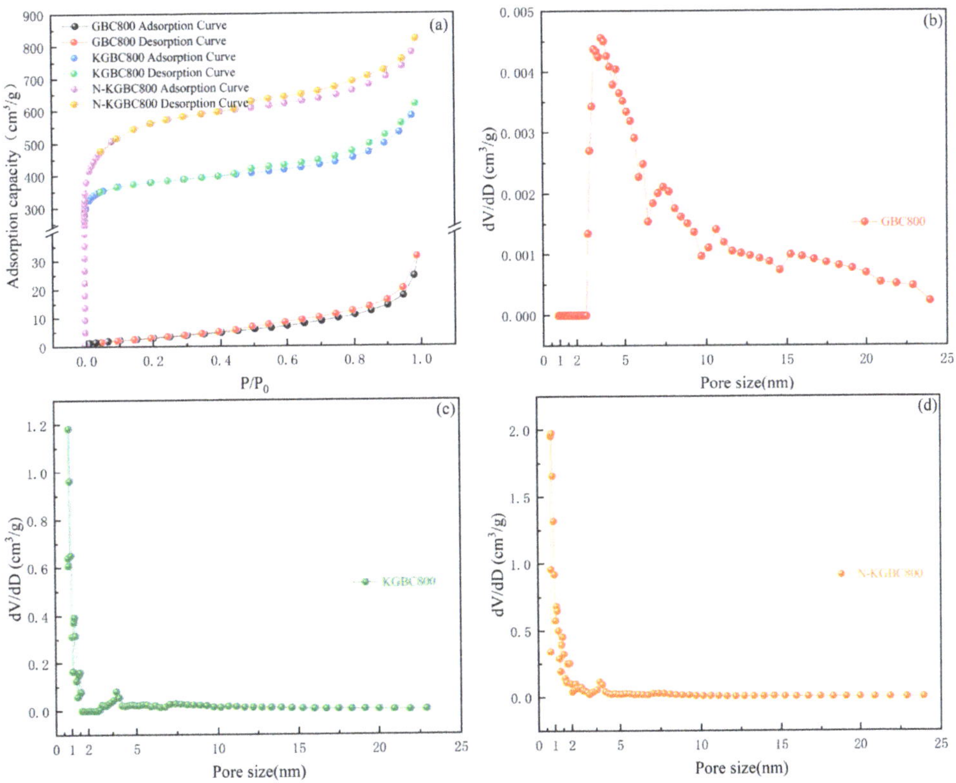

Figure 4. N_2 adsorption–desorption curves and pore size distribution curves of modified rape straw biochar. (**a**) Biochar N_2 adsorption–desorption curve; (**b**) GBC800 pore size distribution graph; (**c**) KGBC800 pore size distribution graph; and (**d**) N-KGBC800 pore size distribution graph.

Table 1. Modified rape straw biochar BET parameters.

Parameters	Unit	GBC800	KGBC800	N-KGBC800
S_{BET}	m^2/g	11.7	1477	2090
V_{Total}	cm^3/g	0.05	0.96	1.27
V_{Mic}	cm^3/g	0.00	0.41	0.36
L_0 [1]	nm	16.56	2.59	2.43

Note: [1] The average pore size.

The pore structure of biochar is mainly divided into three types: macropore with a pore radius greater than 50 nm, micropore with a pore radius less than 2 nm, and mesoporous with a pore radius between 2–50 nm. According to IUPAC [29], GBC800, KGBC800 and N-KGBC800 belong to type IV adsorption isotherms. A significant hysteresis loop can be observed at $p/p_0 > 0.4$, which is due to capillary condensation occurring in mesoporous materials, indicating the presence of mesoporous biochar [30]. By observing (a) in Figure 4, it is found that GBC800 has no obvious inflection point at $p/p_0 < 0.1$, indicating that there are a large number of mesoporous pores on the surface of GBC800, while the number of micropores is very small, which is consistent with the aperture distribution diagram. However, KGBC800 and N-KGBC800 have steep turning points before $p/p_0 < 0.1$, indicating that they contain abundant micropore structures. Combined with (c,d) in Figure 4, it can be seen that micropores occupy a dominant position in their structures. However, according to parameter V_{mic} in Table 1, the micropore volume of KGBC800 accounts for 42.64% of the total pore volume. The micropore volume of N-KGBC800 accounted for 28.61% of the total pore volume, because the number of micropores on the biochar surface did not necessarily have a positive correlation with V_{Mic}/V_{Total}. Although the average pore widths of KGBC800 and N-KGBC800 are 2.59 nm and 2.43 nm, respectively, the presence of a large number of micropores on the surface still indicates that they belong to microporous materials.

3.1.2. Analysis of Chemical Properties of Materials

FT-IR was used to characterize and analyze the functional groups on the surface of biochar, and the characterization results are shown in Figure 5. In the figure, The main characteristic peaks of GBC800, KGBC800, and N-KGBC800 appeared around 3425 cm^{-1}, 2918 cm^{-1}, 2850 cm^{-1}, 1618 cm^{-1}, and 1094 cm^{-1}. However, it is worth noting that a new characteristic stretching peak appears near 1386 cm^{-1} for N-KGBC800. The stretching peak near 3425 cm^{-1} can be attributed to the O–H stretching of phenols, alcohols, and water molecules [31]. Furthermore, the peak intensities of GBC800 and N-KGBC800 are significantly greater than that of KGBC800, which may be related to the N–H bond therein. The stretching peak near 2918 cm^{-1} is produced by the asymmetric stretching vibration of alkyl–CH_2 or the contraction vibration of C–H [32]. The stretching peak near 2850 cm^{-1} is in line with the symmetrical stretching vibration of alkyl–CH_2 [33]. The stretching peak near 1618 cm^{-1} is the C=C or C=O transformed from the aromatic ring in the biomass lignocellulose [34]. N-KGBC800 and GBC800 have a weak absorption peak at 1386 cm^{-1}, mainly due to the C–N single bond formed by the affinity addition reaction between the external nitrogen source or the internal nitrogen source of the biomass and the hydroxyl and epoxy compounds [35]. A wide absorption peak appears near 1094 cm^{-1}, which is the characteristic peak of the aliphatic C–N stretching vibration [16]. The absorption peak of GBC800 at 799 cm^{-1} is caused by the out-of-plane deformation vibration of the aromatic C–H, and the absorption peak near 462 cm^{-1} is the stretching vibration of the Si–O–Si bond of the inorganic salt component [36].

The absorption peaks of the three kinds of biochar at 3425 cm^{-1} were obviously broad and strong; this is the telescopic oscillation of free or bonded –OH on carboxyl, hydroxyl, and carbonyl groups, and the presence of –OH is favorable for the occurrence of ion-exchange adsorption [37]. Zhao Xingchain et al. [38] showed that the intensity of the –OH absorption peak decreases and stabilizes with an increasing pyrolysis temperature, which indicates that the combined water is detached and the hydrogen-bonded hydroxyl group is broken. The magnitude of intensity at 1618 cm^{-1} was in the order of N-KGBC800 > KGBC800 > GBC800, suggesting that the aromatization of biochar was enhanced due to the increase in thermolysis and the introduction of $KHCO_3$. In contrast, the introduction of doped nitrogen intensified the aromatization. A smaller peak at 1386 cm^{-1} for N-KGBC800 indicates that there are still many reactive $-NH_2$ functional groups on its surface [35]. The absorption intensity at 1090 cm^{-1} (C–N) was GBC800 > N-KGBC800 > KGBC800, indicating that the number of nitrogen-containing functional groups in the biochar decreases due to the activation effect of $KHCO_3$ and then increases with the doping of urea, directly proving that nitrogen element has been successfully doped into the modified biochar, which is a direct proof of the successful doping of nitrogen into the modified biochar [21].

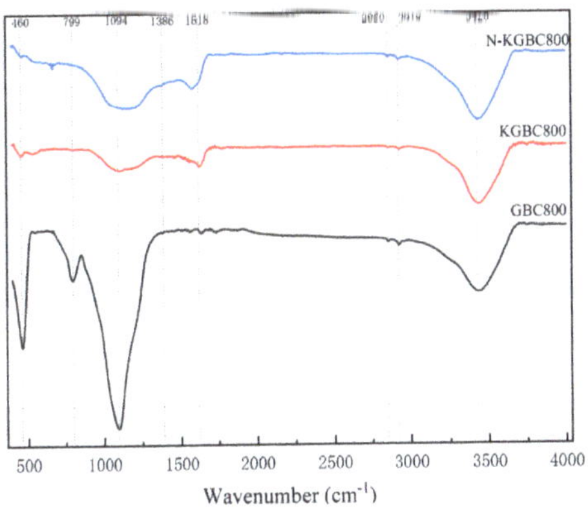

Figure 5. FT-IR plots of different biochar materials.

As can be seen from the proportion of elements on the surface of biochar in Table 2, the elements that make up the surface of biochar are mainly C and O. All the absorption peaks of the three biochars are at the same position. However, the intensities are inconsistent, indicating that the functional groups between GBC800, KGBC800, and N-KGBC800 are similar. According to the results in Figure 6, the obvious O1s peak (523.08 eV), the C1s peak (284.80 eV), and the weaker N1s peak (400.08 eV) can be observed. After the activation and nitrogen doping of the biochar, the O1s peak is significantly lower than that of the unmodified biochar, while the intensity of the C1s peak increases, indicating that after activation by $KHCO_3$, the oxygen-containing functional groups decrease and the degree of carbonization is higher. From the perspective of the elemental composition on the surface of the biochar, the N content of the unmodified biochar is relatively high. However, N decreases sharply after activation, while N doping by urea compensates for the loss of nitrogen-containing functional groups caused by high temperature to a certain extent [21].

Table 2. The elemental composition of different biochar surfaces.

Biochar	Percentage of Surface Elements/%			
	O1s	C1s	N1s	Other Elements
GBC800	25.22	63.41	1.15	10.22
KGBC800	13.82	84.32	0.45	1.41
N-KGBC800	13.37	83.95	1.04	1.64

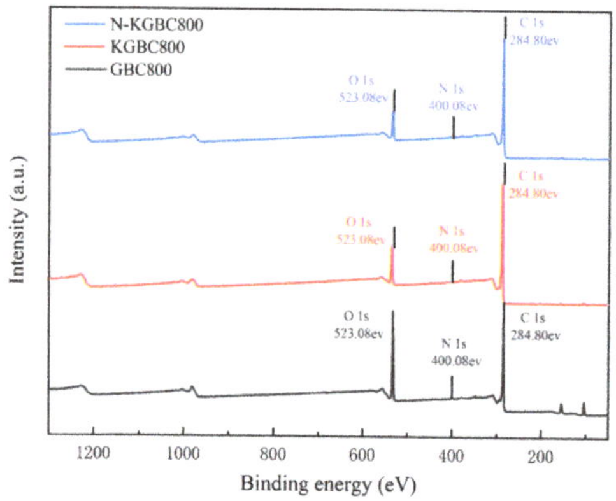

Figure 6. The XPS spectra of different biochar materials.

3.2. *Effect of pH on the Performance of Biochar for TC Adsorption*

The general solution pH of tetracycline is about 3.5, and the adsorption effect of TC with an initial pH in the range of 3~11 was investigated in this experiment. The biochar dosage was 0.15 g/L, the concentration of TC was 100 mg/L, the reaction temperature was 25 °C, and the reaction time was 12 h. The results of the adsorption process are presented in Figure 7.

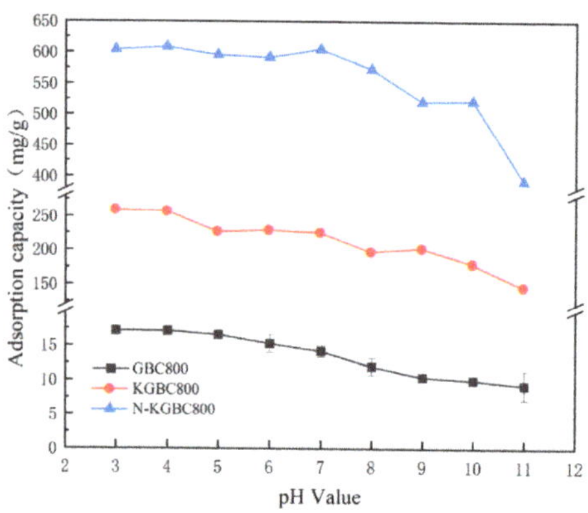

Figure 7. The effect of the pH on the impact of TC adsorption by biochar.

TC is a hydrophilic amphoteric substance, the pK_{a1}, pK_{a2}, and pK_{a3} of which are 3.3, 7.7, and 9.7, respectively. Under different pH conditions, the existing forms of TC are different. When the pH < 3.4, it is mainly TC^{+}; when 3.4 < pH < 7.6, it is mainly TC^{0}; when 7.6 < pH < 9.7, it is mainly TC^{-}; and when the pH > 9.7, it is mainly TC^{-} and TC^{2-}. Biochar has a zero-charge point (pH_{pzc}). When the pH of the solution < pH_{pzc}, the biochar surface is mainly positively charged, and vice versa, it is negatively charged [39].

With an increase in the solution pH, the adsorption of the TC on three biochars showed approximately the same trend, with the maximum adsorption around pH = 3 and then a gradually decreasing trend, consistent with the study by Xiang et al. [40]. The maximum equilibrium adsorption capacities of TC by GBC800, KGBC800, and N-KGBC800 are 16.73 mg/g, 259.33 mg/g, and 609.06 mg/g, respectively. At this time, the protonation degree of the surface functional groups of biochar is relatively high, and TC also exists in the form of TC^{+} or TC^{0} [41]. The adsorption capacity should be weak, but the experimental results show the maximum adsorption capacity. This indicates that electrostatic interaction is not the main factor affecting the adsorption process at this time. When the pH is 4–5 and 7–9, all three kinds of biochar show a certain degree of decline. This is because as the pH increases, the surface of the biochar has negative charges, and TC also exists in the form of TC^0 (3.4 < pH < 7.6), TC^- (7.6 < pH < 9.7), and TC^{2-} (pH > 9.7) in the solution. The electrostatic repulsion between biochar and TC increases, resulting in a significant decrease in adsorption capacity. When the pH > 9, the strong repulsion between negatively charged biochar and TC^-, TC^{2-} makes the adsorption more difficult, resulting in a lower adsorption efficiency of TC. When there is a chemical adsorption process in the adsorption system, when the alkalinity is too strong, the active sites on the surface of the biochar will be passivated, thereby causing a sharp decline in the adsorption capacity of the biochar [42].

3.3. Influence of Biochar Dosage on TC Adsorption Performance

The dosage of the three kinds of biochar (GBC800, KGBC800, and N-KGBC800) was 0.05, 0.10, 0.15, 0.20, 0.25, 0.30, 0.35, and 0.40 g/L, respectively; the concentration of TC was 100 mg/L; pH ≈ 3.5; the reaction temperature was 25 °C; the reaction time was 60 min; and the oscillation frequency was 150 r/min. The experimental results are shown in Figure 8.

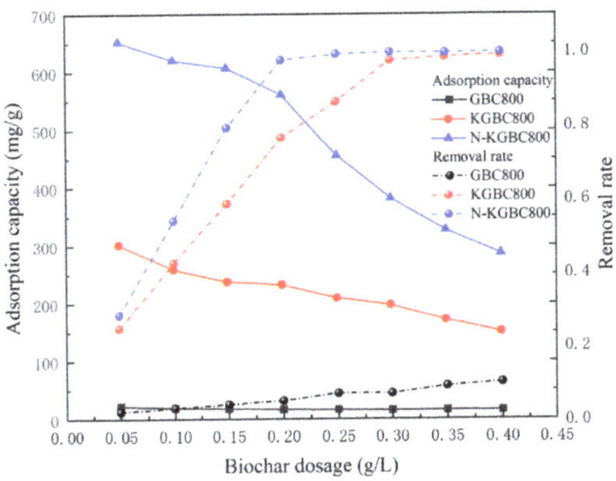

Figure 8. The influence of the biochar dosage on TC-adsorption effects.

An increase in the biochar dosage from 0.05 g/L to 0.4 g/L reduced the adsorption of TC by GBC800, KGBC800, and N-KGBC800, with a corresponding decrease from 21.4 to 14.98 mg/g. The adsorption of TC by GBC800, KGBC800, and N-KGBC800 decreased from 21.4 mg/g to 14.98 mg/g, 302.5 mg/g to 151.0 mg/L, and 652.8 mg/L to 286.8 mg/L, respectively; meanwhile, the removal rate increased from 2% to 10%, 24.7% to 99.05%, and 28.3% to 99.76%, respectively. The dosage of biochar showed a positive correlation with the TC removal rate within a certain range, while it showed a negative correlation with the adsorption amount. Deng Yu et al. [43] also indicated that the TC removal rate gradually increased with biochar addition. In contrast, the adsorption amount showed the opposite trend when preparing corn cob biochar for tetracycline adsorption. The primary rationale for this phenomenon is that an elevated dose of biochar can augment the number of functional groups and the number of efficacious active sites engaged in adsorption, which greatly enhances the possibility of contact between the biochar and TC in solution, resulting in almost close to 100% removal of TC in the solution [44]. In order to ensure the removal effect of TC, the best dosage of KGBC800 is about 0.25~0.3 g/L, and the best dosage of N-KGBC800 is about 0.15~0.2 g/L.

3.4. Adsorption Kinetic Modeling Analysis

Under the conditions of a TC concentration of 10, 50, and 100 mg/L; dosage of adsorbent of 0.15 mg/L; temperature of reaction of 25 °C; and adsorption duration of

720 min, the effects of TC adsorption by modified biochar were studied and kinetically analyzed based on the experimental data of this phase. Three types of kinetics, quasi-primary, quasi-secondary, and Elovich, and the model of intra-particle diffusion (the following equation), have been used to fit the relevant experimental data [45–47]. The kinetic fitting results are shown in Figure 9, with data points and coefficients in Table 3.

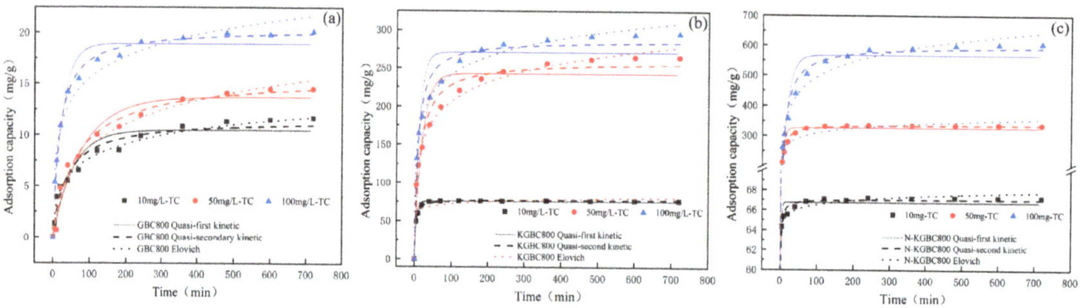

Figure 9. The adsorption kinetics model of TC adsorption by biochar: (**a**) GBC800, (**b**) KGBC800, and (**c**) N-KGBC800.

Table 3. The estimation of kinetic model parameters for TC adsorption on biochar.

Biochar	Kinetic Model	Parameters	Initial TC Concentration		
			10 mg/L	50 mg/L	100 mg/L
GBC800	Quasi-primary kinetic model	K_1/min^{-1}	0.0186	0.0124	0.0383
		$q_{e1,exp}/(\text{mg}\cdot\text{g}^{-1})$	11.3514	14.0823	16.9700
		$q_{e1}/(\text{mg}\cdot\text{g}^{-1})$	10.4794	13.7089	17.9445
		R^2	0.9156	0.9643	0.8851
	Quasi-secondary kinetic model	K_2/min^{-1}	0.0023	0.0009	0.0026
		$q_{e2,exp}/(\text{mg}\cdot\text{g}^{-1})$	11.3514	14.0823	16.9700
		$q_{e2}/(\text{mg}\cdot\text{g}^{-1})$	11.5208	15.6983	18.4010
		R^2	0.9624	0.9836	0.9399
	Elovich kinetic model	$\beta(\text{mg/g})$	0.4773	0.2928	0.3214
		R^2	0.9276	0.9568	0.9345
KGBC800	Quasi-primary kinetic model	K_1/min^{-1}	0.1830	0.0503	0.0799
		$q_{e1,exp}/(\text{mg}\cdot\text{g}^{-1})$	76.5766	264.4726	294.8649
		$q_{e1}/(\text{mg}\cdot\text{g}^{-1})$	75.4745	242.3914	270.7169
		R^2	0.9909	0.9002	0.8927
	Quasi-secondary kinetic model	K_2/min^{-1}	0.0046	0.0002	0.0004
		$q_{e2,exp}/(\text{mg}\cdot\text{g}^{-1})$	76.5766	264.2799	294.8469
		$q_{e2}/(\text{mg}\cdot\text{g}^{-1})$	77.4602	258.8755	286.5655
		R^2	0.9974	0.9700	0.9674
	Elovich kinetic model	$\beta(\text{mg/g})$	0.2281	0.0279	0.02954
		R^2	0.9529	0.9636	0.9508
N-KGBC800	Quasi-primary kinetic model	K_1/min^{-1}	0.6549	0.16286	0.063
		$q_{e1,exp}/(\text{mg}\cdot\text{g}^{-1})$	67.2072	336.0661	604.7147
		$q_{e1}/(\text{mg}\cdot\text{g}^{-1})$	66.7497	326.7898	566.6394
		R^2	0.9987	0.9665	0.9166
	Quasi-secondary kinetic model	K_2/min^{-1}	0.0624	0.0008	0.0001
		$q_{e2,exp}/(\text{mg}\cdot\text{g}^{-1})$	67.2072	336.0661	604.7147
		$q_{e2}/(\text{mg}\cdot\text{g}^{-1})$	67.0695	337.5815	598.7235
		R^2	0.9997	0.9969	0.9772
	Elovich kinetic model	$\beta(\text{mg/g})$	1.5842	0.0427	0.0135
		R^2	0.9796	0.9693	0.9804

Quasi-primary kinetic model:

$$\frac{dq_t}{d_t} = K_1(q_e - q_t) \tag{2}$$

Quasi-secondary kinetic model:

$$q_t = \frac{k_2 t q_e^2}{1 + k_2 t q_e} \tag{3}$$

Elovich kinetic model:

$$q_t = a + b \ln t \tag{4}$$

Intra-particle diffusion model:

$$q_t = K_{ip} t^{0.5} + C \tag{5}$$

q_e—The adsorbed amount of the adsorbent at equilibrium, mg/g.
q_t—The adsorbed amount of the adsorbent at time t, mg/g.
K_1—The quasi-primary kinetic adsorption rate constant, min^{-1}.
t—The adsorption time, min; K_2, the quasi-secondary kinetic modeling rate constant, g/(mg·min).
α—The initial adsorption rate, mg/(g·min).
b—The desorption constant.
K_{ip}—The intra-particle diffusion rate constant, and C is a constant.

As shown in Figure 9, given the fitting situation and the relevant fitting parameters in Table 3, the correlation coefficients R^2 of the quasi-secondary kinetic models for the three biosorption TCs at different initial concentrations were closer to 1. Furthermore, the average difference values between the theoretical saturated adsorption capacity q_{e1} obtained by the quasi-first-order kinetics simulation and the experimentally obtained saturated adsorption capacity $q_{e1,exp}$ for GBC800, KGBC800, and N-KGBC800 are −0.97 mg/g, 15.77 mg/g, and 47.80 mg/g, respectively. The average difference values between the theoretical saturated adsorption capacity q_{e2} obtained by the quasi-second-order kinetics simulation and the experimentally obtained saturated adsorption capacity $q_{e2,exp}$ are −0.43 mg/g, 3.30 mg/g, and 1.537 mg/g, respectively. This indicates that the fitting effect of the quasi-second-order kinetics is better than that of the quasi-first-order kinetics and the Elovich model. When the initial concentration of TC is 100 mg/L, the experimental saturated adsorption capacity of N-KGBC800 on TC is 604.71 mg/g, which shows the excellent adsorption capacity of the former compared with the saturated adsorption capacity of GBC800 16.97 mg/g. Although the overall nitrogen content on the surface of N-KGBC800 is slightly lower than that of GBC800, nitrogenous compound are slightly introduced to the surface of the biochar during the reaction process, and the newly introduced nitrogen-containing functional groups can increase the adsorption capacity, which is consistent with the study by Guo et al. [48]. This phenomenon may be attributed to the promotion of a reaction between nitrogen and oxygen groups on the surface of the biochar, which in turn facilitates the mutual transformation of nitrogen-containing functional groups [49]. The incorporation of nitrogen-containing functional groups diminishes the electronegativity of the C-layer, thereby enhancing its capacity to accept π-electrons and facilitating the formation of π–π bonds and Lewis acid–base interactions with organic molecules during adsorption. This ultimately results in an improvement in the adsorption capacity [50]. During the adsorption process, the uptake of TC by GBC800, KGBC800, and N-KGBC800 was the consequence of a synergistic interaction between physical and chemical adsorption. The chemical adsorption process was identified as the primary rate control step in this process. The adsorption process may be characterized by the sharing of electrons, the formation of valence bonds through electron exchange, and π–π interactions.

In order to gain further insight into the stages of the adsorption process, it is necessary to employ the intra-particle diffusion model for further analysis, and the model parameters are shown in Table 4. As illustrated in Figure 10, the adsorption process of TC by GBC800, KGBC800, and N-KGBC800 can be delineated into three linear phases. The initial stage is the outward transfer of TC molecules from the solution to the surface of the biochar. The second phase is the diffusion of these molecules from the surface into the internal pores. The subsequent phase is the equilibrium stage, during which chemical interactions occur between the TC molecules and the adsorption sites on the biochar. The model-fitting results indicate that the fitted curve of the intra-particle diffusion model consists of three linear cross-sections, which suggests that the adsorption rate is influenced by intra-particle diffusion. However, this does not represent the sole rate-controlling step, as other mechanisms, including liquid film diffusion, electrostatic interactions, hydrogen bonding, and interactions, are also involved in controlling the adsorption rate [49,51]. Given the constants C in Table 5, it can be observed that $C_{i,3} > C_{i,2} > C_{i,1}$ exists in all three steps, suggesting that liquid film diffusion may also be an influencing factor limiting the adsorption rate [52]. Prior to reaching adsorption equilibrium, TC molecules are transported from the solution to the surface of the biochar. This occurs initially via external mass transfer and liquid film diffusion, and subsequently, the molecules enter the inner pore structure and undergo adsorption on the internal surface of the biochar, a process that is controlled by intraparticle diffusion [39,53,54].

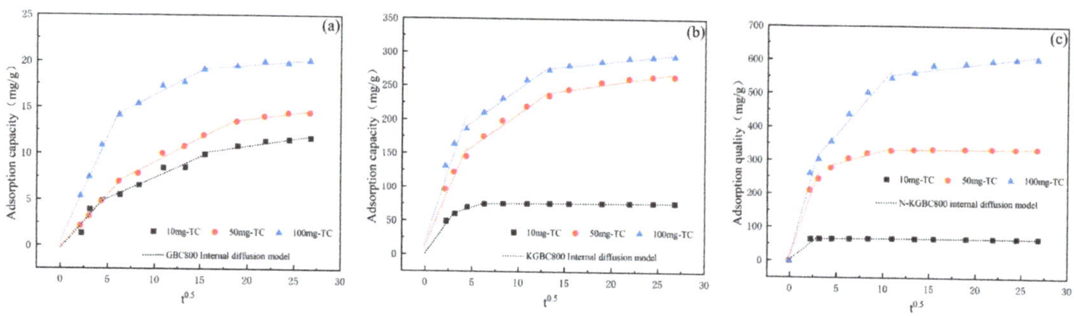

Figure 10. The intraparticle diffusion model of TC adsorption by biochar: (**a**) GBC800, (**b**) KGBC800, and (**c**) N-KGBC800.

Table 4. The parameter estimation of the internal diffusion model for TC adsorption by biochar.

Biochar	Parameters	Initial TC Concentration		
		10 mg/L	50 mg/L	100 mg/L
GBC800	$K_{GBC,1}$ [mg/(g·min)]	1.1472	1.1146	2.2700
	$C_{GBC,1}$ (mg/g)	−0.2934	−0.1857	0.2270
	R_1^2	0.8761	0.9948	0.9946
	$K_{GBC,2}$ [mg/(g·min)]	0.4552	0.5209	0.5179
	$C_{GBC,2}$ (mg/g)	2.8389	3.8104	11.0938
	R_2^2	0.9588	0.9835	0.9625
	$K_{GBC,3}$ [mg/(g·min)]	0.1546	0.1309	0.0856
	$C_{GBC,3}$ (mg/g)	7.7316	11.1160	17.8559
	R_3^2	0.8967	0.8797	0.8352

Table 4. *Cont.*

Biochar	Parameters	Initial TC Concentration		
		10 mg/L	50 mg/L	100 mg/L
KGBC800	$K_{KGBC,1}$ [mg/(g·min)]	19.4498	33.3154	42.9313
	$C_{KGBC,1}$ (mg/g)	1.2772	8.9329	14.5346
	${R_1}^2$	0.9702	0.9425	0.9098
	$K_{KGBC,2}$ [mg/(g·min)]	4.5743	9.8530	9.8794
	$C_{KGBC,2}$ (mg/g)	46.8038	109.2741	146.3773
	${R_2}^2$	0.8133	0.9556	0.9805
	$K_{KGBC,3}$ [mg/(g·min)]	0.0784	2.1578	1.5079
	$C_{KGBC,3}$ (mg/g)	74.5739	210.8439	256.5668
	${R_3}^2$	0.8758	0.8816	0.9202
N-KGBC800	$K_{N\text{-}KGBC,1}$ [mg/(g·min)]	22.1605	80.5541	99.6993
	$C_{N\text{-}KGBC,1}$ (mg/g)	3.3464	6.9789	8.5941
	${R_1}^2$	0.8522	0.9488	0.9493
	$K_{N\text{-}KGBC,2}$ [mg/(g·min)]	0.2413	8.0325	31.8177
	$C_{N\text{-}KGBC,2}$ (mg/g)	64.6160	249.5722	217.5784
	${R_2}^2$	0.9359	0.8337	0.9533
	$K_{N\text{-}KGBC,3}$ [mg/(g·min)]	0.0132	0.2524	3.4903
	$C_{N\text{-}KGBC,3}$ (mg/g)	66.8827	329.9344	517.9382
	${R_3}^2$	0.8348	0.8519	0.8541

Table 5. The estimation of isothermal parameters for TC adsorption by biochar.

Biochar	Model	Parameters	Reaction Temperature (T)		
			288 K	298 K	308 K
GBC800	Langmuir	q_m (mg/g)	16.1679	20.8750	24.1704
		K_L (L/mg)	1.0074	0.6959	0.3604
		R_L	0.0098	0.0141	0.0269
		R^2	0.9384	0.9220	0.9425
	Freundlich	K_F [mg·g^{-1}·(L·mg)$^{-1/n}$]	9.4458	10.2429	10.1680
		$1/n$	0.1476	0.1733	0.2011
		R^2	0.9235	0.9197	0.8894
	Temkin	b_T (kJ/mol)	1.2296	0.9503	0.7765
		K_T (L/mg)	90.1630	38.0677	15.2946
		R^2	0.9250	0.9768	0.9371
KGBC800	Langmuir	q_m (mg/g)	268.4533	285.3526	306.5144
		K_L (L/mg)	1.2706	1.4866	1.3825
		R_L	0.0078	0.0066	0.0071
		R^2	0.9886	0.9727	0.9823
	Freundlich	K_F [mg·g^{-1}·(L·mg)$^{-1/n}$]	149.3361	158.7649	167.5674
		$1/n$	0.1638	0.1691	0.1739
		R^2	0.8550	0.8649	0.8711
	Temkin	b_T (kJ/mol)	0.0717	0.0680	0.0648
		K_T (L/mg)	69.5907	65.3959	60.3399
		R^2	0.9377	0.9456	0.9523
N-KGBC800	Langmuir	q_m (mg/g)	630.5727	660.6366	688.3172
		K_L (L/mg)	1.0369	0.6756	0.6062
		R_L	0.0095	0.0145	0.0162
		R^2	0.9973	0.9917	0.9868
	Freundlich	K_F [mg·g^{-1}·(L·mg)$^{-1/n}$]	301.1075	300.6277	302.3327
		$1/n$	0.1987	0.2198	0.2306
		R^2	0.8163	0.8801	0.8787
	Temkin	b_T (kJ/mol)	0.0262	0.0261	0.0253
		K_T (L/mg)	25.6450	22.1160	18.4779
		R^2	0.9219	0.9549	0.9534

3.5. Adsorption Isotherm Modeling Analysis

The experimental data of TC adsorption in the solution for GBC800, KGBC800, and N-KGBC800 were isothermally fitted using the Langmuir model, Freundlich model, and Temkin model (as follows) in turn at temperatures of 288 K, 298 K, and 308 K, with a biochar dosage of 0. 15 mg/L and an adsorption time of 720 min. The isothermal adsorption fitting curves are shown in Figure 11, and the fitting parameters are shown in Table 5.

The Langmuir isothermal adsorption model:

$$\frac{C_e}{q_e} = \frac{C_e}{q_m} + \frac{1}{q_m K_L} \tag{6}$$

Freundlich isothermal adsorption model:

$$q_e = k_f \times C_e^{\frac{1}{n}} \tag{7}$$

The Temkin isothermal adsorption model:

$$q_m = \frac{RT}{b_T} \ln K_T + \frac{RT}{b_T} \ln C_e \tag{8}$$

q_e—The adsorption capacity of the adsorbent at the adsorption equilibrium, mg/g.
q_m—The theoretical saturated adsorption capacity of the adsorbent for pollutants, mg/g.
C_e—The pollutant concentration in the solution at the adsorption equilibrium, mg/L.
K_L—The Langmuir adsorption constant, L/mg, indicating the adsorption strength of the adsorbent for the adsorbate.
K_f—The Freundlich adsorption constant, (mg/g)·(L/mg)$^{1/n}$, indicating the adsorption capacity of the adsorbent, which is related to the properties and dosage of the adsorbent and the reaction temperature.
n—The nonlinear adsorption index, reflecting the heterogeneity of the adsorption sites on the surface of the adsorbent.
b_T and K_T are the constants of the Temkin model; R is the universal gas constant, J/(mol·K); and T is the Kelvin temperature, K.

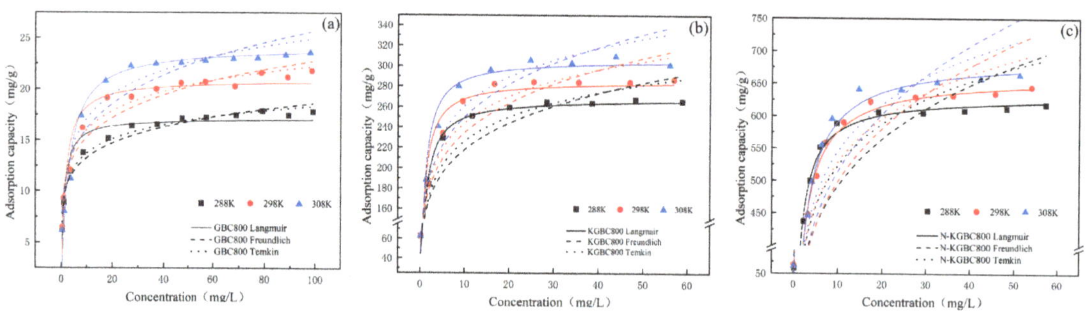

Figure 11. The isothermal modeling of TC adsorption by biochar: (**a**) GBC800, (**b**) KGBC800, and (**c**) N-KGBC800.

The fitting results indicate that the process of TC adsorption by GBC800, KGBC800, and N-KGBC800 at different temperatures is best described by the Langmuir model, followed by the Temkin and Freundlich models. This suggests that the adsorption process is biased toward monolayer adsorption, whereby a homogeneous monolayer of TC molecules is formed on the surface of the biochar, with chemical adsorption dominating the process [55]. The adsorption reaction is a process whereby the adsorbate continuously migrates from the main body of the aqueous phase to the solid adsorbent phase. In contrast, the concentration gradient between the two phases provides the main driving force for this migration. The adsorbent surface is energetically homogeneous; the adsorption sites are equipotent, there is no interaction between TC molecules, and the final equilibrium state of the adsorption is dynamic [56]. The fitting results of the Langmuir isothermal model indicate that the theoretical maximum adsorption capacities of GBC800, KGBC800, and N-KGBC800 at 25 °C are 16.16 mg/g, 268.45 mg/g, and 630.57 mg/g, respectively. The experimentally determined values are 16.82 mg/g, 265.48 mg/g, and 617.91 mg/g, respectively, which are basically consistent. This further confirms the applicability of the Langmuir model.

The correlation coefficient (R^2) of the Temkin isothermal model was found to be greater than 0.9 for all cases, indicating that the adsorption process of TC by GBC800, KGBC800, and N-KGBC800 was predominantly chemisorption-driven and that strong intermolecular forces played a pivotal role in the TC adsorption process [54]. b_T is an index of the reactive chemical interactions; the KGBC800 and N-KGBC800 b_T values are much lower than those of GBC800, indicating that the former two play a weaker role than the latter in the whole adsorption process by chemisorption [57]. This phenomenon may be attributed to the surface structure of GBC800 biochar. The lower specific surface area and less developed porosity of GBC800 result in a diminished role for physical adsorption in the TC-adsorption process. Consequently, the functional groups of the carbon surface become the primary means of contact between GBC800 and TC throughout the adsorption process.

3.6. Thermodynamic Analysis of Adsorption

A comprehensive analysis of the adsorption behavior of modified biochar on TC requires the thermodynamic characterization of the aforementioned process by the three biochars at three distinct temperatures: 298 K (25 °C), 308 K (35 °C), and 318 K (45 °C). The equations used are shown in (9)–(11), the fitting results are indicated by Figure 12, and the thermodynamic parameters are listed in Table 6.

$$\Delta G^{\theta} = -RT \times \ln k \tag{9}$$

$$\ln k = \frac{\Delta S^{\theta}}{R} - \frac{\Delta H^{\theta}}{RT} \tag{10}$$

$$\Delta G^{\theta} = \Delta H^{\theta} - T\Delta S^{\theta} \tag{11}$$

ΔG^{θ}—The Gibbs free energy, kJ/mol.
ΔH^{θ}—The enthalpy, kJ/mol.
ΔS^{θ}—The entropy, kJ/(mol·K).
T—The absolute temperature, K.
K—The Langmuir adsorption equilibrium constant, L/mg.
R—The universal gas constant, J/(mol·K), R = 8.314 × 10^{-3}, KJ/(mol·K).

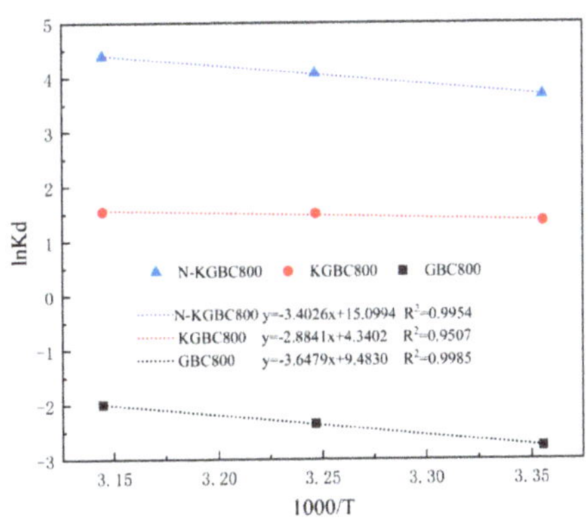

Figure 12. The thermodynamic modeling of TC adsorption on biochar.

Table 6. The estimation of thermodynamic parameters for TC adsorption on biochar.

Biochar	Parameters	Temperature (K)		
		298	308	318
GBC800	ΔG_0 (kJ/mol)	6.8480	6.0157	5.2736
	ΔH_0 (kJ/mol)		30.2920	
	ΔS_0 (J/mol·K)		0.0787	
	R^2		0.9989	
KGBC800	ΔG_0 (kJ/mol)	−3.3619	−3.8494	−4.0779
	ΔH_0 (kJ/mol)		27.2631	
	ΔS_0 (J/mol·K)		0.0358	
	R^2		0.9782	
N-KGBC800	ΔG_0 (kJ/mol)	−9.0964	−10.4260	−11.6034
	ΔH_0 (kJ/mol)		28.232	
	ΔS_0 (J/mol·K)		0.1253	
	R^2		0.9998	

The enthalpy changes (ΔH) of TC adsorption of GBC800, KGBC800, and N-KGBC800 were 30.2920 kJ/mol, 27.2631 kJ/mol, and 28.232 kJ/mol, respectively, and their enthalpy changes were positive, indicating that the adsorption process is a thermally active reaction and that it is more favorable for adsorption at higher temperatures. This is in agreement with the results of the adsorption isotherm fitting. In addition, according to the results of Garba et al. [58], the enthalpy change ΔH being lower than 20 kJ/mol adsorption is mainly a physical reaction. Vice versa, it is dominated by chemical action, and the adsorption behaviors of GBC800, KGBC800, and N-KGBC800 are all greater than 20 kJ/mol, indicating that the adsorption behaviors of all three to TC in water are heat-absorbing reactions and dominated by chemical adsorption. Significant adsorption between adsorbent and adsorbate can occur when the free energy of the adsorption ΔG value is negative [59]. The adsorption of TC on the surface of the adsorbent was observed to be spontaneous in all cases, and the absolute value of ΔH was found to increase with the rise in temperature. This indicates that the spontaneous tendency of the process of adsorption of TC by N-KGBC800 and KGBC800 increases with the rise in ambient temperature and that the temperature rise is beneficial for the adsorption process. The entropy changes (ΔS) of GBC800, KGBC800, and N-KGBC800 in the adsorption of TC were 0.0296 J/(mol·K), 0.0357 J/(mol·K), and 0.1008 J/(mol·K), respectively. These values indicate that the adsorption process was less reversible. The adsorption process can be described as a dynamic equilibrium between adsorption and desorption, with a positive adsorption value that is significantly higher than the desorption value [60]. Furthermore, as the adsorption process progresses, the solid–liquid interface disorder gradually increases, stabilizing the adsorption process.

3.7. Adsorption Mechanism Analysis

In order to better explore the adsorption mechanism of modified biochar on TC, the biochar after adsorption and adsorption saturation was placed in 0.05 mol/L NaOH solution for analysis for 3 h and rinsed with deionized water repeatedly until pH $\approx$ 7. After being desorbed, separated, and dried, the biochar was obtained. The BET and FT-IR characterization of the biochar were performed to compare the changes of pore structure and surface functional groups before and after the adsorption of TC by biochar.

As illustrated in Figure 13, the BET characterization of biochar following the adsorption of tetracycline yielded the following results. The number of micropores in the adsorbed biochar was significantly reduced, and there was no significant change in mesopores. When comparing Tables 1 and 7, the specific surface areas of GBC800, KGBC800, and N-KGBC800 after adsorption are 9.634 m^2/g, 1035.101 m^2/g, and 784.946 m^2/g, respectively, and the total pore volumes are 0.03658 cm^3/g, 0.8192 cm^3/g, and 0.6179 cm^3/g, respectively. Compared with before adsorption, the specific surface area of GBC800 decreased by 18.19%, and the total pore volume decreased by 12.67%; the specific surface area of KGBC800 decreased by 29.96%, and the total pore volume decreased by 14.58%; the specific surface area of N-KGBC800 decreased by 62.45%, and the total pore volume decreased by 51.42%. Kowalczyk et al. [61] proposed that carbon materials with pore diameters of approximately 1.4–1.6 nanometers may cause the molecular-sieving effect of TC. Therefore, the reduction in the surface area and pore volume of KGBC800 and N-KGBC800 mainly occurs when the pore diameter is smaller than the dynamic diameter of TC, which may be due to the blocking of pores by TC molecules. Generally speaking, pores larger than 1–2 times the diameter of the adsorbate molecule show good adsorption performance [25]. The dynamic diameter of TC molecules is approximately 1.27 nm, and the average pore diameters of GBC800 are 16.56 nm. Moreover, the BET and SEM of GBC800 characterization results show that its pores are poorly developed. Therefore, the adsorption capacity of GBC800 for TC is weak. The average pore diameters of KGBC800 and N-KGBC800 are 2.5956 nm and 2.4325 nm, respectively. Tetracycline molecules can easily enter the pore structures of the two biochars, and their microporous structures can be fully utilized as adsorption sites for tetracycline. Thus, one of the main mechanisms for the removal of tetracycline from water by modified biochar is pore-filling.

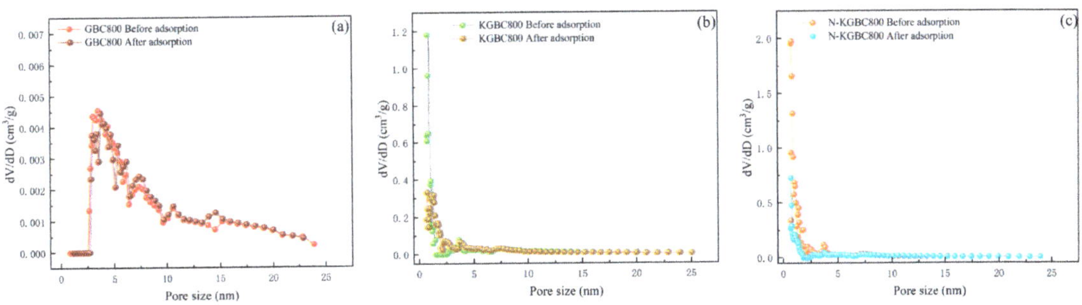

Figure 13. The distribution of front and rear absorption holes: (**a**) GBC800, (**b**) KGBC800, and (**c**) N-KGBC800.

Table 7. BET parameters after TC adsorption on biochar.

Parameters	Unit	GBC800	KGBC800	N-KGBC800
S_{BET}	m^2/g	9.6	1035	785
V_{Total}	cm^3/g	0.04	0.82	0.62
V_{Mic}	cm^3/g	0.00	0.17	0.19
L_0 [1]	nm	13.50	3.16	3.15

Note: [1] The average pore size.

Comparing the FT-IR patterns of GBC800, KGBC800, and N-KGBC800 before and after adsorption (Figure 14), the data demonstrate that the absorption peaks of all three types of biochars at approximately 3425 cm^{-1} are all shifted to the right and exhibit a reduction in peak intensity. This suggests that the oxygen-containing functional groups present in the biochar were involved in the adsorption reaction, potentially interacting with the hydroxyl, amino, carboxyl, and other functional groups on the TC molecule to form hydrogen bonds or n–π interactions. It is similarly conceivable that the nitrogen-containing functional groups on the surface of the biochar may have undergone a chemical bonding process with the TC, which could have resulted in a change in the position of its telescopic vibration [62–64]. The C=C and C=O (near 1618 cm^{-1}) absorption peaks of N-KGBC800 are significantly shifted, indicating that there may be a π–π stacking interaction between the aryl ring structure on N-KGBC800 and the TC [65]. At C-N (1094 cm^{-1}), the peak intensities before and after the adsorption of GBC800 and N-KGBC800 changed significantly, while KGBC800 remained almost unchanged, which may be attributed to the Lewis acid–base interactions between nitrogen-containing functional groups on the surface of the biochar and the TC molecules. Furthermore, the amino group, which is formed by the deprotonation of tetracycline, can undergo a reaction with the benzene ring present in biochar, resulting in the formation of a cationic–π interaction [66]. When the solution pH is higher than the zero-point charge of biochar, the oxygen-containing functional groups on the surface of biochar undergo deprotonation, thereby acquiring a negative charge. This promotes the electrostatic interaction between biochar and tetracycline [67,68].

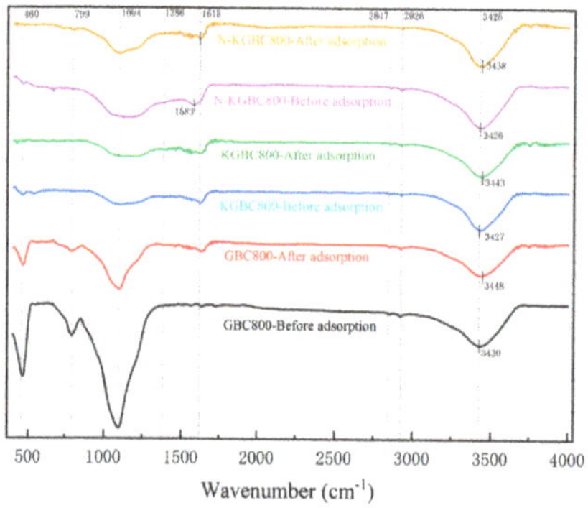

Figure 14. A comparison of FT-IR changes before and after TC adsorption on biochar.

Based on the above-mentioned discussion, the adsorption mechanisms of GBC800, KGBC800, and N-KGBC800 for TC in the adsorption are shown in Figure 15. The adsorption processes of the three biochars for TC are basically the same, and they are all the results of the combined effect of physical and chemical actions. The difference lies in the different intensities of different actions. Specifically, TC in the solution rapidly diffuses to the surface of the biochar, resulting in liquid film diffusion, and continuously fills the macropores, mesopores, and micropores of the biochar. The TC molecules entering the pores form electrostatic interactions with the charged groups on the surface of the biochar, interact with oxygen-containing functional groups such as –OH to form hydrogen bonds or n–π interactions, have Lewis acid–base interactions with nitrogen-containing functional groups, and form π–π stacking or cation–π interactions between the aromatic ring structure of the carbon itself and TC.

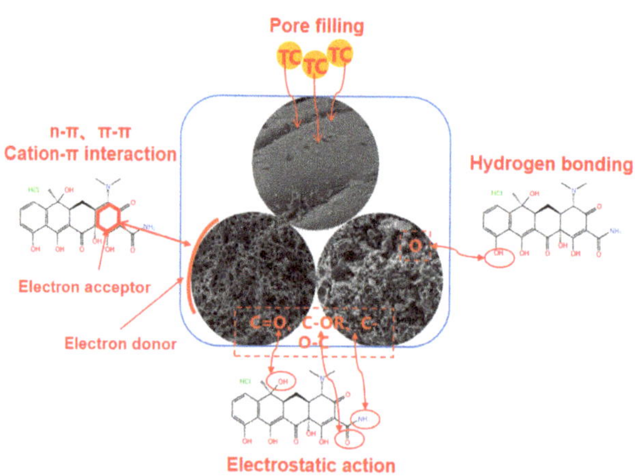

Figure 15. An adsorption mechanism diagram of TC adsorption on modified biochar.

3.8. Experimental Adsorption of Tetracycline on Biochar Fixed-Bed and Adsorption Model Analysis

After modification, the adsorption and removal effects of rapeseed straw biochar (KGBC800 and N-KGBC800) on TC in water were greatly improved compared to those of unmodified GBC800 of rapeseed straw, and they can effectively remove TC in water. Adsorption can be divided into batch and fixed-bed adsorption according to different operation modes. In practical applications, batch adsorption has many limitations, such as influent water, reaction site, operation management, etc. In the fixed-bed adsorption process, the length of the effective mass-transfer zone moves along the liquid flow to the entire bed depth until the fixed bed is fully utilized [69]. Compared with batch adsorption, the fixed-bed reactor is more economical, operable, and has a continuous processing capacity, thus becoming the mainstream wastewater treatment method. This chapter investigates the dynamic adsorption behaviors of three TC on biochars in pure water and salt water and uses the Thomas model and Yoon–Nelson mass-transfer model to conduct further analysis of dynamic adsorption.

3.8.1. Biochar Fixed-Bed Adsorption of Tetracycline Experiment

The penetration point in the dynamic adsorption process is defined as $C_t/C_0 = 0.05$, indicating that the TC effective concentration is approximately 5% of the initial concentration. The adsorption saturation point is $C_t/C_0 = 0.95$, representing an effective TC concentration of approximately 95% of the initial concentration. As shown in Figure 16, the penetration points of GBC800, KGBC800, and N-KGBC800 in the dynamic adsorption process under pure water conditions were 10 min (when the adsorption penetration point < 10 min, it was set as 10 min), 70 min, and 270 min, respectively, and the adsorption penetration points under brine conditions were 10 min, 40 min, and 240 min, respectively. As the adsorption sites of biochar in the adsorption column were gradually consumed, the concentration of TC in the effluent gradually increased and finally reached saturation. The saturation points of the dynamic adsorption process of GBC800, KGBC800, and N-KGBC800 were 60 min, 210 min, and 460 min in pure water, and the adsorption penetration points were 40 min, 190 min, and 270 min in brine, respectively. The adsorption penetration points under saline conditions were 40 min, 190 min, and 390 min, respectively.

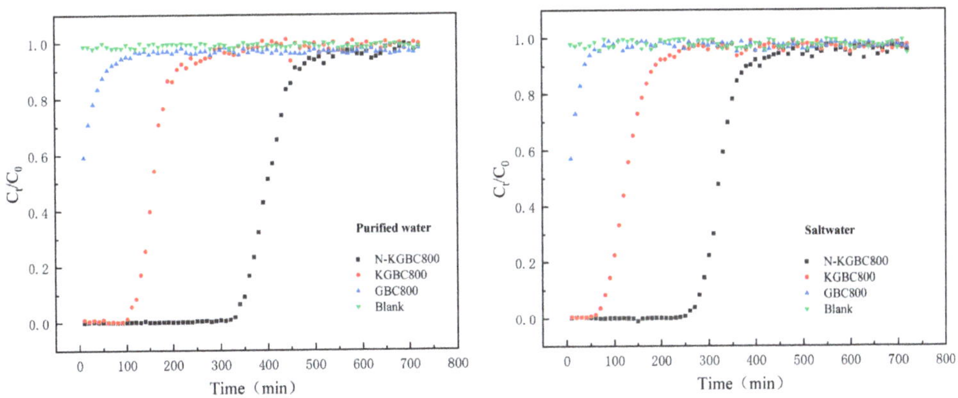

Figure 16. The adsorption effect of biochar fixed bed on TC under different conditions.

The adsorption properties of GBC800, KGBC800, and N-KGBC800 showed obvious differences in the process of dynamic adsorption of TC. The saturated adsorption capacities of GBC800, KGBC800, and N-KGBC800 were calculated by integral method and were 15.11 mg/g, 238.37 mg/g, and 601.03 mg/g, respectively, under pure water conditions. Under saline conditions, the results were 11.63 mg/g, 211.93 mg/g and 573.42 mg/g, respectively. A range of inorganic ions present in brine exert a discernible inhibitory effect on the adsorption performance of the fixed bed of biochar. Moreover, a considerable quantity of Na^+ in the solution has been found to exert a certain inhibitory effect on the adsorption of tetracycline. Furthermore, the adsorption of tetracycline can be significantly disrupted by the presence of Mg^{2+} and Ca^{2+}. The presence of an electrostatic effect between Ca^{2+} and the adsorbent of biochar, on the one hand, and the presence of Ca^{2+} with the adsorbent of biochar, on the other hand, results in competition between Ca^{2+} and TC for hydroxyl groups on the surface of biochar. This leads to a reduction in the number of adsorption sites on the surface of the adsorbent [70]. Zhao et al. [71] demonstrated that divalent cations exhibit higher ionic strengths, resulting in a greater salting-out effect in solution than the crowding-out effect. Additionally, the higher covalency of divalent cations reduces the number of adsorption sites available on the surface of the adsorbent, consequently leading to a decline in the adsorption capacity for TC. In contrast, the anions chloride (Cl^-) and sulfate ($SO_4{}^{2-}$) have been observed to exert a negligible effect on the TC-adsorption capacity of biochar [56].

3.8.2. Biochar Fixed-Bed Adsorption Modeling

The Thomas model is a descriptive and evaluative tool that can be used to analyze the penetration curve of pollutants in a fixed bed, as well as to assess the adsorption capacity of the adsorbent in relation to the pollutants. It can also be employed to estimate the equilibrium adsorption amount of the adsorbent and the adsorption rate constants [72–74]. The Yoon–Nelson adsorption model predicts the trend of the whole adsorption curve and fits the typical penetration curves with an S-shape very well, mainly predicting 50% of the penetration time. The pertinent curves are illustrated in Figure 17, and the pertinent data are enumerated in Table 8.

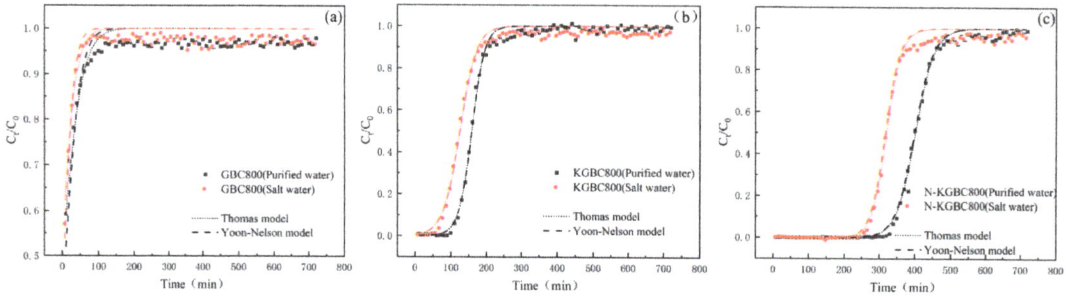

Figure 17. The fitting results of the Thomas model and Yoon–Nelson model. (**a**) GBC800, (**b**) KGBC800 and (**c**) N-KGBC800.

Table 8. The fitting parameters estimation of the Thomas model and the Yoon–Nelson model for the adsorption of TC by a biochar fixed bed.

Biochar	Conditional	Thomas Model Parameters				Yoon–Nelson Model Parameters		
		K_T (mL/(mg·min))	q_e (mg/g)	q_{exp} (mg/g)	R^2	K_Y (L/min)	τ (min)	R^2
GBC800	Purified water	0.3664	16.7136	15.1134	0.8842	0.0482	8.3346	0.8711
	Salted water	0.5024	15.4731	11.6302	0.8513	0.0716	7.8948	0.8321
KGBC800	Purified water	0.4325	262.0138	238.3792	0.9949	0.0424	158.2122	0.9872
	Salted water	0.5212	230.5852	211.9348	0.9897	0.0540	128.0997	0.9818
N-KGBC800	Purified water	0.3721	599.7145	601.0326	0.9973	0.0379	399.2547	0.9971
	Salted water	0.4690	585.2991	573.4281	0.9938	0.0479	321.3415	0.9932

The fitting coefficients R^2 for TC adsorption under pure water conditions were 0.8842, 0.9949, and 0.9973 for GBC800, KGBC800, and N-KGBC800, respectively, and 0.8513, 0.9897, and 0.9938 for the adsorption of TC under brine conditions, respectively, which shows that the Thomas model was better able to describe, to a certain extent, the fixed-bed adsorption process of TC by KGBC800 and N-KGBC800. At the same time, it is slightly less effective in describing GBC800. In the Thomas model, there is an inverse relationship between the magnitude of K_T obtained for the same adsorbent under different reaction conditions and the equilibrium adsorption amount q_e. In contrast, the K_T between different adsorbents reflects the strength of the adsorbent to the adsorbate per unit of time [75]. Observing the K_T values obtained from the fitting, K_T (pure water) < K_T (brine) existed for all three biochars, indicating that the equilibrium adsorption of TC by the same biochar in brine was lower than that in pure water conditions, and the maximum adsorption q_e calculated from the fitting proved exactly the same. The maximum adsorption capacities of GBC800 predicted by the Thomas model under pure water and saltwater conditions were 16.71 mg/g and 15.47 mg/g, respectively. The errors between them and the actual measured values of 15.11 mg/g and 11.63 mg/g were 10.2% and 21.8%, respectively. Both errors were greater than 10%, which might be because the adsorption capacity of GBC800 for TC was limited and the high concentration of TC caused certain errors in the experimental results. The maximum predicted adsorption capacities of KGBC800 in pure water and saltwater were 262.01 mg/g and 230.58 mg/g, respectively. The errors between them and the experimentally calculated values of 238.37 mg/g and 211.93 mg/g were 9.02% and 8.08%, respectively. The maximum predicted adsorption capacities of N-KGBC800 in pure water and saltwater were 599.71 mg/g and 581.29 mg/g, respectively. The errors between them and the experimentally calculated values of 601.03 mg/g and 573.42 mg/g were 0.21% and 1.3%, respectively. The errors between the predicted and calculated values of the adsorption capacities of KGBC800 and N-KGBC800 for TC were both less than 10%, indicating that the Thomas model can describe the dynamic adsorption process of TC by the modified biochar in this study, and the estimated saturated adsorption capacity has a high degree of feasibility.

GBC800, KGBC800, and N-KGBC800 have the Yoon–Nelson model coefficient R^2 of absorption of TC under pure and salty water conditions, as shown in Table 8, of 0.8711, 0.9872, 0.9971, and 0.8321, respectively, and 0.9818, 0.9932, and R^2 for KGBC 800 and KGBC 800 are greater than 0.98, respectively. The parameters in the Yoon–Nelson model reflect the time required to penetrate 50% of the adhesive. GBC800 is calculated at 8.33 min and 7.89 min, respectively, in pure and saltwater conditions, whereas 50% of the absorption time is measured in 10 min. KGBC800's predicted values of 158 min and 128 min, respectively, are almost consistent with actual 50% absorbent penetration times of 150 to 160 min and 120 to 130 min. In the Yoon–Nelson model, the N-KGBC800 is 399 min and 321 min in pure and saltwater conditions, respectively, and is consistent with the values measured in actual experiments of 390 to 400 min and 320 to 330 min. The above experimental results suggest that the Yoon–Nelson adsorption model can simulate the process of this modified biocarbon dynamic absorption TC. However, because the model simplifies the actual conditions and ignores the structural characteristics of the adsorbent and the geometry of the fixed bed, there is a certain reasonable error.

4. Conclusions

(1) Using rape straw as the raw material and potassium bicarbonate ($KHCO_3$) and urea ($CO(NH_2)_2$) as the modification reagents, they were mixed in a certain proportion to form the raw pyrolysis materials. The unmodified biochar GBC800, $KHCO_3$-modified biochar KGBC800, and ($KHCO_3$)/($CO(NH_2)_2$) co-modified biochar N-KGBC800 were, respectively, prepared by the one-pot method at 800 °C. The surface of GBC800 is smooth and dense, without an obvious pore structure, and the specific surface area and total pore volume are small. After $KHCO_3$ activation, the surface of biochar KGBC800 showed an obvious coral-like three-dimensional carbon skeleton, the number of micropores and

specific surface area were significantly increased, and the adsorption performance of TC was slightly improved. However, with the addition of $KHCO_3$, the N in biochar is sharply reduced, and nitrogen-containing functional groups can be effectively introduced after urea co-pyrolysis. N-KGBC800 prepared by $KHCO_3/CO(NH_2)_2$ co-modification has a coral-like three-dimensional carbon skeleton similar to KGBC800. There are more carbon clusters, and the carbon layers have significant changes and interlayer gaps, showing a multistage porous structure. The adsorption performance of the three biochars was the best when the pH was 3~4, and the adsorption equilibrium was reached within 720 min.

(2) The kinetic models of tetracycline adsorption by the three biochar types show that they conform to the quasi-second-order kinetic model, and the adsorption process results from both physical and chemical effects. The adsorption isotherm model showed that the adsorption of tetracycline by the three biochar types was consistent with the Langmuir isotherm model, preferring monolayer adsorption, and the intermolecular force played a key role in the adsorption process. The adsorption thermodynamics showed that the adsorption of tetracycline by the three biochar was endothermic. By comparing the BET and FT-IR characterization results of GBC800, KGBC800, and N-KGBC800 before and after adsorption, the adsorption mechanisms include pore filling, electrostatic interaction, hydrogen bonds, $n-\pi$ interaction, Lewis acid–base interaction with nitrogen-containing functional groups, and $\pi-\pi$ stacking or cation $-\pi$ interaction between the aromatic ring structure of the carbon itself and TC.

(3) In the biochar fixed-bed adsorption experiment, the biochar with stronger adsorption capacity in the static adsorption performed better in the fixed-bed adsorption, and the inorganic ions in brine inhibit or interfere with the adsorption process to a certain degree, resulting in a decrease in adsorption capacity. Overall, the Thomas model and Yoon–Nelson model can describe the adsorption of tetracycline by KGBC800 and an N-KGBC800 biochar fixed bed.

Author Contributions: Conceptualization, Z.Z. and C.T.; methodology, C.T.; software, H.W.; validation, H.W. and P.G.; formal analysis, C.T., M.Z. and Z.Z.; resources, W.X. and Y.C.; data curation, M.Z.; writing—original draft preparation, C.T.; writing—review and editing, C.T.; visualization, Z.Z. and P.G.; supervision, Y.C.; project administration, Y.C. and W.X.; funding acquisition, Y.C. and W.X. All authors have read and agreed to the published version of the manuscript.

Funding: This research was funded by the open fund of the Engineering & Technology Center of Groundwater Pollution Control for Environmental Protection in Sichuan (2024-01), the open fund of the Sichuan Provincial Engineering Research Center of City Solid Waste Energy and Building Materials Conversion and Utilization Technology (No. GF2023YB002), and the Chengdu Institute of Technology 2023 Provincial Da Chuang Project (S202311116036).

Data Availability Statement: Data are available upon request from the authors.

Conflicts of Interest: Author Yiyang Chen was employed by the company Sichuan Ecology and Environment Industry Group Co., Ltd. The remaining authors declare that the research was conducted in the absence of any commercial or financial relationships that could be construed as a potential conflict of interest.

References

1. Li, J.; Wang, Z. The road to agricultural power for small farmers in large countries: Restraint and cracking. *Southwest Financ.* **2023**, *12*, 41–53.
2. Huang, L.; Zhu, Y.; Wang, Q.; Zhu, A.; Liu, Z.; Wang, Y.; Allen, D.T.; Li, L. Assessment of the effects of straw burning bans in China: Emissions, air quality, and health impacts. *Sci. Total Environ.* **2021**, *789*, 147935. [CrossRef] [PubMed]
3. Sun, Y.; Li, X.; Hu, X.; Xu, X.; Yang, W.; Sun, R.; Xue, Y.; Xu, Z. The situation, utilization status and development suggestion of rape straw resources in China. *Chin. J. Oil Crops* **2024**, 1–8. [CrossRef]
4. Li, L.; Chen, B.; Yan, G.; Gao, G.; Xu, K.; Xie, T.; Zhang, F.; Wu, X. Proposed Strategies and Current Progress of Research and Utilization of Oilseed Rape Germplasm in China. *J. Plant Genet. Resour.* **2020**, *21*, 1–19.
5. Doula, M.K.; Sarris, A.; Hliaoutakis, A.; Kydonakis, A.; Papadopoulos, N.S.; Argyriou, L. Building a strategy for soil protection at local and regional scale-the case of agricultural wastes landspreading. *Environ. Monit. Assess.* **2016**, *188*, 141. [CrossRef]

6. Li, G.; Feng, Y.; Zhu, W.; Zhang, X. Enhanced adsorptive performance of tetracycline antibiotics on lanthanum modified diatomite. *Korean J. Chem. Eng.* **2015**, *32*, 2109–2115. [CrossRef]
7. Shao, Y.; Wang, Y.; Yuan, Y.; Xie, Y. A systematic review on antibiotics misuse in livestock and aquaculture and regulation implications in China. *Sci. Total Environ.* **2021**, *798*, 149205. [CrossRef]
8. 2022 National Fishery Economic Statistical Bulletin. *Chin. Aquat. Prod.* **2023**, *8*, 7–8.
9. Zuo, R.; Peng, J.; Li, R.; Liu, S.; Wang, Y.; Wen, Q.; Hou, H.; Lu, M.; Li, Y. Progress in environmental behavior and ecotoxicology of tetracycline veterinary drugs. *Adv. Vet. Med.* **2018**, *39*, 98–101. [CrossRef]
10. Zhu, Z.; Li, L.; Xu, M.; Zhang, Q.; Hong, J. The rhomboidal flaky Fe-Mn catalyst activated persulfate for degradation of tetracycline. *Chin. Environ. Sci.* **2021**, *41*, 5142–5152. [CrossRef]
11. Wang, Z.; Bai, M.; Li, X.; Weng, L.; Ye, H. Research progress of microbial degradation of tetracycline antibiotics. *J. Agric. Environ. Sci.* **2022**, *41*, 2779–2786.
12. Zou, S.-J.; Chen, Y.-F.; Zhang, Y.; Wang, X.-F.; You, N.; Fan, H.-T. A hybrid sorbent of α-iron oxide/reduced graphene oxide: Studies for adsorptive removal of tetracycline antibiotics. *J. Alloys Compd.* **2021**, *863*, 158475. [CrossRef]
13. Chen, K.; Huang, D. Application of various adsorbent materials in removal of tetracycline antibiotic residues in water. *J. Guangxi Med. Univ.* **2018**, *35*, 1169–1172. [CrossRef]
14. Du, C.; Wu, T.; Hu, T.; Liu, Y.; Deng, J. Preparation of biochar by ball milling and its adsorption of methylene blue. *J. Sichuan Univ. Arts Sci.* **2023**, *33*, 21–25.
15. Zhang, T.; Yang, Y.; Li, W.; Zhou, Y.; Wu, L.; Cheng, Z. Adsorption of tetracycline by alkali-modified biochar. *J. Yichun Univ.* **2022**, *44*, 6–12+47.
16. Sun, Y.; Zheng, L.; Zheng, X.; Xiao, D.; Yang, Y.; Zhang, Z.; Ai, B.; Sheng, Z. Adsorption of Sulfonamides in Aqueous Solution on Reusable Coconut-Shell Biochar Modified by Alkaline Activation and Magnetization. *Front. Chem.* **2022**, *9*, 814647. [CrossRef]
17. Che, H.; Wei, G.; Fan, Z.; Zhu, Y.; Zhang, L.; Wei, Z.; Huang, X.; Wei, L. Super facile one-step synthesis of sugarcane bagasse derived N-doped porous biochar for adsorption of ciprofloxacin. *J. Environ. Manag.* **2023**, *335*, 117566. [CrossRef]
18. Lu, S.; Huang, X.; Tang, M.; Peng, Y.; Wang, S.; Makwarimba, C.P. Synthesis of N-doped hierarchical porous carbon with excellent toluene adsorption properties and its activation mechanism. *Environ. Pollut.* **2021**, *284*, 117113. [CrossRef]
19. Li, H.; Tang, M.; Huang, X.; Wang, L.; Liu, Q.; Lu, S. An efficient biochar adsorbent for CO_2 capture: Combined experimental and theoretical study on the promotion mechanism of N-doping. *Chem. Eng. J.* **2023**, *466*, 143095. [CrossRef]
20. Chen, X.; Oh, W.-D.; Hu, Z.-T.; Sun, Y.-M.; Webster, R.D.; Li, S.-Z.; Lim, T.-T. Enhancing sulfacetamide degradation by peroxymonosulfate activation with N-doped graphene produced through delicately-controlled nitrogen functionalization via tweaking thermal annealing processes. *Appl. Catal. B-Environ.* **2018**, *225*, 243–257. [CrossRef]
21. Cheng, Y.; Wang, B.; Shen, J.; Yan, P.; Kang, J.; Wang, W.; Bi, L.; Zhu, X.; Li, Y.; Wang, S.; et al. Preparation of novel N-doped biochar and its high adsorption capacity for atrazine based on π–π electron donor-acceptor interaction. *J. Hazard. Mater.* **2022**, *432*, 128757. [CrossRef] [PubMed]
22. Obanla, O.R.; Hestekin, J.A.; Ojewumi, M.E.; Bousrih, I.; Fawole, M.C. Enhancing rubber (*Hevea brasiliensis*) seed shell biochar through acid-base modification for effective phenol removal from aqueous environments. *Results Eng.* **2023**, *20*, 15. [CrossRef]
23. Huang, Y.; Li, K.; Yao, W.; Gao, Y. Factors affecting the pore structure and adsorption performance of rice straw carbon prepared by potassium bicarbonate activation. *Environ. Chem.* **2018**, *37*, 569–575.
24. Tian, W.; Zhang, H.; Sun, H.; Tade, M.O.; Wang, S. Template-free synthesis of N-doped carbon with pillared-layered pores as bifunctional materials for supercapacitor and environmental applications. *Carbon* **2017**, *118*, 98–105. [CrossRef]
25. Li, Y.; Xing, B.; Wang, X.; Wang, K.; Zhu, L.; Wang, S. Nitrogen-Doped Hierarchical Porous Biochar Derived from Corn Stalks for Phenol-Enhanced Adsorption. *Energy Fuels* **2019**, *33*, 12459–12468. [CrossRef]
26. Yim, S.D.; Kim, S.J.; Baik, J.H.; Nam, I.S.; Mok, Y.S.; Lee, J.H.; Cho, B.K.; Oh, S.H. Decomposition of urea into NH_3 for the SCR process. *Ind. Eng. Chem. Res.* **2004**, *43*, 4856–4863. [CrossRef]
27. Li, K.; Chen, W.; Yang, H.; Chen, Y.; Xia, S.; Xia, M.; Tu, X.; Chen, H. Mechanism of biomass activation and ammonia modification for nitrogen-doped porous carbon materials. *Bioresour. Technol.* **2019**, *280*, 260–268. [CrossRef]
28. Chen, W.; Yang, H.P.; Chen, Y.Q.; Chen, X.; Fang, Y.; Chen, H.P. Biomass pyrolysis for nitrogen-containing liquid chemicals and nitrogen-doped carbon materials. *J. Anal. Appl. Pyrolysis* **2016**, *120*, 186–193. [CrossRef]
29. Al-Ghouti, M.A.; Da'ana, D.A. Guidelines for the use and interpretation of adsorption isotherm models: A review. *J. Hazard. Mater.* **2020**, *393*, 122383. [CrossRef]
30. Duan, X.; Sun, H.; Kang, J.; Wang, Y.; Indrawirawan, S.; Wang, S. Insights into Heterogeneous Catalysis of Persulfate Activation on Dimensional-Structured Nanocarbons. *ACS Catal.* **2015**, *5*, 4629–4636. [CrossRef]
31. Chen, J.; Yang, J.; Hu, G.; Hu, X.; Li, Z.; Shen, S.; Radosz, M.; Fan, M. Enhanced CO_2 Capture Capacity of Nitrogen-Doped Biomass-Derived Porous Carbons. *ACS Sustain. Chem. Eng.* **2016**, *4*, 1439–1445. [CrossRef]
32. Ying, Z.W.; Chen, X.W.; Li, H.; Liu, X.Q.; Zhang, C.; Zhang, J.; Yi, G.F. Efficient Adsorption of Methylene Blue by Porous Biochar Derived from Soybean Dreg Using a One-Pot Synthesis Method. *Molecules* **2021**, *26*, 661. [CrossRef]
33. Huang, W.; Zhang, X.P.; Zhang, J.Q.; Tang, Y.X.; Luo, Z. Preparation of magnetically modified sheep manure-derived ZVI-biochar and characterization of its activation over monosulfate for AO7 degradation. *J. Environ. Sci. Eng.* **2022**, *42*, 196–208. [CrossRef]
34. Yu, H.X.; Gu, L.; Chen, L.; Wen, H.F.; Zhang, D.F.; Tao, H. Activation of grapefruit derived biochar by its peel extracts and its performance for tetracycline removal. *Bioresour. Technol.* **2020**, *316*, 123971. [CrossRef] [PubMed]

35. Li, S.C.; Du, G.Y.; Hu, Z.H.; Zhang, H.T. Preparation of nitrogen-doped biochar and its adsorption characteristics on heavy metal lead in water. *New Chem. Mater.* **2023**, *51*, 214–220. [CrossRef]
36. Kang, Y.X.; Kang, X.; Li, D.P.; Sun, Z.Y.; Song, X.S.; Zhao, X.X. Adsorption properties of pigeon feces biochar prepared at different pyrolysis temperatures on methicillin in water. *J. Donghua Univ.* **2021**, *47*, 100–108. [CrossRef]
37. Zhang, S.; Yang, X.; Liu, L.; Ju, M.; Zheng, K. Adsorption Behavior of Selective Recognition Functionalized Biochar to Cd(II) in Wastewater. *Materials* **2018**, *11*, 299. [CrossRef]
38. Zhao, X.L.; Yang, L.N.; Wang, Y.; Dong, N. Adsorption removal of imidacloprid from water by biochar from tobacco stalks. *Appl. Chem.* **2024**, *53*, 308–313+318. [CrossRef]
39. Jang, H.M.; Kan, E. A novel hay-derived biochar for removal of tetracyclines in water. *Bioresour. Technol.* **2019**, *274*, 162–172. [CrossRef]
40. Xiang, Y.J.; Xu, Z.Y.; Wei, Y.Y.; Zhou, Y.Y.; Yang, X.; Yang, Y.; Yang, J.; Zhang, J.C.; Luo, L.; Zhou, Z. Carbon-based materials as adsorbent for antibiotics removal: Mechanisms and influencing factors. *J. Environ. Manag.* **2019**, *237*, 128–138. [CrossRef]
41. Chen, J.; Li, X.; Li, B.; Sun, X.; Huang, W.; Lu, Y. Study on adsorption performance of rape straw biochar to water methylene blue. *Water Treat. Technol.* **2022**, *48*, 37–42+47. [CrossRef]
42. Li, H.; Hu, J.; Meng, Y.; Su, J.; Wang, X. An investigation into the rapid removal of tetracycline using multilayered graphene-phase biochar derived from waste chicken feather. *Sci. Total Environ.* **2017**, *603*, 39–48. [CrossRef] [PubMed]
43. Deng, Y.; She, A.L.; Zhang, Y.; Ni, F.Q.; Ao, T.Q.; Chen, W.Q. Effect of pyrolysis temperature on carbon structure, ash content and adsorption of tetracycline on corn kernel biochar. *J. Agric. Environ. Sci.* **2024**, 1–18.
44. Liu, S.; Li, J.; Xu, S.; Wang, M.; Zhang, Y.; Xue, X. A modified method for enhancing adsorption capability of banana pseudostem biochar towards methylene blue at low temperature. *Bioresour. Technol.* **2019**, *282*, 48–55. [CrossRef]
45. Guo, X.; Wang, J. A general kinetic model for adsorption: Theoretical analysis and modeling. *J. Mol. Liq.* **2019**, *288*, 111100. [CrossRef]
46. Wang, J.; Guo, X. Adsorption kinetic models: Physical meanings, applications, and solving methods. *J. Hazard. Mater.* **2020**, *390*, 122156. [CrossRef]
47. Ho, Y.S.; Wase, D.A.J.; Forster, C.F. Removal of lead ions from aqueous solution using sphagnum moss peat as adsorbent. *Water SA* **1996**, *22*, 219–224.
48. Liu, H.; Xu, G.; Li, G. Preparation of porous biochar based on pharmaceutical sludge activated by NaOH and its application in the adsorption of tetracycline. *J. Colloid Interface Sci.* **2021**, *587*, 271–278. [CrossRef]
49. Yu, W.; Lian, F.; Cui, G.; Liu, Z. N-doping effectively enhances the adsorption capacity of biochar for heavy metal ions from aqueous solution. *Chemosphere* **2018**, *193*, 8–16. [CrossRef]
50. Lian, F.; Cui, G.; Liu, Z.; Duo, L.; Zhang, G.; Xing, B. One-step synthesis of a novel N-doped microporous biochar derived from crop straws with high dye adsorption capacity. *J. Environ. Manag.* **2016**, *176*, 61–68. [CrossRef]
51. Hao, Z.; Wang, C.; Yan, Z.; Jiang, H.; Xu, H. Magnetic particles modification of coconut shell-derived activated carbon and biochar for effective removal of phenol from water. *Chemosphere* **2018**, *211*, 962–969. [CrossRef]
52. Liu, L.; Deng, G.; Shi, X. Adsorption characteristics and mechanism of p-nitrophenol by pine sawdust biochar samples produced at different pyrolysis temperatures. *Sci. Rep.* **2020**, *10*, 5149. [CrossRef]
53. Fierro, V.; Torne-Fernandez, V.; Montane, D.; Celzard, A. Adsorption of phenol onto activated carbons having different textural and surface properties. *Microporous Mesoporous Mater.* **2008**, *111*, 276–284. [CrossRef]
54. Zhang, P.; Li, Y.; Cao, Y.; Han, L. Characteristics of tetracycline adsorption by cow manure biochar prepared at different pyrolysis temperatures. *Bioresour. Technol.* **2019**, *285*, 121348. [CrossRef]
55. Cui, J.T.; Zhang, M.Y.; Mi, M.; Zhao, Y.M.; Jin, Z.W.; Wong, M.H.; Shan, S.D.; Ping, L.F. Effects of swine manure and straw biochars on fluorine adsorption-desorption in soils. *PLoS ONE* **2024**, *19*, 21. [CrossRef] [PubMed]
56. Dai, J.; Meng, X.; Zhang, Y.; Huang, Y. Effects of modification and magnetization of rice straw derived biochar on adsorption of tetracycline from water. *Bioresour. Technol.* **2020**, *311*, 123455. [CrossRef] [PubMed]
57. Nandhakumar, V.; Elavarasan, A.; Chandramohan, M.; Balasuramanian, N. Kinetics and isotherm studies on the adsorption of hexavalent chromium onto phosphoric acid activated *Mimusops elengi* leaves carbon. *Int. J. Curr. Res. Chem. Pharm. Sci.* **2017**, *4*, 67–76. [CrossRef]
58. Garba, Z.N.; Lawan, I.; Zhou, W.; Zhang, M.; Wang, L.; Yuan, Z. Microcrystalline cellulose (MCC) based materials as emerging adsorbents for the removal of dyes and heavy metals—A review. *Sci. Total Environ.* **2020**, *717*, 135070. [CrossRef] [PubMed]
59. Srivastava, V.C.; Swamy, M.M.; Mall, I.D.; Prasad, B.; Mishra, I.M. Adsorptive removal of phenol by bagasse fly ash and activated carbon: Equilibrium, kinetics and thermodynamics. *Colloids Surf. A Physicochem. Eng. Asp.* **2006**, *272*, 89–104. [CrossRef]
60. Sun, A.H.; Bian, S.Y.; Li, L.Z.; Guo, Z.J.; Li, W.J.; Li, J.H.; Xu, S.Y.; Liu, P.D. Preparation of highly adsorptive biochar by sequential iron impregnation under refluxing and pyrolysis at low temperature for removal of tetracycline. *Environ. Pollut.* **2024**, *348*, 10. [CrossRef]
61. Kamali, M.; Appels, L.; Kwon, E.E.; Aminabhavi, T.M.; Dewil, R. Biochar in water and wastewater treatment-a sustainability assessment. *Chem. Eng. J.* **2021**, *420*, 129946. [CrossRef]
62. Chen, Y.; Liu, J.; Zeng, Q.; Liang, Z.; Ye, X.; Lv, Y.; Liu, M. Preparation of *Eucommia ulmoides* lignin-based high-performance biochar containing sulfonic group: Synergistic pyrolysis mechanism and tetracycline hydrochloride adsorption. *Bioresour. Technol.* **2021**, *329*, 124856. [CrossRef]

63. Yu, K.; Chen, Y.X.; Wang, C.B.; Sun, D.; Shen, Q. Adsorption performance of hydrothermal bamboo biochar on tetracycline hydrochloride. *J. Huzhou Norm. Univ.* **2021**, *43*, 54–58.
64. Sewu, D.D.; Jung, H.; Kim, S.S.; Lee, D.S.; Woo, S.H. Decolorization of cationic and anionic dye-laden wastewater by steam-activated biochar produced at an industrial-scale from spent mushroom substrate. *Bioresour. Technol.* **2019**, *277*, 77–86. [CrossRef] [PubMed]
65. Ndoun, M.C.; Elliott, H.A.; Preisendanz, H.E.; Williams, C.F.; Knopf, A.; Watson, J.E. Adsorption of pharmaceuticals from aqueous solutions using biochar derived from cotton gin waste and guayule bagasse. *Biochar* **2021**, *3*, 89–104. [CrossRef]
66. He, L.; Liu, F.-f.; Zhao, M.; Qi, Z.; Sun, X.; Afzal, M.Z.; Sun, X.; Li, Y.; Hao, J.; Wang, S. Electronic-property dependent interactions between tetracycline and graphene nanomaterials in aqueous solution. *J. Environ. Sci.* **2018**, *66*, 286–294. [CrossRef]
67. Zheng, H.; Wang, Z.; Zhao, J.; Herbert, S.; Xing, B. Sorption of antibiotic sulfamethoxazole varies with biochars produced at different temperatures. *Environ. Pollut.* **2013**, *181*, 60–67. [CrossRef] [PubMed]
68. Zhang, D.; He, Q.; Hu, X.; Zhang, K.; Chen, C.; Xue, Y. Enhanced adsorption for the removal of tetracycline hydrochloride (TC) using ball-milled biochar derived from crayfish shell. *Colloids Surf. A Physicochem. Eng. Asp.* **2021**, *615*, 126254. [CrossRef]
69. Xu, Z.; Cai, J.-G.; Pan, B.-C. Mathematically modeling fixed-bed adsorption in aqueous systems. *J. Zhejiang Univ.-Sci. A* **2013**, *14*, 155–176. [CrossRef]
70. Lin, J.; Zhan, Y.; Wang, H.; Chu, M.; Wang, C.; He, Y.; Wang, X. Effect of calcium ion on phosphate adsorption onto hydrous zirconium oxide. *Chem. Eng. J.* **2017**, *309*, 118–129. [CrossRef]
71. Zhao, J.; Gao, F.; Sun, Y.; Fang, W.; Li, X.; Dai, Y. New use for biochar derived from bovine manure for tetracycline removal. *J. Environ. Chem. Eng.* **2021**, *9*, 105585. [CrossRef]
72. Kumar, P.A.; Chakraborty, S. Fixed-bed column study for hexavalent chromium removal and recovery by short-chain polyaniline synthesized on jute fiber. *J. Hazard. Mater.* **2009**, *162*, 1086–1098. [CrossRef] [PubMed]
73. Dolatyari, L.; Yaftian, M.R.; Rostamnia, S. Adsorption of Th(IV) and U(VI) on functionalized SBA-15 mesoporous silica materials using fixed bed column method; breakthrough curves prediction and modeling. *Sep. Sci. Technol.* **2018**, *53*, 1282–1294. [CrossRef]
74. Hasan, S.H.; Ranjan, D.; Talat, M. Agro-industrial waste 'wheat bran' for the biosorptive remediation of selenium through continuous up-flow fixed-bed column. *J. Hazard. Mater.* **2010**, *181*, 1134–1142. [CrossRef]
75. Wang, L.; Feng, J.J.; Tao, M.L. Dynamic adsorption of dyes by quaternary ammonium ionized acrylic fibers. *Chem. Ind. Eng.* **2020**, *37*, 73–79. [CrossRef]

water

MDPI

Article

Characteristics of Nitrogen Removal and Functional Gene Transcription of Heterotrophic Nitrification-Aerobic Denitrification Strain, *Acinetobacter* sp. JQ1004

Liangang Hou [1,*], Feng Huang [1], Zhengwei Pan [1], Wei Chen [1,3] and Xiujie Wang [2,*]

[1] Water & Environmental Protection Department, China Construction First Group Construction & Development Co., Ltd., Beijing 100102, China
[2] The College of Environmental and Chemical Engineering, Jiangsu University of Science and Technology, Zhenjiang 212100, China
[3] China Construction First Group Corporation Limited, Beijing 100161, China
* Correspondence: houliangang@chinaonebuild.com (L.H.); wangxiujie@just.edu.cn (X.W.)

Abstract: In this study, the heterotrophic nitrification–aerobic denitrification strain JQ1004 was investigated in terms of its nitrogen removal mechanism and kinetic properties, laying the foundation for its application in the field of wastewater treatment. Nitrogen balance analysis revealed that the final metabolic product was N_2, and approximately 54.61% of N was converted into cellular structure through assimilation. According to the fitting of the Compertz model, the maximum degradation rates of ammonia and nitrate were 7.93 mg/(L·h) and 4.08 mg/(L·h), respectively. A weakly alkaline environment was conducive to N removal, and the sensitivity of functional genes to acidic environments was *amoA* > *nirS* > *narG*. An appropriate increase in dissolved oxygen significantly enhanced heterotrophic nitrification activity, and notably, the denitrification-related functional gene *narG* exhibited greater tolerance to dissolved oxygen compared to *nirS*. The transcription level of *amoA* was significantly higher than that of *narG* or *nirS*, confirming that there might have been direct ammonia oxidation metabolic pathways ($NH_4^+ \rightarrow NH_2OH \rightarrow N_2$) besides the complete nitrification and denitrification pathway. The annotation of nitrogen assimilation-related functional genes (including *gltB*, *gltD*, *glnA*, *nasA*, *nirB*, *narK*, *nrtP*, *cynT*, and *gdhA* genes) in the whole-genome sequencing analysis further confirmed the high assimilation nitrogen activity of the HN-AD strain.

Keywords: wastewater; nitrogen balance; kinetics; *amoA*; whole genome

1. Introduction

With the rapid growth of the global population and the development of the social economy, a large amount of nitrogen-containing wastewater is being released, seriously damaging the aquatic environment ecosystem [1]. Excessive nitrogen discharge not only leads to the eutrophication of water bodies and hypoxia in animals or plants, but also poses a threat to human health [2]. Microbial biotechnology is one of the most common methods for nitrogen removal in wastewater treatment. However, conventional nitrifiers or denitrifiers require separate aerobic or anaerobic environments, respectively, for bacterial growth and metabolism [3,4]. This has led to limitations of traditional biotechnology for nitrogen removal, further causing low treatment efficiency and high infrastructure investment [5].

Recent research has focused on heterotrophic nitrification and aerobic denitrification (HNAD), which can effectively overcome the shortcomings of traditional biological nitrogen removal technologies. The HNAD process is considered to convert NH_4^+ into NO_3^- through heterotrophic nitrification reactions under aerobic conditions while further denitrifying NO_3^- or NO_2^- into gaseous nitrogen, achieving a complete nitrogen cycle in a single reactor [6]. At present, an increasing number of HNAD bacteria have been isolated and purified from different natural environments, such as *Vibrio* sp. [7], *Zobellella* sp. [8], and *Pseudomonas* sp. [9]. Compared with traditional nitrifying bacteria, HNAD bacteria have several advantages, such as a fast growth rate, high nitrogen removal efficiency, strong environmental tolerance, and the ability to nitrify and denitrify simultaneously in the same aerobic environments [10]. The HNAD process has broad application prospects in the field of wastewater treatment due to the simplification of nitrogen conversion processes and the effective improvement of nitrogen removal efficiency [3,11].

Currently, the study of HNAD bacteria mainly focuses on bacterial isolation and identification, nitrogen removal efficiency, nitrogenous gas composition, and nitrogen metabolic mechanisms [12,13]. There are fewer studies on the kinetics and molecular mechanism of HNAD, so it is significant to search for a universal HNAD bacterium. The aim of this study is to elucidate the characteristics of the growth dynamics of HNAD bacteria, including the rules of nitrogen transformation and the differences in gene transcription under various environmental conditions. In the present study, the biomass proliferation pattern of HNAD strain JQ1004, isolated in previous work [14], was simulated using the Logistic model. Nitrogen balance analysis was performed to determine the nitrogen conversion mechanism. The effects of environmental conditions on the strain JQ1004 were investigated based on the performance of nitrogen removal and the expression levels of functional genes. The whole genome was further analyzed to study the characteristics of cellular molecules and the potential degrading ability of pollutants in wastewater.

2. Materials and Methods

2.1. Media and Strains

The HNAD strain *Acinetobacter* sp. JQ1004 was reported in the published research with the GenBank ID: MF033517.1 [14]. The strain was obtained using the methods of gradient dilution and plate scratching. The components of the heterotrophic nitrification medium (HM) and denitrification medium (DM) referred to the description in the reported work, with some modifications [15]. 1 mol/L HCl or NaOH solution was used to adjust the pH. All culture media were required to be sterilized at 121 °C for 15 min using high-pressure steam before use.

2.2. Nitrogen Balance Analysis

The analysis of nitrogen balance by strain JQ1004 was based on the published work [15]. Firstly, 200 mL of liquid HM containing 50 mg/L ammonia was put into a 500 mL serum bottle, and an O_2/He gas mixture (2 L/min, for 10 min) was injected to ensure complete air displacement. The bottle was sealed and sterilized at 121 °C for 30 min. Bacterial suspensions were inoculated into the serum vials at a 2% (*v*/*v*) inoculation ratio. The inoculation process ensured that the vials remained sealed to avoid mixing with external nitrogen sources. The bacterial mixture was cultured at 140 rpm and 30 °C. Gas samples were taken with a 1000 μL gas-tight needle, and the sampling volume was 200 μL. N_2O and N_2 were determined by gas chromatography. Cells were centrifuged and freeze-dried to determine the intracellular nitrogen using an element analyzer, and the supernatant was taken to detect NH_4^+, NO_2^-, NO_3^-, and total nitrogen (TN).

2.3. Kinetics of Nitrogen Degradation and Cell Growth

The bacterial suspension was inoculated into the liquid HM or DM to conduct experiments on the degradation of different nitrogen compounds by the strain JQ1004. The mixed solution was incubated continuously at 140 rpm and 30 °C for 48 h. N concentration changes were monitored through timed sampling. The degradation kinetics model for different nitrogen compounds could be fitted using a modified Compertz model [16]:

$$S = S_0\left\{1 - \exp\left[-\exp\left(\frac{eR_m}{S_0}(t_0 - t) + 1\right)\right]\right\} \quad (1)$$

where R_m is the maximum degrading rate (mg/(L·h)), S is the N concentration (mg/L) at a certain time, S_0 is the initial N concentration (mg/L), t_0 is the lag time (h), t is the culture time (h), and e is a mathematical constant (Euler's number, approximately equal to 2.71828). The Logistic model [17] was employed to fit the cell growth curves when using ammonia as the nitrogen source. The model equations were:

$$y(t) = \frac{a}{1 + be^{-kt}} \tag{2}$$

where a, b, and k are the constants of the growth equation and e is the mathematical constant.

2.4. The Effect of Environmental Conditions on Nitrogen Removal

To investigate the characteristics of nitrogen removal by the strain JQ1004 under different environmental conditions, sodium succinate and ammonium sulfate were used as carbon and nitrogen sources, respectively, with the initial ammonia concentration set at 100 mg/L and a C/N ratio of 7.5. The control culture conditions were set as pH 7.0, 30 °C, and 140 rpm. For different pH experiments, the initial pH was set to 5.5, 6.5, 7.0, 7.5, or 8.5, respectively. For different rotational speed experiments, the speed was set to 120, 140, 160, 180, or 200 rpm, respectively. For different temperature experiments, the temperature was set to 20 °C, 25 °C, 30 °C, or 35 °C. Except for the variables being tested, all other cultivation conditions were the same as those of the control group. Samples were taken at regular intervals for the determination of NH_4^+ and OD_{600}, and the cultured cells were collected for the determination of the relative expression of functional genes.

2.5. Functional Genes Analysis

Functional genes involved in the nitrogen removal process were amplified using polymerase chain reaction (PCR), and their transcription levels were tested by quantitative PCR (Q-PCR). The primer pairs for each functional gene used in PCR are shown in Table 1. The lengths of the PCR-amplified fragments were detected via 2% agarose gel electrophoresis. The sequences were submitted to the BLAST program at the National Center for Biotechnology Information (NCBI) for online comparison, and representative sequences with high homology were downloaded for analysis based on the comparison results. Total RNA was extracted using an RNAprep Pure Cell/Bacteria Kit (Tiangen Biochemistry, Beijing, China) for Q-PCR. The 16S rDNA gene was used as an internal reference gene to correct for differences in the cDNA content between different samples [18]. Q-PCR was also performed on the target and internal reference genes of each sample using SYBR Green I. The data were analyzed using the $2^{-\Delta\Delta CT}$ method [19].

Table 1. The PCR amplification primers list.

Functional Genes	Primer Sequence (5′-3′)	Product Length (bp)
amoA	AAGGATTGGCCATTGCTCTG GTGGACCTAAAATCCAAGCATT	606
nirS	TAAAAGTTCACACACAAAAAGCAACGC ACAAGTACTGCACCCAGTAATTTGG	517
narG	AATCGCAGATCAATTCCAAGCG TCAGCTTCAGTCTGACTAGATTCTAGT	537

2.6. Whole Genome Sequencing and Analysis

The DNA sample was extracted using the magnetic bead method for subsequent gene library construction. The NEB Next® Ultra™ DNA Library Prep Kit (Shanghai, China, Sangon) was used for the construction of sequencing libraries, and 2% agarose gel was used for electrophoresis to check the size of the library. Samples were subjected to paired-end sequencing using the Illumina HiSeq XTen. The data obtained from sequencing were trimmed and filtered using Trimmomatic, and genome assembly was performed on the clean data after quality control. The assembled genome was subjected to Denovo prediction of repetitive sequences. The predicted open reading frame (ORF) sequences were converted into protein sequences and compared to the Gene Ontology (GO), Clusters of Orthologous Groups of proteins (COG), or Kyoto Encyclopedia of Genes and Genomes (KEGG) databases to obtain the GO, COG, and KEGG pathway annotations of the genes.

2.7. Analytical Methods

The contents of NH_4^+, NO_3^-, NO_2^-, and TN were detected according to standard methods [20]. The pH and temperature were measured using a portable tester (Munich, German, Water Testing Worldwide). OD_{600} was expressed as the absorbance value of the bacterial suspension at 600 nm.

3. Results and Discussion

3.1. Nitrogen Balance Analysis

From the results of the nitrogen balance analysis in Table 2, it could be seen that the strain JQ1004 converted ammonia into N_2O and N_2 through the HNAD process, in which 12.28% and 22.26% of nitrogen were transformed into N_2O and N_2, respectively, and emitted into the atmosphere. In addition, up to 54.61% of nitrogen was assimilated into the intracellular nitrogen of the biomass. Xu et al. [21] stated that a novel fungal strain, *Fusarium keratoplasticum* FSP1, with the HNAD function, assimilated 64.28% of nitrogen into cell substances and dissimilated 20.41% into gaseous nitrogen. During the nitrogen metabolism process, only a very small amount (<0.1 mg) of nitrate or nitrite was detected, which was negligible within the detection error range, indicating that JQ1004 was able to secrete nitrate reductase and nitrite reductase with high activity [22]. There was a 2.02% ammonia deficiency in the nitrogen removal process, which might have been due to an experimental error.

Table 2. The nitrogen balance of strain JQ1004 during the heterotrophic nitrogen removal process.

Initial TN(mg/L)	Nitrogen Content (mg/L)						Error
	NH_4^+	NO_3^-	NO_2^-	Intracellular N	N_2O	N_2	
50.58 ± 0.74	4.37 ± 0.08	0.04 ± 0.01	0.06 ± 0.02	27.62 ± 0.57	6.21 ± 0.14	11.26 ± 0.94	2.02%

Note: Error% = 100 × (Initial TN content − Final TN content)/Initial TN content.

3.2. Kinetic Analysis of Nitrogen Degradation and Cell Growth

As shown in Figure 1a, the modified Compertz model aptly described the degradation patterns of different nitrogen sources by strain JQ1004. According to Table 3, the correlation coefficients (R^2) of the fitted curves for the degradation of ammonia and nitrate were 0.997 and 0.985, respectively. From the fitting results, it could be seen that strain JQ1004 had an obvious delay period in degrading nitrogen pollutants, which was 5.91 h or 12.74 h, respectively, for metabolizing ammonia or nitrate. These results indicated that the strain JQ1004 showed a preference for degrading ammonia, which was in accordance with previously published work [23]. When using different nitrogen sources, the average degradation rate of ammonia was significantly higher than that of nitrate based on the kinetic parameters. Ammonia was rapidly degraded within 6–24 h, and the ammonia removal efficiency reached 97.3% at t = 24 h, with the highest degradation rate of R_m = 7.93 mg/(L·h). Ren et al. [24] studied the ammonia removal efficiency of *Acinetobacter junii* YB under different C/N conditions and found that the maximum ammonia removal rate ranged from 4.04 to 10.09 mg/(L·h). Similarly, nitrate entered a quick consumption period at 13–35 h, with a nitrate removal rate of 82.3% at t = 36 h and a maximum nitrate degradation rate of R_m = 4.08 mg/(L·h). The strain JQ1004 had higher degradation activity in metabolizing ammonia than nitrate, possibly due to higher enzyme activity during ammonia oxidation compared to denitrification [25].

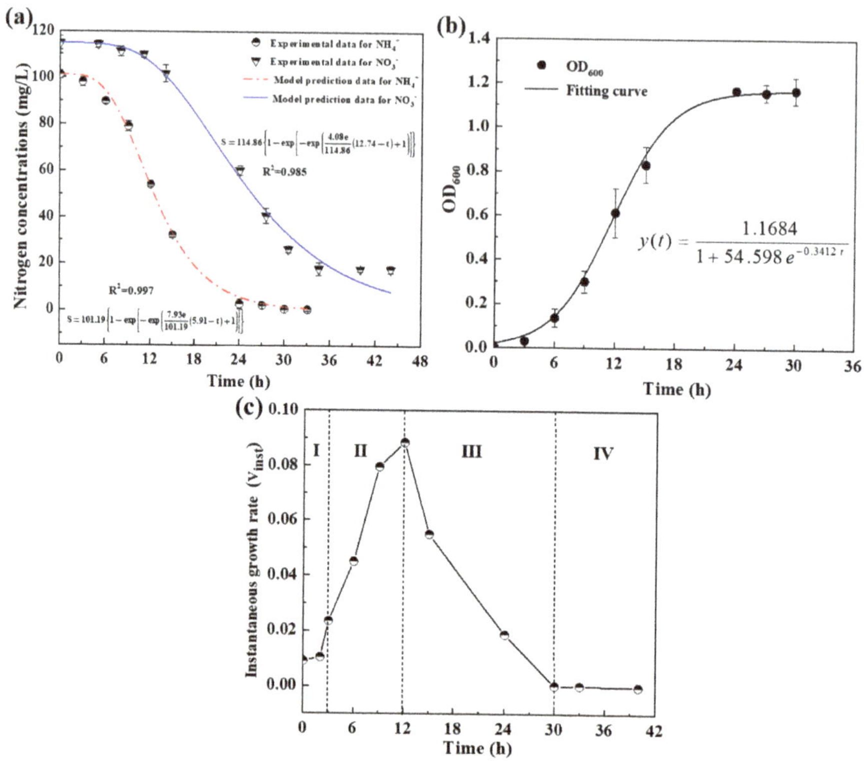

Figure 1. The nitrogen degradation curve (**a**), growth curve (**b**), and its first derivative (**c**) of heterotrophic nitrifying strain JQ1004.

Table 3. The kinetic parameters of degradation of different nitrogen by strain JQ1004.

Kinetic Parameters	Ammonia	Nitrate
S_0 (mg/L)	101.19	114.86
t_0 (h)	5.91	12.74
R_m [mg/(L·h)]	7.93	4.08
R^2	0.997	0.985

The Logistic equation, modeled as a typical S-curve, can be used to describe the effect of increasing cell concentration on the microbial self-metabolism rate during growth and proliferation [26]. As shown in Figure 1b, the equation was able to effectively fit the actual experimental data of cell growth with an R^2 of 0.99. From the results of the regression analysis, it can be seen that the maximum specific growth rate of strain JQ1004 using ammonia as a nitrogen source was 0.34 h^{-1}. Furthermore, the biomass of strain JQ1004 reached the maximum OD_{600} = 1.16 A at 24 h, according to the fitting curve. The growth curves were first-order derived using the Origin 2018 software to obtain the instantaneous growth rate, V_{inst}. The instantaneous growth rate was calculated as:

$$V_{inst} = \frac{dy}{dt} = \lim_{x \to 0} \frac{\Delta y}{\Delta t_i} = \lim_{x \to 0} \frac{f(t + \Delta t) - f(t)}{\Delta t} \tag{3}$$

where the proliferation rate was taken into account when $\Delta t \to 0$, which could more effectively describe cell proliferation than the average growth rate. As shown in Figure 1c, the proliferation process of strain JQ1004 was obviously divided into four stages. Stage I was a slow-growth hysteresis period which lasted for about 0–3 h. In this stage, microorganisms mainly adapted to the environment by adjusting their own metabolic patterns and produced functional enzymes that could be used for subsequent metabolism. Subsequently, the strain entered the logarithmic growth period (stage II), during which substrate was consumed and utilized in large quantities to synthesize microbial cellular matter. Bacterial biomass grew rapidly, and the duration of this stage was 3–12 h. After the logarithmic period, the bacterial growth gradually entered a period of slow growth due to the large amount of substrate consumption, and the duration of this stage (Stage III) was 12–30 h. Finally, the strain entered the stable period (Stage IV), in which the growth rate was close to zero, and the biomass no longer increased or even showed a slight decline due to cellular demise and lysis.

3.3. Amplification and Expression of Functional Genes

3.3.1. PCR Amplification of Functional Genes

To further investigate the nitrogen removal mechanism, functional genes related to nitrogen metabolism, including *amoA*, *narG*, *nirS*, and *nirK*, were amplified by PCR. Ammonia monooxygenase (AMO), encoded by the *amoA* gene, catalyzed the conversion of ammonia into hydroxylamine, which was the beginning of the nitrification reaction [27]. Generally, the *amoA* gene is used as a marker functional gene for identifying nitrifying bacteria, as it is widely found among autotrophic or heterotrophic nitrifying bacteria [28]. As shown in Figure 2a, an *amoA* gene fragment with a length of 606 bp was amplified from the genome of strain JQ1004. The sequences were submitted to NCBI for BLAST online comparison, and the results showed that the gene was homologous to the strain *Acinetobacter pittii* AB17H194 (CP040911.1), with up to 99% homology. Nitrate reduction ($NO_3^- \rightarrow NO_2^-$) in the aerobic denitrification process was mainly completed by membrane-bound nitrate reductase (*narG*) or periplasmic nitrate reductase (*napA*), which were located in the cell membrane or periplasmic space [29]. According to the characteristics of aerobic denitrification exhibited by JQ1004, the *napA* gene, as the signature gene, was theoretically present in its genome. However, as shown in Figure 2b, only the *narG* gene was detected, with a fragment length of 537 bp, due to the limitations of PCR technology. Furthermore, the *nirS* gene, coding for the nitrite reductase containing cytochrome cd1, was obtained with a fragment length of 381 bp. The results of the homology comparison showed that the *nirS* gene had the highest similarity (97.64%) with the strain *Acinetobacter haemolyticus* XH900 (CP018260.1). The successful acquisition of the *nirS* gene further verified the denitrification ability of strain JQ1004 at the molecular level [30].

3.3.2. The Patterns of Functional Genes Expression

The transcriptional patterns of three genes—*amoA*, *narG*, and *nirS*—during nitrogen removal by strain JQ1004 are shown in Figure 2d. The findings indicated that the *amoA* gene was expressed first, and AMO was synthesized and secreted with high activity, leading to ammonia being degraded rapidly. The transcription level of the *amoA* gene reached a maximum value of 1.98×10^4 copies at 12 h, and then decreased with ammonia consumption. There was an obvious delay in the expression of the *narG* and *nirS* genes, and their expression reached a maximum value of about 1.81×10^4 copies and 1.69×10^4 copies, respectively, at 16 h. It has been shown that nitrite reductase is an inducible enzyme requiring the presence of nitrate or nitrite to induce its expression [31]. In addition, the transcription level of the *amoA* gene was significantly higher than that of the *narG* and *nirS* genes, yet strain JQ1004 did not produce accumulation of intermediates such as nitrate or nitrite, suggesting that the strain had other metabolic pathways in addition to the complete nitrification and denitrification pathways.

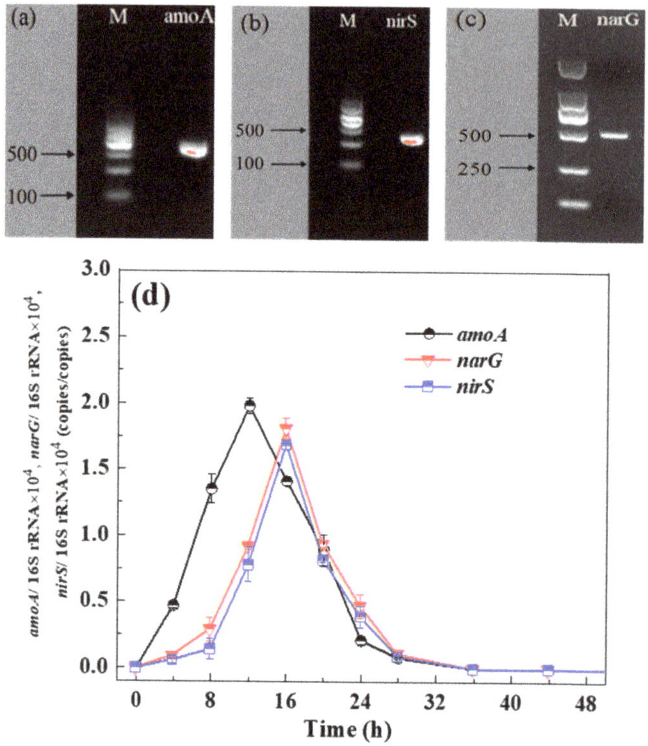

Figure 2. The PCR amplification results (**a–c**) and expression levels (**d**) of different functional genes (*amoA*, *narG*, and *nirS*) for nitrogen removal.

3.4. Effect of Environmental Conditions

3.4.1. Different pH

Figure 3a shows the nitrogen removal ability and growth characteristics of strain JQ1004 under different pH conditions, with sodium succinate as the carbon source. After 24 h of continuous culture, the average TN degrading efficiencies were up to 71.44%, 83.28%, 90.22%, 91.87%, and 88.12% at pH values of 5.5, 6.5, 7.0, 7.5, and 8.5, respectively, and the maximum specific growth rates were 0.24 h^{-1}, 0.28 h^{-1}, 0.30 h^{-1}, 0.34 h^{-1}, and 0.30 h^{-1}, respectively. It was concluded that a weakly alkaline environment (such as pH = 7.5) promoted ammonia degradation and the growth rate of the strain, while a too-high or too-low pH environment inhibited nitrogen removal and bacterial growth. There are two main reasons for the increase in nitrogen removal ability when the pH is weakly alkaline. On the one hand, according to Loh et al. [32], aerobic bacteria tend to change their metabolic mechanisms to adapt to environmental variations. When the pH increases due to the alkalinity consumption of the nitrifying reaction, microorganisms will accelerate their nitrification activity in order to produce more acid to lower the pH, thus keeping the cells in an optimal pH environment. On the other hand, the metabolic substrate of AMO is NH_3 rather than NH_4^+, and an alkaline environment will accelerate NH_3 production, which will improve the nitrification activity and further promote TN removal efficiency [33]. Figure 3b shows the changes in the relative expression of *amoA*, *nirS*, and *narG* in the strain JQ1004 under different pH conditions. The results indicated that the alkaline environment was beneficial for the expression of the *amoA*, *nirS*, and *narG* genes. Compared with the control, *amoA*, *nirS*, and *narG* expression increased by 34%, 23%, and 24%, respectively, at pH 7.5, while *amoA*, *nirS*, and *narG* expression increased by 36%, 21%, and 19%, respectively, at pH 8.5. As mentioned above, an alkaline environment promoted the activity of AMO and induced more NH_4^+ to be converted to NO_2^- or NO_3^-, which further enhanced the expression of the *narG* and *nirS* genes. In addition, all acidic environments inhibited the expression of functional genes. Compared to the neutral environment, the expression of *amoA*, *nirS*, and *narG* decreased by 37%, 26%, and 18%, respectively, at pH 5.5. These results revealed that the sensitivities of these three genes to an acidic environment were in the order of *amoA* > *nirS* > *narG*.

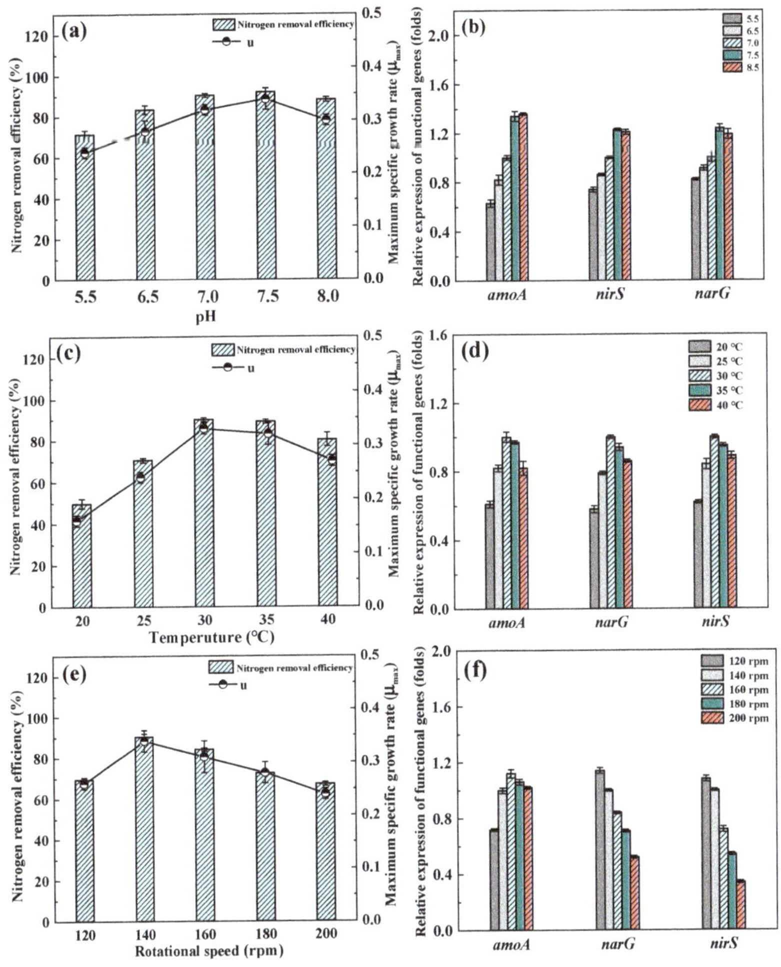

Figure 3. The effect of environmental conditions on nitrogen removal and gene expression of strain JQ1004: (**a**,**b**) different pH; (**c**,**d**) different temperatures; (**e**,**f**) different rotational speeds.

3.4.2. Different Temperatures

As shown in Figure 3c, the temperature significantly affected nitrogen removal and the corresponding gene expression of strain JQ1004. The highest growth and nitrogen removal efficiency were observed at 30 °C, with a maximum specific growth rate of 0.33 h^{-1} and an average TN degrading efficiency of 90.08% at 24 h, respectively. When the temperature increased to 35 °C, the growth and nitrogen degradation were not overly affected. The maximum specific growth rate and average TN removal efficiency were 0.32 h^{-1} and 89.11% at 24 h, respectively. It could be seen that the optimal temperature range for strain JQ1004 was 30–35 °C, which was consistent with the results of previous studies, i.e., the optimal temperature for the metabolic growth of heterotrophic nitrifying bacteria was 30–37 °C [33,34]. Furthermore, the maximum specific growth rates were 0.16 h^{-1}, 0.24 h^{-1}, and 0.27 h^{-1}, with average TN removal efficiencies of 49.82%, 70.53%, and 80.41%, when the temperature was 20 °C, 25 °C, and 40 °C, respectively. The findings suggested that low temperatures resulted in a slow microbial growth rate and weak metabolic vitality, further leading to a decrease in the rate of nitrogen removal. When the temperature was too high, it easily led to the denaturation of nucleic acids or proteins, reducing the activity of related metabolic enzymes and affecting bacterial growth and reproduction [25]. Figure 3d demonstrates that temperatures which were either too high or too low inhibited the expression of *amoA*, *nirS*, and *narG*. Compared with 30 °C, the transcriptional levels of the *amoA* gene were reduced by 39%, 18%, 3%, and 18% at 20 °C, 25 °C, 35 °C, and 40 °C, respectively. The transcriptional levels of the *nirS* gene were decreased by 38%, 16%, 5%, and 11%, while those of the *narG* gene were decreased by 42%, 21%, 6%, and 14%, respectively. It can be concluded that the different functional enzymes involved in the nitrogen cycle of strain JQ1004 were sensitive to temperature changes, and the enzyme activities were decreased by both high and low temperatures.

3.4.3. Different Rotational Speeds

Figure 3e shows that the maximum specific growth rate of strain JQ1004 was 0.26 h^{-1} at 120 rpm, and the average TN removal efficiency was only 69.41% after 24 h. When the shaking speed increased to 140 rpm, the bacterial growth and metabolism rate were accelerated with the rising DO content. The maximum specific growth rate was 0.34 h^{-1}, and the average TN removal efficiency was 90.36% after 24 h. When the speed successively increased to 160 rpm, 180 rpm, and 200 rpm, the specific growth rates of the strain in 24 h decreased to 0.31 h^{-1}, 0.28 h^{-1}, and 0.24 h^{-1}, respectively, and the corresponding TN removal efficiencies reduced to 84.19%, 72.58%, and 67.16%. The reasons for this phenomenon might lie in the increase in dissolved oxygen (DO) caused by the ascending rotational speed, promoting the nitrification activity within a certain concentration range and further enhancing the nitrogen removal performance. However, an excessively high DO concentration could inhibit the denitrification process, thereby reducing the TN removal efficiency. According to the coupling mechanism proposed by Wehrfritz et al. [35], O_2, as one of the essential substrates for HNAD bacteria, was involved in nitrogen conversion under the catalysis of AMO and synergistic respiration with nitrate or nitrite. Although HNAD bacteria mostly grow under aerobic conditions, nitrite reductase in the denitrification process is sensitive to oxygen, and too high a DO concentration will inhibit its activity [36]. On the contrary, a shortage of DO will also limit the nitrification reaction, so controlling the DO at the optimal concentration is especially important for microbial growth and the nitrogen degradation process. As shown in Figure 3f, compared with the control group, the expression of the *amoA* gene decreased by 28% in the 120 rpm group and increased by 12%, 6%, and 2% in the 160, 180, and 200 rpm groups, respectively. The results demonstrated that the reduction in DO concentration inhibited the expression of the *amoA* gene and thus reduced the heterotrophic nitrification activity of the strain. Shoda and Ishikawa [37] studied the effect of DO on the HNAD strain *Alcaligenes faecalis* strain No. 4, stating that the microorganism still retained a high heterotrophic nitrification capacity when the DO concentration was 2 mg/L. The nitrogen removal ability of the HNAD strain decreased significantly once the DO concentration fell below 1 mg/L. In addition, the expressions of functional genes related to the denitrification process were inhibited to different degrees due to the increase in rotational speed. Compared with the control group, the expression of the *narG* gene decreased by 16%, 29%, and 48% at 160 rpm, 180 rpm, and 200 rpm, respectively, while the expression of the *nirS* gene decreased by 28%, 46%, and 64%. The findings confirm that denitrification genes have different degrees of sensitivity to DO, and that the *narG* gene is more tolerant than the *nirS* gene to DO.

In summary, *Acinetobacter* sp. JQ1004 is capable of nitrogen transformation through the heterotrophic nitrification–aerobic denitrification process. The overexpression of *amoA* genes compared to denitrification-related genes suggests the potential existence of a direct ammonia oxidation pathway ($NH_4^+ \rightarrow NH_2OH \rightarrow N_2$). This direct ammonia oxidation reaction utilizes ammonia as an electron donor and O_2 as an electron acceptor, transforming ammonia directly to N_2 via NH_2OH as an intermediate [38]. Unlike the common complete nitrification–denitrification process, the direct ammonia oxidation process significantly reduces the consumption of electrons, demonstrating immense application potential in the field of biological nitrogen removal [39]. Kinetic parameters related to the growth and metabolism of HN-AD bacteria, such as the growth rate and substrate consumption rate, can be obtained through the Logistic model and the modified Compertz model, providing a theoretical basis for practical applications. These kinetic models can predict the metabolic patterns of the strain towards pollutants under different conditions, enabling the optimization of reactor design and operational parameters and ultimately enhancing treatment efficiency.

3.5. Analysis of the Whole Genome

3.5.1. Molecular Characterization of the Genome

The genome of strain JQ1004 was sequenced using the Illumina Hiseq™ sequencing platform, and the total number of reads obtained after quality control of the raw data was 9.19×10^6. The average sequence length was 146.2 bp, of which the high-quality data standardized by Q20 were 1.33×10^9 bp, accounting for 99.1% of the total number of bases. The genome was assembled to obtain a genome size of 1.34×10^9 bp and divided into 35 contigs with a GC content of 40.1%. The assembled results were subjected to gene element prediction and preliminary annotation, while Denovo prediction of repetitive sequences was carried out at the same time. A total of 3291 genes were predicted and involved in encoding proteins. Among them, the numbers of genes containing bases ≥500 bp and ≥1000 bp were 2408 and 1171, respectively, and the numbers of tRNAs and rRNAs were 62 and 8, respectively.

3.5.2. Functional Genes Annotation

As shown in Figure 4a, the bacterial genome was subjected to functional annotation using the COG database. There were 3201 functional genes predicted by ORF, of which 2351 genes could be annotated in the COG database, accounting for 73.5% of the total number of genes. Genes could be annotated in the COG database under the categories of "Energy production and conversion", "Amino acid transport and metabolism", "Translation, ribosomal structure and biogenesis", "Transcription", "Cell wall/membrane/envelope biogenesis", and "Inorganic ion transport and metabolism", which accounted for 6.2%, 7.2%, 7.3%, 7.9%, 6.1%, and 6.2%, respectively. In addition, there were 256 genes (approximately 10.9%) in strain JQ1004 with uncertain protein functions, which need to be further investigated. The GO database was utilized to enrich the genes of strain JQ1004 from three aspects: biological process, cellular component, and molecular function. A total of 2057 genes were annotated in the GO database, accounting for about 64.3% of the total functional genes. As shown in Figure 4b, most of the functional genes were enriched in the "metabolic process", "cellular process", or "single-organism process" of the biological pathway; "cell", "cell part", "membrane", or "membrane part" of the cellular components; and "catalytic activity", "transporter activity", or "binding" of the molecular functions related to catalytic binding. The genome was enriched and analyzed in terms of pathways using the KEGG database. A total of 1672 genes, accounting for about 52.2%, were annotated in the KEGG database. As shown in Figure 4c, the strain JQ1004 had a large number of genes enriched in the "Amino acid metabolism", "Carbohydrate metabolism", and "Energy metabolism" pathways, and the most enriched pathway was the "Amino acid metabolism" pathway, with a total of 242 enriched genes.

Analysis of the metabolic pathways showed that the strain JQ1004 contained not only general essential metabolic pathways, such as carbon metabolism (ko01200), the tricarboxylic acid cycle (TCA cycle, ko00020), glycolysis/gluconeogenesis metabolic pathways (ko00010), and pyruvate metabolism (ko00620), but also nitrogen metabolism (ko00910), sulfur metabolism (ko00920), fatty acid metabolism (ko01212), methane metabolism (ko00680), selenium metabolism (ko00450), propionic acid metabolism (ko00640), butyric acid metabolism (ko00650), glyoxylate and dibasic acid metabolism (ko00630), fructose and mannose metabolism (ko00051), starch and sucrose metabolism (ko00500), and other basic metabolic pathways. Among them, the genes that could be annotated to the nitrogen metabolic pathway were mainly the *gltB*, *gltD*, *glnA*, *nasA*, *nirB*, *narK*, *nrtP*, *cynT*, and *gdhA* genes. In addition to the ability to metabolize ammonia, nitrate, and other nitrogen compounds, the results of functional annotation showed that the strain JQ1004 was able to degrade aromatic compounds, limonene, pinene, aminobenzoic acid, 2-carbonylsalicylic acid, benzoic acid, geraniol, caprolactam, bisphenol, chloroalkanes, styrene, ethylbenzene, nitrotoluene, xylenes, atrazine, dioxin, naphthalene, etc. This indicates that the bacterium has great potential for application in wastewater treatment.

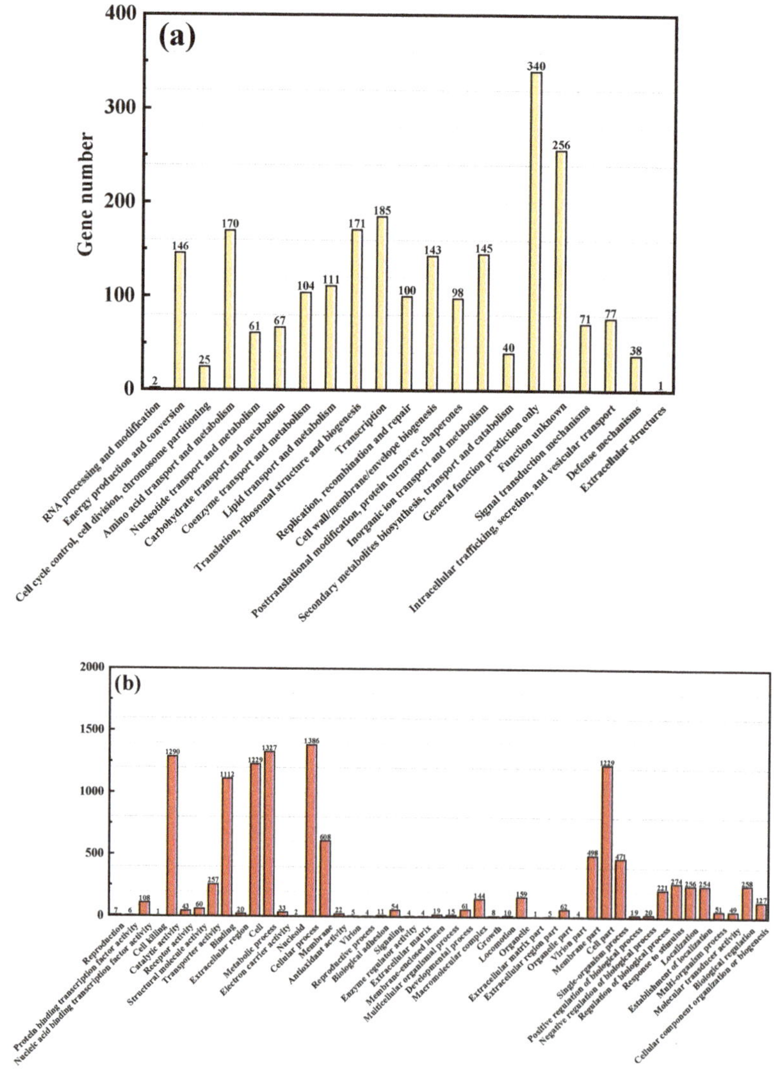

Figure 4. *Cont.*

(c)

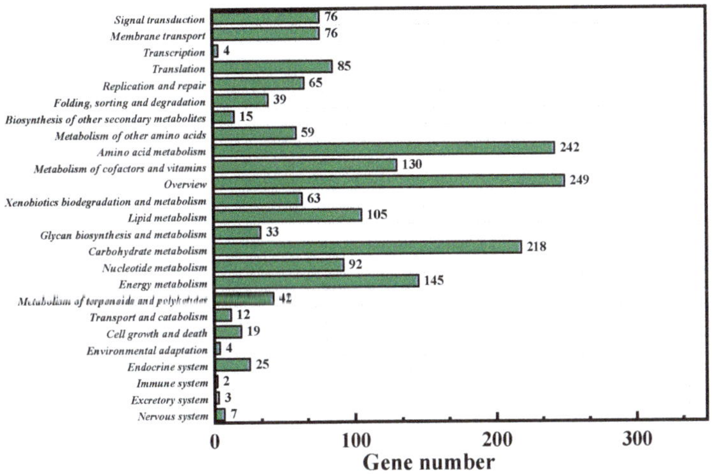

Figure 4. The functional annotation classifications of strain JQ1004 in the COG (**a**), GO (**b**), and KEGG (**c**) databases.

4. Conclusions

The present study mainly investigated the nitrogen removal characteristics of the strain JQ1004 under different environmental conditions, as well as its whole genome. Nitrogen balance analysis showed that about 12.3% and 22.3% of nitrogen was converted to N_2O and N_2, respectively. Up to 54.6% of nitrogen was converted to intracellular nitrogen by assimilation. The *amoA*, *narG*, and *nirS* genes were successfully expressed, and the transcription of denitrification-related genes lagged obviously behind *amoA* genes. A weakly alkaline environment promoted the expression of the *amoA*, *narG*, and *nirS* genes, while environments with too-high or too-low pH values inhibited nitrogen removal and cell growth. The optimal temperature was 30–35 °C. Increasing the rotational speed was able to improve the heterotrophic nitrification ability, but suppressed the expression of genes related to the denitrification process. The total number of functional genes predicted by the ORF was 3201, among which 2351 genes could be annotated to the COG database, accounting for 73.45%. About 2057 genes were annotated to the GO database, accounting for about 64.26%, while 1672 genes (52.23%) were annotated to the KEGG database. Among these, the most enriched pathway was "Amino acid metabolism", collecting approximately 242 genes.

Author Contributions: L.H., investigation, methodology, and writing—original draft; F.H., investigation, methodology, and project administration; Z.P., investigation, conceptualization, and project administration; W.C., investigation, conceptualization, and project administration; X.W., formal analysis and writing—review and editing. All authors have read and agreed to the published version of the manuscript.

Funding: This research was funded by the Science and Technology R&D Plan of China Construction First Group Co., Ltd. (KJYF-2022-9) and the Jiangsu Province Natural Science Fund (BK20210896).

Data Availability Statement: Some or all data, models, and codes that support the findings of this study are available from the corresponding author upon reasonable request.

Conflicts of Interest: Authors Liangang Hou, Feng Huang, Zhengwei Pan were employed by the company China Construction First Group Construction & Development Co., Ltd. Author Wei Chen was employed by the company China Construction First Group Construction & Development Co., Ltd. and China Construction First Group Corporation Limited. The remaining authors declare that the research was conducted in the absence of any commercial or financial relationships that could be construed as a potential conflict of interest.

References

1. Shu, H.; Sun, H.; Huang, W.; Zhao, Y.; Ma, Y.; Chen, W.; Sun, Y.; Chen, X.; Zhong, P.; Yang, H.; et al. Nitrogen removal characteristics and potential application of the heterotrophic nitrifying-aerobic denitrifying bacteria *Pseudomonas mendocina* S16 and *Enterobacter cloacae* DS'5 isolated from aquaculture wastewater ponds. *Bioresour. Technol.* **2022**, *345*, 126541. [CrossRef] [PubMed]
2. Joo, H.; Mitsuyo, H.; Makoto, S. Characteristics of ammonium removal by heterotrophic nitrification-aerobic denitrification by *Alcaligenes faecalis* No. 4. *J. Biosci. Bioeng.* **2005**, *100*, 184–191. [CrossRef] [PubMed]
3. Wang, X.; Hou, H.; Liu, P.; Hou, L.; Yang, T.; Dai, H.; Li, J. Acceleration of nitrogen removal performance in a biofilm reactor augmented with *Pseudomonas* sp. using polycaprolactone as carbon source for treating low carbon to nitrogen wastewater. *Bioresour. Technol.* **2023**, *386*, 129507. [CrossRef]
4. Hou, L.; Li, J.; Liu, Y. Microbial communities variation analysis of denitrifying bacteria immobilized particles. *Process Biochem.* **2019**, *87*, 151–156. [CrossRef]
5. Gu, X.; Leng, J.; Zhu, J.; Zhang, K.; Zhao, J.; Wu, P.; Xing, Q.; Tang, K.; Li, X.; Hu, B. Influence mechanism of C/N ratio on heterotrophic nitrification-aerobic denitrification process. *Bioresour. Technol.* **2022**, *343*, 126116. [CrossRef] [PubMed]
6. Zhao, B.; Tian, M.; An, Q.; Ye, J.; Guo, J. Characteristics of a heterotrophic nitrogen removal bacterium and its potential application on treatment of ammonium-rich wastewater. *Bioresour. Technol.* **2017**, *226*, 46–54. [CrossRef] [PubMed]
7. Ren, J.; Di, F. Sensitivity analysis strategy to assess the salting-out problem during CO_2 geological storage process. *Shock Vib.* **2021**, *2021*, 2809467. [CrossRef]
8. Fu, G.; Zhao, L.; Huangshen, L.; Wu, J. Isolation and identification of a salt-tolerant aerobic denitrifying bacterial strain and its application to saline wastewater treatment in constructed wetlands. *Bioresour. Technol.* **2019**, *290*, 121725. [CrossRef] [PubMed]
9. Xie, F.; Myat, T.; Wang, H. Simultaneous heterotrophic nitrification and aerobic denitrification by a novel isolated *Pseudomonas mendocina* X49. *Bioresour. Technol.* **2021**, *319*, 124198. [CrossRef]
10. Yao, S.; Ni, J.; Ma, T.; Li, C. Heterotrophic nitrification and aerobic denitrification at low temperature by a newly isolated bacterium, *Acinetobacter* sp. HA2. *Bioresour. Technol.* **2013**, *139*, 80–86. [CrossRef]
11. Xi, H.; Zhou, X.; Arslan, M.; Luo, Z.; Wei, J.; Wu, Z.; El-Din, M. Heterotrophic nitrification and aerobic denitrification process: Promising but a long way to go in the wastewater treatment. *Sci. Total Environ.* **2022**, *805*, 150212. [CrossRef] [PubMed]
12. Song, T.; Zhang, X.; Li, J.; Wu, X.; Feng, H.; Dong, W. A review of research progress of heterotrophic nitrification and aerobic denitrification microorganisms (HNADMs). *Sci. Total Environ.* **2021**, *801*, 149319. [CrossRef] [PubMed]
13. Zhang, Q.; Liu, H.; Liang, S.; Chen, W.; Tan, S.; Yang, C.; Qin, S.; Long, K. Comparison of moving bed biofilm reactor and bio-contact oxidation reactor start-up with heterotrophic nitrification-aerobic denitrification bacteria and activated sludge inoculation under high ammonia nitrogen conditions. *Bioresour. Technol.* **2024**, *395*, 130408. [CrossRef] [PubMed]
14. Wang, X.; Wang, W.; Zhang, Y.; Zhang, J.; Li, J.; Wang, S.; Chen, G. Isolation and characterization of *Acinetobacter* sp. JQ1004 and evaluation of its inhibitory kinetics by free ammonia. *Desalin. Water Treat.* **2019**, *147*, 316–325. [CrossRef]
15. Wang, X.; Wang, W.; Zhang, Y.; Sun, Z.; Zhang, J.; Chen, G.; Li, J. Simultaneous nitrification and denitrification by a novel isolated *Pseudomonas* sp. JQ-H3 using polycaprolactone as carbon source. *Bioresour. Technol.* **2019**, *288*, 121506. [CrossRef]
16. Li, J.; Gu, J.; Pan, L. Transformation of dimethyl phthalate, dimethyl isophthalate and dimethyl terephthalate by *Rhodococcus rubber Sa* and modeling the processes using the modified Gompertz model. *Int. Biodeter. Biodegr.* **2005**, *55*, 223–232. [CrossRef]
17. Liu, H.; Zeng, W.; Fan, Z.; Li, J.; Zhan, M.; Peng, Y. Effect of iron on enhanced nitrogen removal from wastewater by sulfur autotrophic denitrification coupled to heterotrophic denitrification under different substrate ratios. *Chem. Eng. J.* **2021**, *421*, 129828. [CrossRef]
18. Chai, Y.; Huang, C.; Sui, M.; Yin, Y.; Sun, N.; Chen, Y.; Liao, Z.; Sun, X.; Shen, W.; Tang, S. Fe-loaded alginate hydrogel beads activating peroxymonosulfate for enhancing anaerobic fermentation of waste activated sludge: Performance and potential mechanism. *J. Environ. Manag.* **2023**, *341*, 118079. [CrossRef]
19. Zhou, X.; Wang, Y.; Tan, X.; Sheng, Y.; Li, Y.; Zhang, Q.; Xu, J.; Shi, Z. Genomics and nitrogen metabolic characteristics of a novel heterotrophic nitrifying-aerobic denitrifying bacterium *Acinetobacter oleivorans* AHP123. *Bioresour. Technol.* **2023**, *375*, 128822. [CrossRef]
20. APHA; AWWA; WEF. *Standards Methods for the Examination of Water and Wastewater*; American Public Health Association: Washington, DC, USA, 2005.
21. Xu, M.; Cui, Y.; Huang, M.; Sui, Y. Simultaneous inorganic nitrogen and phosphate removal by aerobic-heterotrophic fungus *Fusarium keratoplasticum* FSP1: Performance, pathway and application. *Bioresour. Technol.* **2024**, *393*, 130141. [CrossRef]
22. Chen, M.; He, T.; Wu, Q.; Zhang, M.; He, K. Enhanced heterotrophic nitrification and aerobic denitrification performance of *Glutamicibacter arilaitensis* EM-H8 with different carbon sources. *Chemosphere* **2023**, *323*, 138266. [CrossRef]
23. Chen, X.; Li, S.; Zhang, W.; Li, S.; Gu, Y.; Ouyang, L. A newly isolated *Rhodococcus* sp. S2 from landfill leachate capable of heterotrophic nitrification and aerobic denitrification. *Water* **2024**, *16*, 431. [CrossRef]
24. Ren, Y.; Yang, L.; Liang, X. The characteristics of a novel heterotrophic nitrifying and aerobic denitrifying bacterium, *Acinetobacter junii* YB. *Bioresour. Technol.* **2014**, *171*, 1–9. [CrossRef]

25. Ke, X.; Liu, C.; Tang, S.; Guo, T.; Pan, L.; Xue, Y.; Zheng, Y. Characterization of *Acinetobacter* indicus ZJB20129 for heterotrophic nitrification and aerobic denitrification isolated from an urban sewage treatment plant. *Bioresour. Technol.* **2022**, *347*, 126423. [CrossRef] [PubMed]
26. Gupta, S.; Ansari, F.; Nasr, M.; Rawat, I.; Nayunigari, K.; Bux, F. Cultivation of *Chlorella sorokiniana* and *Scenedesmus obliquus* in wastewater: Fuzzy intelligence for evaluation of growth parameters and metabolites extraction. *J. Clean. Prod.* **2017**, *147*, 419–430. [CrossRef]
27. Huang, X.; Yi, X.; He, T.; Jia, H.; Feng, M.; Xiang, S.; Wang, S.; Ni, J.; Xie, D.; Li, Z. Ammonium transformed into nitrous oxide via nitric oxide by *Pseudomonas putida* Y-9 under aerobic conditions without hydroxylamine as intermediate. *Bioresour. Technol.* **2019**, *277*, 87–93. [CrossRef]
28. Chen, P.; Wang, J.; Lv, J.; Wang, Q.; Zhang, C.; Zhao, W.; Li, S. Nitrogen removal by *Rhodococcus* sp. SY24 under linear alkylbenzene sulphonate stress: Carbon source metabolism activity, kinetics, and optimum culture conditions. *Bioresour. Technol.* **2023**, *368*, 128348. [CrossRef] [PubMed]
29. Zhao, B.; Yang, C.; Pan, T.; An, Q.; Guo, J. Characterization of an aerobic denitrifier *Pseudomonas stutzeri* strain XL-2 to achieve efficient nitrate removal. *Bioresour. Technol.* **2018**, *250*, 564–573. [CrossRef] [PubMed]
30. Braker, G.; Zhou, J.; Wu, L.; Devol, A.; Tiedje, J. Nitrite reductase genes (*nirK* and *nirS*) as functional markers to investigate diversity of denitrifying bacteria in pacific northwest marine sediment communities. *Appl. Environ. Microb.* **2000**, *66*, 2096–2104. [CrossRef]
31. Härtig, E.; Walter, Z. Kinetics of *nirS* expression (cytochromecd1 nitrite reductase) in *Pseudomonas stutzeri* during the transition from aerobic respiration to denitrification: Evidence for a denitrification specific nitrate and nitrite responsive regulatory system. *J. Bacteriol.* **1999**, *181*, 161–166. [CrossRef]
32. Loh, X. The effect of pH on the hydrolytic degradation of poly(ε-caprolactone)-block-poly(ethylene glycol) copolymers. *J. Appl. Polym. Sci.* **2013**, *127*, 2046–2056. [CrossRef]
33. Chen, Q.; Ni, J. Ammonium removal by *Agrobacterium* sp. LAD9 capable of heterotrophic nitrification-aerobic denitrification. *J. Biosci. Bioeng.* **2012**, *113*, 619–623. [CrossRef] [PubMed]
34. Zhang, Q.; Ying, L.; Ai, G.; Miao, L.; Zheng, H.; Liu, Z. The characteristics of a novel heterotrophic nitrification-aerobic denitrification bacterium, *Bacillus methylotrophicus* strain L7. *Bioresour. Technol.* **2012**, *108*, 35–44. [CrossRef] [PubMed]
35. Wehrfritz, J.; Reilly, A.; Spiro, S.; Richardson, D. Purification of hydroxylamine oxidase from *Thiosphaera pantotropha*. Identification of electron acceptors that couple heterotrophic nitrification to aerobic denitrification. *FEBS Lett.* **1993**, *335*, 246–250. [CrossRef] [PubMed]
36. Ka, J.; Urbance, J.; Ye, W.; Ahn, T.; Tiedje, J. Diversity of oxygen and N-oxide regulation of nitrite reductases in denitrifying bacteria. *FEMS Microbiol. Lett.* **1997**, *156*, 55–60. [CrossRef] [PubMed]
37. Shoda, M.; Yoichi, I. Heterotrophic nitrification and aerobic denitrification of high-strength ammonium in anaerobically digested sludge by *Alcaligenes faecalis* strain No. 4. *J. Biosci. Bioeng.* **2014**, *117*, 737–741. [CrossRef] [PubMed]
38. Pous, N.; Bañeras, L.; Corvini, P.; Liu, S.; Puig, S. Direct ammonium oxidation to nitrogen gas (Dirammox) in *Alcaligenes* strain HO-1: The electrode role. *Environ. Sci. Ecotechnol.* **2023**, *15*, 100253. [CrossRef]
39. Pan, Y.; Liu, D. Tapping the potential of wastewater treatment with Direct ammonia oxidation (Dirammox). *Environ. Sci. Technol.* **2023**, *57*, 7106–7108. [CrossRef]

water

MDPI

Article

Efficient H_2O_2 Production and Activation by Air Diffusion Cathode Combined with Ultraviolet for Lake Water Treatment: A Long-Term Evaluation

Hui Hong [1], Shiwei Xie [1], Aoxue Qiu [1], Yuming Yao [1], Wenzhe Jiang [1], Jilei Li [1], Zihao Wan [1], Shengjie Xiang [1], Chenyi Xi [1], Jingxian Xiao [1] and Na Zhang [2,3,*]

[1] School of Urban Construction, Wuhan University of Science and Technology, Wuhan 430065, China; honghui0910@163.com (H.H.); xieshiwei@wust.edu.cn (S.X.); 1370557615@wust.edu.cn (A.Q.); 15623365126@163.com (Y.Y.); jwz20020702@163.com (W.J.); 15623706700@163.com (J.L.); m15027080735@163.com (Z.W.); xsj2qw@163.com (S.X.); 18186068011@163.com (C.X.); xjx021220@163.com (J.X.)

[2] Hebei Key Laboratory of Wetland Ecology and Conservation, Hengshui University, 1088 Heping West Road, Hengshui 053000, China

[3] State Key Laboratory of Environmental Criteria and Risk Assessment, Chinese Research Academy of Environmental Sciences, Beijing 100012, China

* Correspondence: zhangna118wood@163.com

Abstract: This study utilizes a natural air diffusion cathode (ADC) and an ultraviolet lamp to construct a UV/H_2O_2 reactor for the in situ synthesis and activation of H_2O_2 and evaluates its potential application in practical lake water treatment. The results indicate that the reactor exhibits stable treatment performance during a continuous flow experiment of 80 h. The air diffusion cathode maintains an H_2O_2 concentration of above 350 mg·L^{-1} in sodium sulfate electrolyte and shows no decreasing trend. Under the condition of approximately 59% H_2O_2 utilization, the removal rates of COD and TOC are 37.6% and 40.0%, respectively; the rate of reduction of A_{254} is 64.3%; while the total bacterial count removal rate reaches 100%. Large organic molecules in surface water are degraded to small organic molecules and mineralized to inorganic minor molecules. It effectively ameliorates the problem of organic pollution of surface water and effectively kills bacteria and improves the microbiological safety of the water body. Therefore, the UV/H_2O_2 system developed in this study, based on electrochemically produced H_2O_2, is an effective method for treating micro-polluted surface water.

Keywords: UV/H_2O_2 technology; air diffused cathode; oxidation of organic matter; lake water

1. Introduction

Lakes, as integral ecosystems and essential freshwater resources, are instrumental in the purification and conservation of water, the modulation of climate dynamics, and the preservation of ecological equilibrium and biodiversity. They are even used as domestic water in some water-scarce areas [1,2]. As human society experiences accelerated growth and development, the ecological health and sustainability of lakes are increasingly under severe threat [3]. Due to rapid industrialization and the spread of intensive agricultural practices, eutrophication and pollution have become widespread in many inland water bodies [4]. Eutrophication precipitates the proliferation of algal and cyanobacterial blooms, culminating in the turbidity and toxicity of water bodies, which in turn endangers biodiversity and compromises the safety of human populations [5]. In addition to the challenges posed by algal growth, lake waters are increasingly found to contain a diverse array of emerging organic contaminants, including pesticides, pharmaceuticals, and personal care products, which are being continuously detected [6,7]. Although the detected concentrations of these organic pollutants are low, their recalcitrant nature and propensity for bioaccumulation necessitate the implementation of advanced treatment methods to ensure their effective degradation, thereby safeguarding water security for humans [8].

Conventional water treatment technologies exhibit constrained efficacy in the remediation of nascent organic contaminants. Moreover, certain process technologies, such as chlorination for disinfection, can engender hazardous by-products like organohalides, thereby precipitating ancillary pollution [9]. In the face of these challenges, advanced oxidation processes are regarded as a promising solution, offering the advantages of high efficiency, versatility, and low secondary pollution [10–12]. The potent oxidizing free radicals produced by this process ultimately mineralize chemically recalcitrant organic pollutants present in water to CO_2, H_2O, and additional inorganic minor molecules, or they facilitate the transformation of more toxic pollutants into low-toxicity, readily biodegradable, small-molecule organic compounds [13,14]. Among the advanced oxidation technologies, ultraviolet/hydrogen peroxide (UV/H_2O_2) technology has received widespread attention because of its high oxidizing capacity, few toxic by-products, and easy operation [15–17]. However, H_2O_2 needs to be continuously dosed in the UV/H_2O_2 technology, and as an explosive hazardous chemical, the transportation and storage of high concentrations of H_2O_2 are risky [18].

The electrochemical reduction of oxygen to synthesize H_2O_2—a methodology that is continually advancing toward maturity—has garnered substantial interest within the domain of water-treatment technology. This technique affords significant environmental advantages, permits meticulous regulation of concentration, and obviates the potential risks related to the storage and transportation of H_2O_2 [19,20]. Some studies have shown that the use of electrochemical in situ generation of H_2O_2 combined with UV for the treatment of wastewater with specific pollutants can achieve good performance [21]. However, there is a lack of empirical data on the application of this technology to the treatment of real water bodies. The composition of real water is complex, and the metal cations (Ca^{2+}, Mg^{2+}, etc.) in the water will accumulate at the cathode under the action of an electric field and generate precipitation on the cathode surface, leading to the deterioration of the electrode and a decrease in the H_2O_2 production rate [22,23]. Meanwhile, the stability of the system needs to be verified by long-term experiments.

To address the above problems, this paper utilizes a homemade air diffusion cathode with a UV lamp to construct a UV/H_2O_2 system, which allows the oxygen in the air to naturally diffuse through the gas diffusion layer to the reaction sites on the catalyst layer for H_2O_2 electrogenesis. To avoid the cathodic scaling problem, we added a bypass of H_2O_2 electrosynthesis in a sodium sulfate electrolyte. In this study, an actual lake water sample was used as the treatment object to carry out a long continuous-flow experiment. During the experiment, several indicators, including the potential of hydrogen (pH), H_2O_2 concentration, total organic carbon (TOC), absorbance at 254 nm (A_{254}), chemical oxygen demand (COD), fluorescent substances, and total bacterial counts were monitored and evaluated to comprehensively evaluate the performance of this treatment technology. The results of this paper can verify the feasibility and effectiveness of the UV/H_2O_2 technology of electrochemical in situ generation of H_2O_2 in the treatment of surface water and provide data support for the further promotion and application of the technology. It also allows for local water treatment in areas where clean water is scarce, providing a new way to access clean water.

2. Materials and Methods

2.1. Materials

Conductive carbon black BP2000 was purchased from Cabot Co. (Boston, MA, USA). The 60% polytetrafluoroethylene emulsion (PTFE) and titanium sulfate ($TiSO_4$) were purchased from Aladdin, Inc. (Shanghai, China). Ethanol (C_2H_5OH), manganese dioxide (MnO_2), and sodium sulfate (Na_2SO_4) were purchased from Sinopharm Chemical Reagent Co., Ltd., (Shanghai, China). Catalase (100,000 u/g) was purchased from Shandong Longo Enzyme Preparation Co., Ltd. (Linyi, China). All of the above materials are analytically pure and above and were used as is. Dimensionally stable electrode (DSA, Ti/RuIr) was purchased from Baoji Longsheng Nonferrous Metals Co., Ltd. (Baoji, China)

2.2. Preparation of CB–PTFE Air Diffusion Cathodes

The air diffusion electrode consists of a gas diffusion layer (in contact with air), a catalytic layer (in contact with water), and a current collection layer (connected to the power supply). The conductive carbon black was dispersed in anhydrous ethanol and stirred well and was ultrasonically stirred and mixed for 30 min, while 60% PTFE emulsion was added slowly and dropwise. The mass ratios of carbon black to PTFE in the gas diffusion layer and the catalytic layer were 1:3 and 2:1, respectively. The resulting suspension was dried in a water bath at 80 °C to remove excess ethanol and pressed under a roller press to make a sheet of about 0.2 mm. The 0.2 mm gas diffusion layer and catalytic layer flakes were placed on both sides of a stainless-steel mesh and pressed together to form 0.4 mm flakes, which were left to stand for 24 h and then fired at 340 °C for 1 h.

2.3. Experimental Methods

The experimental water for this study was obtained from Huangjia Lake in Wuhan City, and the main water quality parameters of the lake are shown in Table 1. Samples were taken from the lake at 16:00 every day and kept still in the laboratory. The supernatant was used at 08:00 the next day for the continuous experiments. As shown in Figure 1, the experimental setup was mainly composed of an H_2O_2 generation chamber, raw water electrolysis chamber, excess H_2O_2 treatment chamber, and UV reactor. All three chambers had internal dimensions of 2 cm × 2 cm × 16 cm. The raw water was delivered into the electrolysis chamber by a peristaltic pump (LM60B-RZ1030-4, Nanjing Runze Fluid Control Equipment Co., Ltd., Nanjing, China) at 8 mL·min^{-1}. The electrolysis chamber was equipped with two-dimensionally stable electrodes (DSA, Ti/RuIr) at a distance of 2 cm. A solution of 50 mM Na_2SO_4 was pumped at a rate of 2 mL·min^{-1} into the H_2O_2 generation chamber, which consisted of an air diffusion cathode and a DSA anode at a spacing of 2 cm. The gas diffusion layer of the air diffusion cathode was exposed to natural air without an additional gas supply. Then O_2 entering from the gaseous diffusion layer was reduced by $2e^-$ transfer on the catalytic layer to H_2O_2 in a Na_2SO_4 solution. The resulting H_2O_2 solution was then mixed with the effluent from the electrolysis chamber and fed at a flow rate of 10 mL·min^{-1} into a UV reactor fitted with a low-pressure mercury lamp (254 nm, 10 watts). The volume of the UV reactor was 250 mL and the residence time was 25 min. The effluent of the UV reactor went into the excess H_2O_2 treatment chamber, which was filled with MnO_2 solid particles. The constant current was supplied by DC power (GPS-X303/C, Taiwan Goodwell Electronics Co., Ltd., New Taipei City, Taiwan).

Table 1. Main water quality parameters of Huangjia Lake.

Parameters	Cl^-	SO_4^{2-}	Ca^{2+}	Mg^{2+}	COD	TOC
Concentration (mg·L^{-1})	38.9 ± 0.99	63.4 ± 1.12	40.8 ± 0.82	7.6 ± 0.15	37.0 ± 5.21	7.35 ± 1.36

2.4. Methods of Analysis

pH was measured by pH meter (Multi 3630 IDS, WTW Group, Wuppertal, Germany). H_2O_2 was measured by a UV–visible spectrophotometer (AOE UV-1800, Shanghai Aoyi Instrument Co., Ltd., Shanghai, China) at wavelength 405 nm after color development by titanium sulfate [24]. Total organic carbon (TOC) was determined by a total organic carbon analyzer (HTY-CT000M, Hangzhou Tailin Biotechnology Equipment Co., Ltd., Hangzhou, China). A_{254} was determined by a UV–visible spectrophotometer. The samples for COD analysis were pretreated by catalase to remove the interference of H_2O_2 [25] and then determined by the potassium dichromate method. The enumeration of total bacteria was conducted utilizing the plate culture technique, wherein 0.5 mL water samples were inoculated onto Petri dishes containing agar to achieve an even spread of the sample across the medium's surface. Following this, the dishes were subjected to an incubation period of 24 h at a temperature of 37 °C to allow for the visualization and subsequent counting of bacterial colonies that emerge. Fluorescent organic matter was detected by a fluorescence spectrophotometer (F-7100, Hitachi High-Technologies Co., Tokyo, Japan), and three-dimensional fluorescence spectrograms were drawn to analyze the fluorescent organic matter. The morphologies of the catalytic layer of the air diffusion electrode were examined on a field-emission SEM (ZEISS Gemini SEM 300, Jena, Germany).

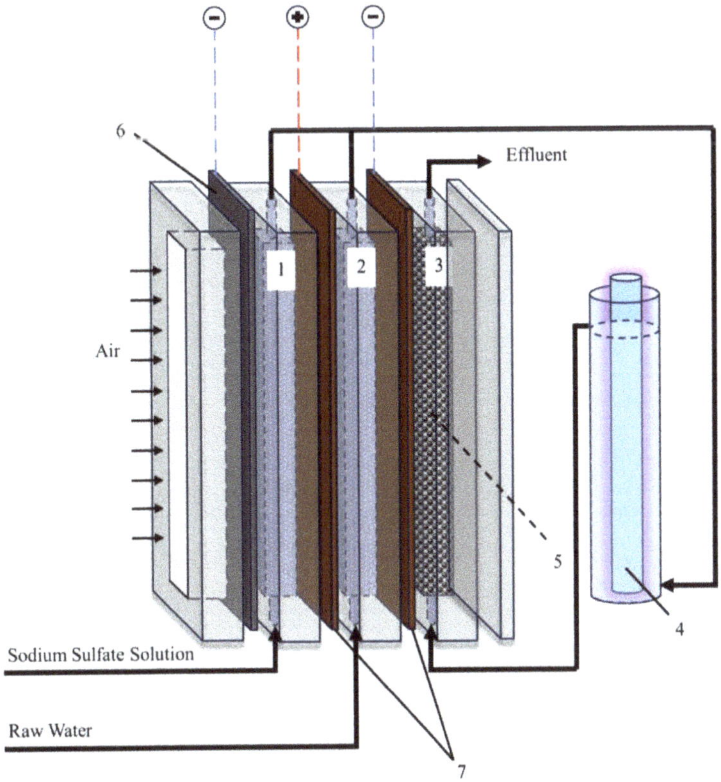

Figure 1. Schematic diagram of the experimental setup (1: H_2O_2 generation chamber; 2: Raw water electrolysis chamber; 3: Excess H_2O_2 treatment chamber; 4: UV reactor; 5: Manganese dioxide; 6: Air diffusion electrode; 7: DSA electrode).

The current efficiency (CE) was calculated using Equation (1) [26],

$$CE = \frac{nCqF}{I} \times 100\% \tag{1}$$

where n is the number of electrons transferred from oxygen reduction to H_2O_2 (2), C represents the H_2O_2 concentration ($mol \cdot L^{-1}$), q is the flow rate ($L \cdot s^{-1}$), F is Faraday's constant (96,486 $C \cdot mol^{-1}$), and I is current intensity (A).

The electric energy consumption (EEC, $kWh \cdot kg^{-1}$) was calculated with Equation (2) [27],

$$EEC = \frac{UI}{Cq} \tag{2}$$

where U and I are the applied voltage (V) and current (A), respectively; C is the concentration of H_2O_2 ($g \cdot L^{-1}$); and q is the flow rate ($L \cdot h^{-1}$).

The unit electrical energy ($kWh \cdot m^{-3}$) to treat a unit volume of lake water was calculated from Equation (3),

$$UEE = \frac{P}{q} \quad (3)$$

where p is the total power (w), which is jointly composed of the electrolysis chamber, the H_2O_2 generation chamber, and the UV reactor power, and q is the flow rate ($L \cdot h^{-1}$).

3. Results and Discussion

3.1. Performance of H_2O_2 Production by the Air Diffusion Cathode

Sodium sulfate solution at a concentration of 50 mM was used as the electrolyte. Then, the effluent concentration of H_2O_2 under different conditions was determined by varying the magnitude of the current density and flow rate, other performance indicators were calculated, and the corresponding results were plotted (Figure 2).

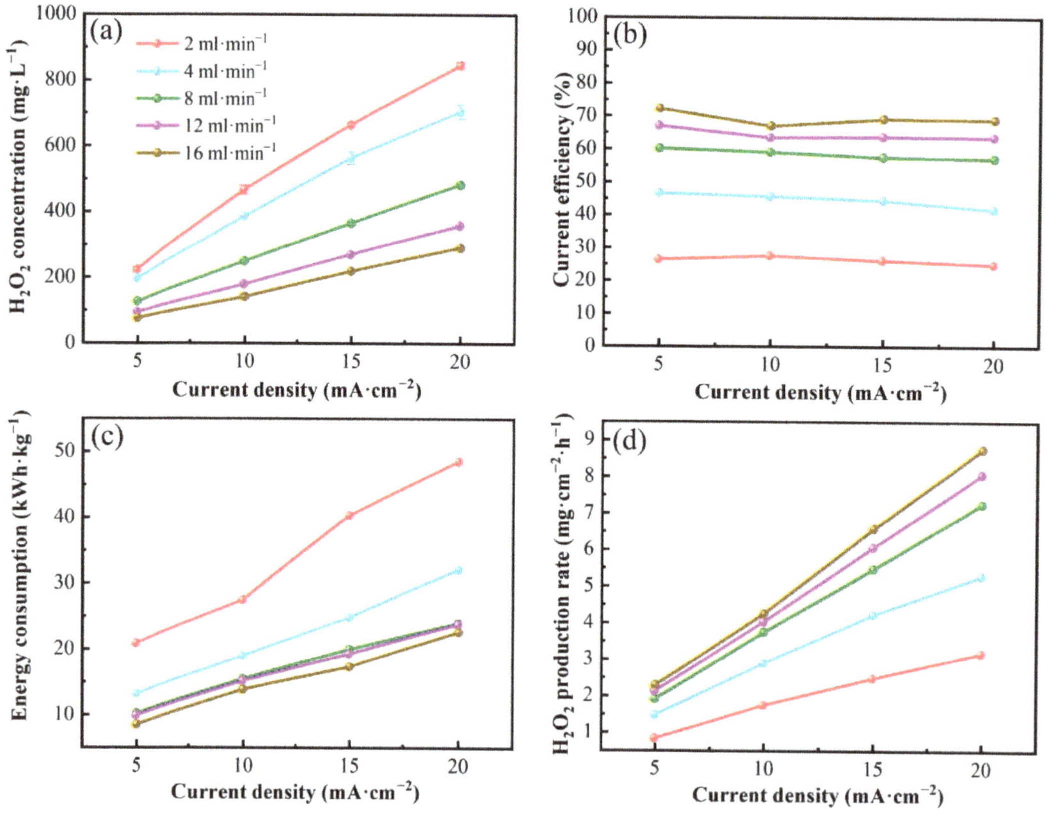

Figure 2. (**a**) H_2O_2 concentration, (**b**) current efficiency, (**c**) energy consumption per unit, (**d**) H_2O_2 generation rate versus flow rate and current density. All experiments were performed in 50 mM sodium sulfate solution.

It was observed that the H_2O_2 effluent concentration, current efficiency, energy consumption per unit, and rate of production were directly related to the magnitude of the applied current density and flow rate. For example, at a current density of 10 $mA \cdot cm^{-2}$, the current efficiency decreased from 67.03% to 26.29% as the flow rate decreased from 16 $mL \cdot min^{-1}$ to 2 $mL \cdot min^{-1}$. At a constant flow rate, the current efficiency showed a slow decreasing trend with increasing current density. At a flow rate of 16 $mL \cdot min^{-1}$, the H_2O_2

production rate increased from 2.29 mg· $cm^{-2} \cdot h^{-1}$ to 8.75 mg· $cm^{-2} \cdot h^{-1}$ as the current density was increased from 5 mA·cm^{-2} to 20 mA·cm^{-2}, showing a linear relationship between the applied current density and the H_2O_2 production rate, which suggests that it is possible to adjust the applied current density to control the production rate. In addition, the unit energy consumption also increased with increasing current density, from 8.51 kWh·kg^{-1} at 5 mA·cm^{-2} to 22.62 kWh·kg^{-1} at 20 mA·cm^{-2}. As the current density increased, the current efficiency decreased, and the energy consumption of the cell gradually increased, which could be due to the accelerated rate of self-decomposition of H_2O_2 [21]. The results showed that the electrochemical in situ H_2O_2 generation reactor in this study was able to provide sufficient H_2O_2 for surface water treatment within an acceptable range of current efficiency and energy consumption. The effluent concentration of H_2O_2 increased accordingly with increasing current density and decreasing flow rate; however, this trend was accompanied by a decrease in current efficiency and an increase in the unit energy consumption. The focus of this study is to perform a comprehensive evaluation of UV/H_2O_2 systems to balance energy consumption and electrolyte cost while pursuing high H_2O_2 yields.

Considering that the device acts on lake water, it was also necessary to verify the H_2O_2 production performance of the air diffusion electrode in real lake water. The actual lake water was injected into the H_2O_2 generation chamber as the electrolyte, and the electrolysis process was carried out for 4 and 8 consecutive hours. The H_2O_2 concentration of the effluent water was then detected. The results are presented in Figure 3a. After 4 h of continuous electrolysis, the H_2O_2 concentration of the effluent water was 11.1 mg·L^{-1}, and the current efficiency was 16.8%. After 8 h of electrolysis, the H_2O_2 effluent concentration was 4.76 mg·L^{-1}, and the current efficiency was 7.2%. Compare (b) and (c) in Figure 3: a layer of white scale, which might be induced by precipitation of Ca^{2+} and Mg^{2+} in the lake water, covered the surface of the ADC after 8 h operation (Equations (4) and (5)). A similar problem has occurred in previous studies, wherein scale formed on ADC led to the performance deterioration of the electrode [22,23]. To avoid degradation of electrode performance due to deterioration of the electrode surface, subsequent experiments will produce H_2O_2 by electrolysis in a simulated electrolyte. The method of mixing the resulting H_2O_2 solution with lake water was used for subsequent lake water treatment experiments.

$$2H_2O + 2e^- \rightarrow 2OH^- + H_2 \uparrow \tag{4}$$

$$Ca^{2+}\left(Mg^{2+}\right) + 2OH^- \rightarrow Ca(OH)_2 \downarrow [Mg(OH)_2] \downarrow \tag{5}$$

Considering that H_2O_2 is a strong oxidizing agent, it is worth exploring the effect of its individual action on water quality. The H_2O_2 solution generated by the electrochemical reduction method was mixed with lake water that had been standing for some time, and after one hour of reaction, the H_2O_2 content and other water quality indicators were tested. As shown in Table 2, the concentration of H_2O_2 after 1 h reaction was almost unchanged, leading to a utilization rate of almost zero, and other water quality indexes were also basically unchanged. This observation indicated that H_2O_2 alone could not oxidize the organic pollutants in the lake water. Therefore, in this study, a quick mixing of raw water with H_2O_2 solution before entering the UV reactor for advanced oxidation reaction was adopted.

Table 2. Parameter changes during 1 h reaction of H_2O_2 with lake water.

Reaction Time	H_2O_2 (mg·L^{-1})	COD (mg·L^{-1})	A_{254}
0 h	63.93 ± 2.20	42 ± 2.83	0.104 ± 0.001
1 h	62.20 ± 0.73	41 ± 4.24	0.097 ± 0.001

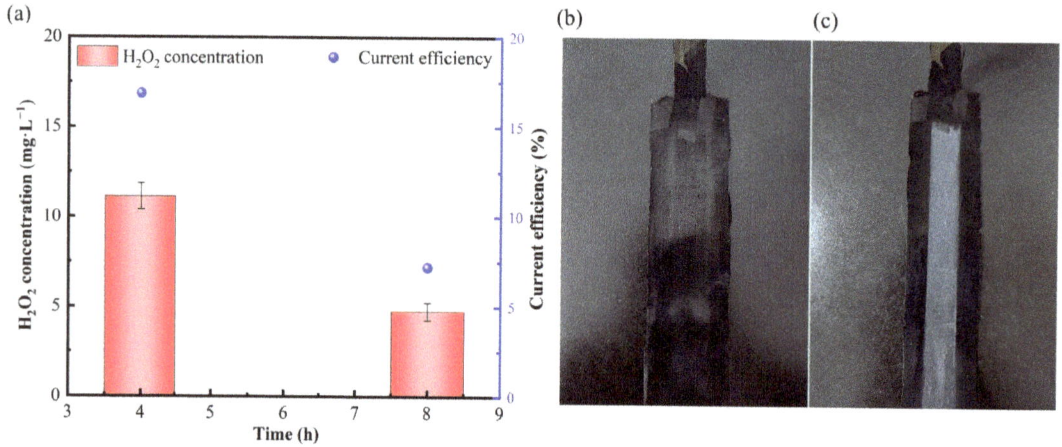

Figure 3. (**a**) Variation in H_2O_2 concentration and current efficiency with electrolysis time (current of 50 mA, flow rate of 8 mL·min^{-1}), (**b**) air diffusion electrode before electrolysis, (**c**) air diffusion electrode after 8 h of electrolysis.

The complex composition of natural water bodies and the scattering and high optical density effects caused by metal cations (Ca^{2+}, Mg^{2+}, etc.), inorganic anions (Cl^-, etc.), natural organic matter (humic acid, fulvic acid, etc.), and other constituents in the water bodies, all lead to a reduction in the efficiency of UV-activated H_2O_2 in generating free radicals [15,28]. In this study, it is proposed to use an appropriate increase in the initial concentration of H_2O_2 in the raw water to reduce the impact of the above problems. Therefore, on the premise of comprehensive consideration of various factors such as energy consumption, treatment effect, device convenience, and water quality environment, and to ensure that the concentration of H_2O_2 effluent from the H_2O_2 generation chamber meets the experimental requirements, for this study we decided that the H_2O_2 generation chamber would conduct the subsequent experiments under the conditions of a current density of 10 mA·cm^{-2} and a flow rate of 2 mL·min^{-1}. The subsequent experiments were carried out at a current size of 30 mA and a voltage of 13 V in the raw water electrolysis chamber. The utilization of H_2O_2 was further explored by monitoring the change in H_2O_2 concentration in the system during the reaction process and assessing the treatment effect of the excess H_2O_2 treatment chamber. Through this comprehensive consideration, this study aims to provide a comprehensive assessment of the production and utilization efficiency of H_2O_2 under optimal experimental conditions and to provide more reliable data support for the experimental results.

According to the results in Figure 4a, the air diffusion cathode performed well in generating H_2O_2 after up to 80 h. The concentration of H_2O_2 effluent from the H_2O_2 generation chamber remained above 350 mg·L^{-1} and did not show a decreasing trend. Through the characterization of the catalytic layer's surface, we can gain insights into its structural and functional properties, which are crucial for understanding its performance in electrochemical reactions. As shown in Figure 4b,c, PTFE and carbon black are uniformly distributed on the catalytic layer. After a long time of electrolysis, the structure of the catalytic layer is still compact and there is no obvious change in the morphology, which is the reason why there is no degradation of the electrode performance. When the effluent from the H_2O_2 generation chamber was mixed with the raw water, the concentration of H_2O_2 in the influent water of the UV was about 78 mg·L^{-1}, the effluent water of the reactor was about 32 mg·L^{-1}, and the utilization rate of H_2O_2 was about 59%. The consumption of H_2O_2 is closely related to the quality of the water and the irradiation time of the UV lamps [29]. To ensure the treatment effect and safety, a post-treatment process was added after the UV reactor effluent to eliminate these potential pitfalls, and the data showed that the H_2O_2 decomposition chamber was able to decompose H_2O_2 effectively with a decomposition efficiency of about 60%. These results further prove the treatment effect and safety of the system.

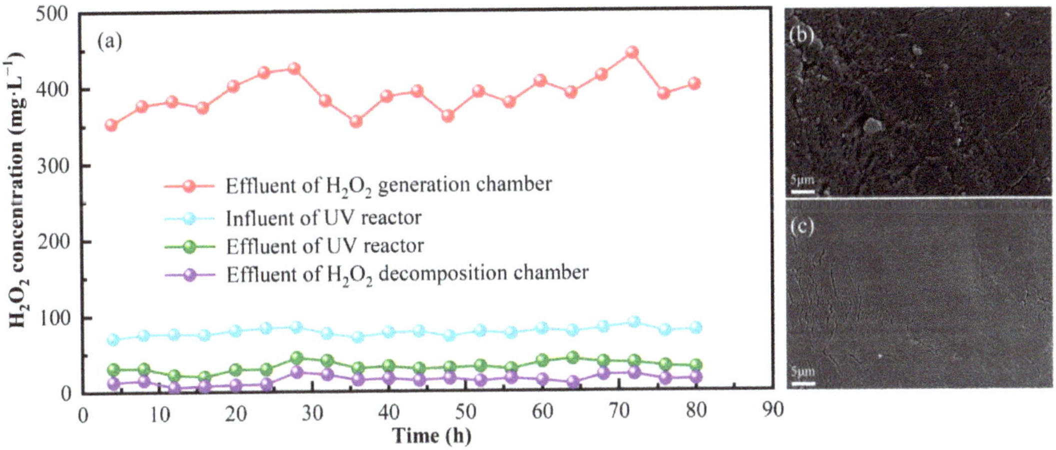

Figure 4. (**a**) Changes in H_2O_2 concentration; SEM images of the catalytic layer of the air diffusion electrode (**b**) before and (**c**) after the long-term experiment.

3.2. Removal of Organic Pollutants and Microorganisms

The pH of the influent water was around 7.65 and the pH of the effluent water from the UV reactor was around 6.88, with a decrease in the effluent pH compared with the influent pH. This phenomenon may be attributed to the fact that some natural organic matter in the influent water is oxidized by ·OH and degraded to weakly acidic products such as CO_2, HCO_3^-, and small-molecule carboxylic acids (Equation (6)) [30]:

$$R + \cdot OH \rightarrow CO_2 + H_2O \tag{6}$$

As an important indicator for assessing the content of organic matter in water, changes in A_{254} directly reflect the degree of degradation of macromolecular organic matter in water. When the A_{254} value changes, it not only indicates a decrease in the total amount of organic matter but also implies a significant change in the structure of organic matter, especially a decrease in aromaticity [31]. According to the results in Figure 5, the UV/H_2O_2 system has a stable reduction effect on A_{254}. The organic content of the effluent water was significantly reduced after the raw water was treated by electrolysis and the UV reactor. The mean A_{254} for the influent water was found to be 0.070, whereas the A_{254} of the effluent water from the electrolysis chamber averaged 0.057, indicating an average reduction of 18.6% in A_{254}. The average value of A_{254} in the effluent water from the UV reactor was 0.025, with an overall average reduction of 64.3% in A_{254}. The electrolytic treatment served as a preliminary purification, while the UV reactor further improved the effluent quality.

The chemical oxygen demand concentration is a measure that reflects the level of contamination by reducing agents in water and is considered one of the significant indicators for evaluating the extent of organic pollution in aquatic environments. As depicted in Figure 6, the system exhibits a pronounced efficacy in the removal of COD, signifying that the process efficiently eliminates the reducing substances present in the water. The mean concentration of COD in the raw water was determined to be 37 $mg \cdot L^{-1}$. After the electrolysis treatment within the electrolytic cell, the mean COD concentration in the effluent water decreased to 31.7 $mg \cdot L^{-1}$, indicating an average removal efficiency of 14.3%. This result shows that the electrolysis process plays a role in the removal of COD, and parts of the reducing substances are directly oxidized on the anode surface, but there is still room for improvement in the removal effect. When the effluent from the electrolysis chamber entered the UV reactor for advanced oxidation, the average value of COD was further reduced to 23.1 mg·L^{-1}, which was 37.6% lower than that of the raw water. The UV/H_2O_2 process is more efficient for COD removal, and the combination of electrolysis and UV reactor has a good synergistic effect on COD removal. The two processes can correspond to different types of pollutants, with easily oxidized pollutants being removed by anodic oxidation and difficult pollutants being removed by oxidation with strong oxidizing radicals generated by advanced oxidation reactions. The electrolysis chamber initially reduces the COD concentration, while the UV reactor further improves the removal efficiency, resulting in a significant improvement of the effluent water quality.

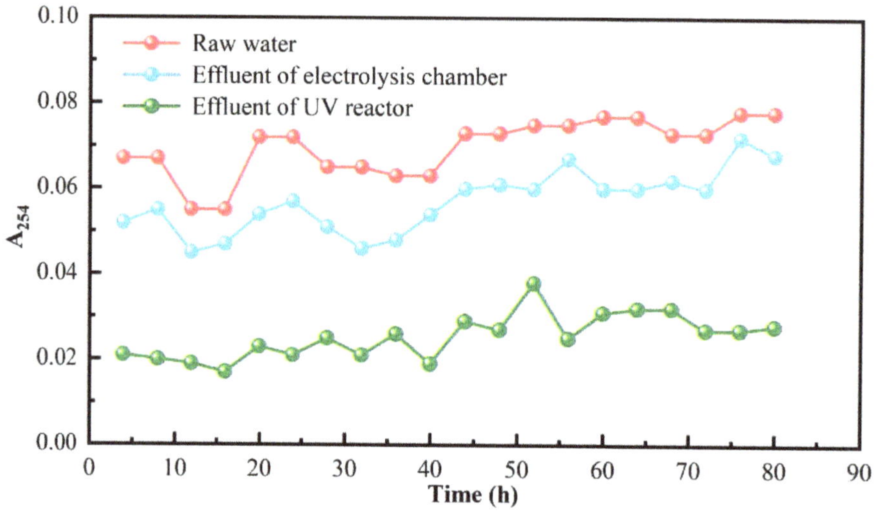

Figure 5. Change in absorbance at 254 nm.

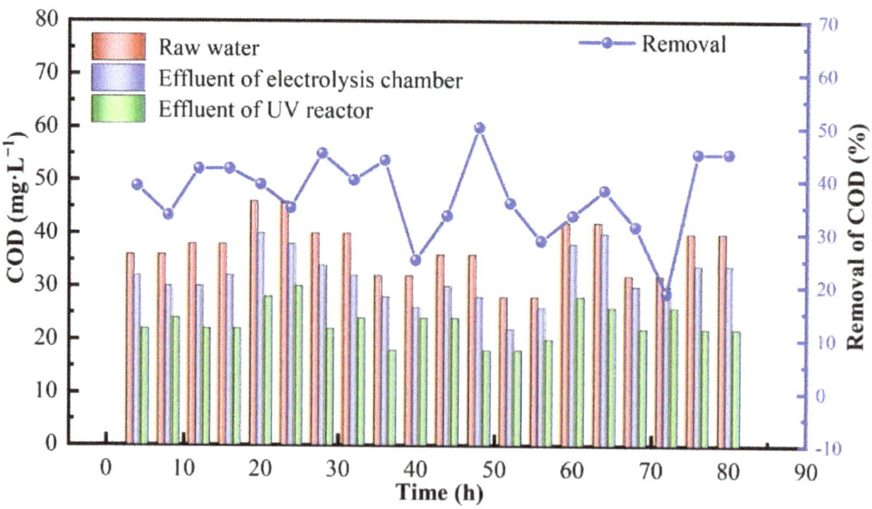

Figure 6. Changes in chemical oxygen demand content.

As illustrated in Figure 7, total organic carbon measurements were conducted on both the raw water and the water samples treated in the electrolysis chamber and UV reactor. This study aimed to assess the effectiveness of these treatments in removing organic pollutants. The average TOC content of the raw water was 7.35 mg·L^{-1}, showing some organic pollution load. After the electrolysis reaction occurred in the electrolysis chamber, the average TOC of the effluent water was reduced to 6.35 mg·L^{-1}, with a removal rate of 13.6%. This change indicates that the electrolysis process has an impact on the removal of organic pollutants from water. The electrolysis process reduces the TOC concentration through oxidation and reduction mechanisms. Some of the organic matter is converted to more degradable forms or is removed directly from the water column. Nevertheless, the TOC concentrations in the discharged water from the electrolysis chamber remained at relatively high levels compared with the raw water samples. This indicates that although the electrolysis process is effective to some extent, it does not completely remove all organic pollutants. Following the wastewater from the electrolysis chamber's entry into the UV reactor and its subsequent participation in an advanced oxidation reaction, the average concentration of TOC in the wastewater was significantly reduced to 4.41 mg·L^{-1}, with a removal rate of 40.0%. This significant decreasing trend indicates that the UV reactor is more effective at removing organic pollutants than the electrolysis process alone. The advanced oxidation reaction was able to attack the organic molecules with a strong oxidant to mineralize them into harmless small molecules, thus achieving efficient TOC removal. Meanwhile, the stability of the effluent TOC concentration indicates that the process has reliable operation.

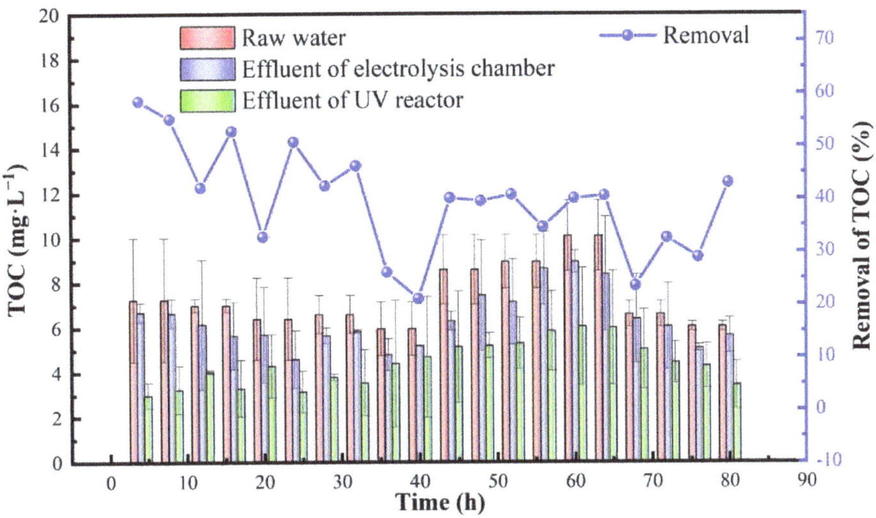

Figure 7. Changes in total organic carbon content.

As shown in Figure 8, (a–c) and (d–f) are the three-dimensional fluorescence spectrograms of the raw water, the effluent from the electrolysis chamber, and the effluent from the UV reactor at different sampling times, respectively. Different fluorescent substances correspond to different wavelength ranges, and blocks I, II, III, IV, and V on the spectrograms represent tyrosine, tryptophan, fulvic acid, dissolved microbial metabolites, and humic acid, respectively. A meticulous comparison of the three-dimensional fluorescence spectrograms of the raw water and the effluent from the electrolysis chamber revealed that the I and II blocks exhibited more pronounced degradation, whereas the changes observed in the other blocks were less pronounced. This finding indicates that the electrolytic treatment is effective in removing some small molecules of organic matter but less so in removing large molecules of organic matter, which suggests that it is not an effective method for eliminating these organic impurities. However, analysis of the three-dimensional fluorescence spectrogram of the effluent from the UV reactor revealed a notable shift in the fluorescence peaks. It can be observed that the three organic compounds tryptophan, fulvic acid, and humic acid, have undergone a complete degradation process, while tyrosine has also been significantly altered. It can be postulated that the disparate outcomes observed in block IV may be attributed to the preferential decomposition of high-molecular-weight fluorescence-absorbing organics, which results in the generation of lower-molecular-weight fluorescence-absorbing organics [32]. The preceding analysis permits the observation of the transformation and decomposition processes of aromatic compounds in water as well as the progressive degradation of macromolecular organics. This is evidenced by a reduction in the number of conjugated bonds in the chain structure and a shift from a linear to a nonlinear ring structure. This implies a reduction in the complexity of the organic molecules and the removal of specific functional groups, such as carbonyl and hydroxyl groups. It also reflects the chemical changes in the organic matter during the treatment process, resulting in improved water quality [33,34]. The results indicate that the advanced oxidation reaction in the UV reactor has a significant impact on the degradation of organic matter. Following the treatment, the larger relative molecular mass of fulvic acid and humic acid organics may be converted into small-molecule organics, with a reduction in the aromaticity of the organic matter in the water.

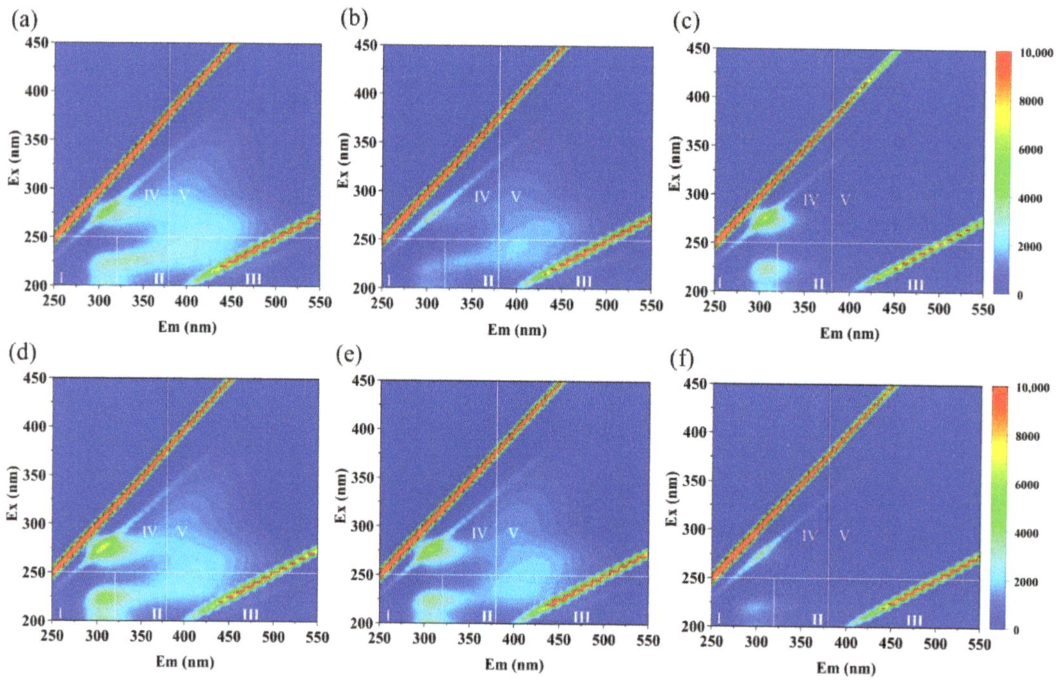

Figure 8. Three-dimensional fluorescence spectra of (**a**) the raw water, (**b**) the effluent from the electrolysis chamber, and (**c**) the effluent from the UV reactor for 40 h of unit operation; (**d**) the raw water, (**e**) the effluent from the electrolysis chamber, and (**f**) the effluent from the UV reactor for 48 h of unit operation.

Total bacterial counts, a key bioindicator, provide valuable insights into the quality of water and the efficacy of organic matter degradation. As illustrated in Figure 9, the total bacterial counts in the water samples of the raw water, the electrolysis chamber effluent, and the UV reactor effluent changed at six time points following a 24 h incubation period.

Firstly, a considerable number of colonies can be observed on the total-bacterial-count plates derived from the raw water samples, which indicates the presence of a certain concentration of bacteria in the water samples. The bacteria in question may have originated from a variety of environmental contaminants, including soil, plant and animal debris, or other water bodies. Following electrolysis treatment, the number of colonies was found to be significantly reduced. This indicates that electrolysis has a significant killing effect on bacteria. Following further treatment with a UV reactor, the number of colonies was found to have decreased to zero. Ultraviolet light has the effect of destroying the DNA structure of bacteria, rendering them incapable of reproduction. At the same time, H_2O_2 produces the strong oxidant ·OH, which further kills the bacteria. The combined effect of this advanced oxidizing reaction was the complete removal of the total number of bacteria. The trend of total bacterial count fully proved the important role of electrolysis and UV reactor in water quality improvement and organic matter degradation.

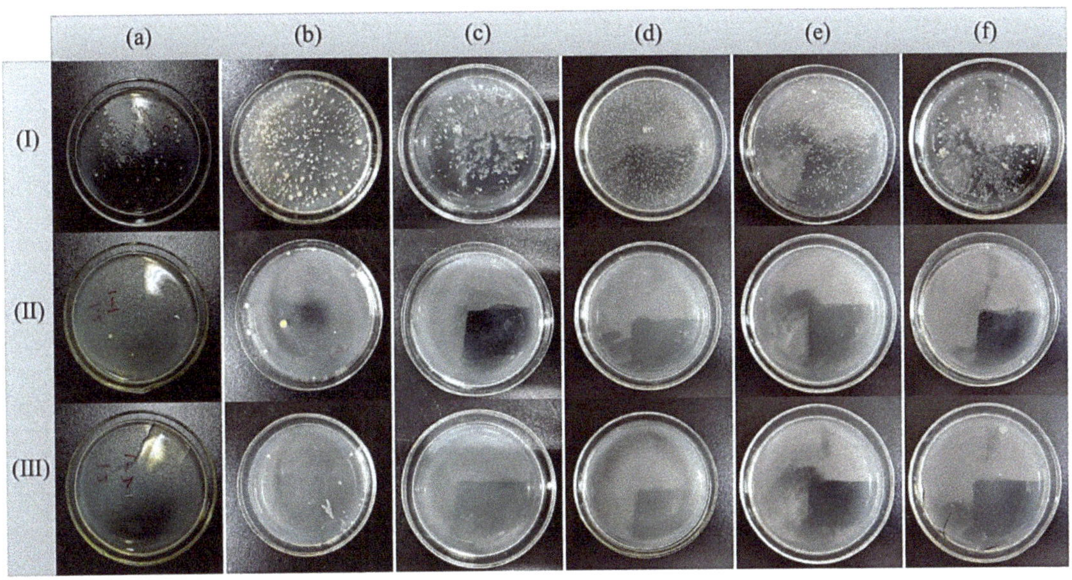

Figure 9. The bacterial colonies formed on the agar plates after incubation with the (**I**) raw water, (**II**) the effluent from the electrolysis chamber, and (**III**) the effluent from the UV reactor after (**a**) 8, (**b**) 16, (**c**) 24, (**d**) 32, (**e**) 40, and (**f**) 48 h of continuous operation of the equipment, respectively. For example, (**II**) (**a**) shows a bacterial culture plate of the effluent from the electrolysis chamber following 8 h of equipment operation.

3.3. *Energy Consumption Analysis*

For any technology, energy is a crucial factor that must be taken into account when evaluating its overall performance. In this paper, actual lake water samples were used as the treatment object, and according to the calculation, the operating cost of the system was 24.8 kWh·m^{-3}, of which the energy consumption of the raw water electrolysis chamber was 0.8 kWh·m^{-3}, that of the production of H_2O_2 was 3.2 kWh·m^{-3}, and that of the ultraviolet lamp was 20.8 kWh·m^{-3}. Some scholars have investigated the removal effect of the UV/H_2O_2 process on a variety of single emerging organic pollutants in a simulated drinking water environment. The results showed that the unit energy consumption of the UV/H_2O_2 process ranged from 0.17 to 2.38 kWh·m^{-3} under the condition of 90% removal of the target [35]. The unit energy consumption of the UV/H_2O_2 process was affected by the operation scale and the water quality [36]: from laboratory research to production tests, the unit energy consumption of the process decreased significantly with the expansion of the test scale, indicating that the expansion of the operation scale improved the energy utilization efficiency. The impact of raw water quality on the process is mainly reflected in the water transmittance and the content of free radical scavengers. The higher the concentration of scavengers (e.g., inorganic anions) and the higher the UV absorption of the raw water, the higher the unit energy consumption.

4. Conclusions

This study was undertaken to design an advanced oxidation process using electrogenerated H_2O_2 and UV irradiation and to evaluate the long-term performance of lake water treatment. The individual H_2O_2 generation chamber exhibited stable and highly efficient H_2O_2 production throughout the operation period of 80 h, whereas H_2O_2 alone could not oxidize the organic pollutants in the lake water. With the catalysis of UV, the system developed in this study showed significant mineralization capacity based on the results of A_{254}, COD, TOC, and fluorescence. In addition, the reactor could effectively kill bacteria and improve the microbiological safety of the water body. The proposed UV/H_2O_2 reactor has the potential for efficient and feasible practical application in surface water treatment. This provides a new and effective means of accessing clean water in areas where it is scarce. However, more effort is needed to explore the performance of the proposed reactor for treating source waters with varied quality and quantity to comprehensively evaluate the applicability for different water requirements.

Author Contributions: Conceptualization, H.H. and S.X. (Shiwei Xie); methodology, H.H. and S.X. (Shiwei Xie); validation, H.H., Y.Y., W.J., J.L., Z.W., S.X. (Shengjie Xiang), C.X. and J.X.; formal analysis, H.H., Y.Y., W.J., J.L., Z.W., S.X. (Shengjie Xiang), C.X. and J.X.; investigation, H.H., Y.Y., W.J., J.L., Z.W., S.X. (Shengjie Xiang), C.X. and J.X.; resources, S.X. (Shiwei Xie) and N.Z.; data curation, H.H. and S.X. (Shiwei Xie); writing—original draft preparation, H.H.; writing—review and editing, H.H., S.X. (Shiwei Xie), A.Q. and N.Z.; supervision, S.X. (Shiwei Xie); project administration, H.H. and S.X. (Shiwei Xie); funding acquisition, N.Z. All authors have read and agreed to the published version of the manuscript.

Funding: This research was funded by the National Natural Science Foundation of China (NSFC, No. 42307311, 51808415), Hebei Natural Science Foundation (No. D2022111001), Central Guiding Local Science and Technology Development Foundation of Hebei Province (No. 236Z7602G), Open Foundation of State Key Laboratory of Environmental Criteria and Risk Assessment, Chinese Research Academy of Environmental Sciences (No. SKLECRA2023OFP02), Science Research Project of Hebei Education Department (No. BJK2022064), Scientific Research Project of Hengshui University (No. 2021GC16), and Hebei Academician Cooperation Key Unit of Hengshui University.

Data Availability Statement: Data will be made available on request.

Conflicts of Interest: The authors declare no conflicts of interest.

References

1. Carpenter, S.R.; Stanley, E.H.; Zanden, M.J.V. State of the World's Freshwater Ecosystems: Physical, Chemical, and Biological Changes. *Annu. Rev. Environ. Resour.* **2011**, *36*, 75–99. [CrossRef]
2. Mammides, C. A global assessment of the human pressure on the world's lakes. *Glob. Environ. Chang.* **2020**, *63*, 102084. [CrossRef]
3. Le, C.; Zha, Y.; Li, Y.; Sun, D.; Lu, H.; Yin, B. Eutrophication of lake waters in China: Cost, causes, and control. *Environ. Manag.* **2010**, *45*, 662–668. [CrossRef] [PubMed]
4. Ran, J.; Xiang, R.; He, J.; Zheng, B. Spatiotemporal variation and driving factors of water quality in Yunnan-Guizhou plateau lakes, China. *J. Contam. Hydrol.* **2023**, *254*, 104141. [CrossRef] [PubMed]
5. Ho, L.T.; Goethals, P.L.M. Opportunities and Challenges for the Sustainability of Lakes and Reservoirs in Relation to the Sustainable Development Goals (SDGs). *Water* **2019**, *11*, 1462. [CrossRef]
6. Meffe, R.; de Bustamante, I. Emerging organic contaminants in surface water and groundwater: A first overview of the situation in Italy. *Sci. Total Environ.* **2014**, *481*, 280–295. [CrossRef] [PubMed]
7. Rodil, R.; Quintana, J.B.; Concha-Graña, E.; López-Mahía, P.; Muniategui-Lorenzo, S.; Prada-Rodríguez, D. Emerging pollutants in sewage, surface and drinking water in Galicia (NW Spain). *Chemosphere* **2012**, *86*, 1040–1049. [CrossRef] [PubMed]
8. Bui, X.T.; Vo, T.P.T.; Ngo, H.H.; Guo, W.S.; Nguyen, T.T. Multicriteria assessment of advanced treatment technologies for micropollutants removal at large-scale applications. *Sci. Total Environ.* **2016**, *563–564*, 1050–1067. [CrossRef]

9. Chu, W.-H.; Gao, N.-Y.; Templeton, M.R.; Yin, D.-Q. Comparison of inclined plate sedimentation and dissolved air flotation for the minimisation of subsequent nitrogenous disinfection by-product formation. *Chemosphere* **2011**, *83*, 647–651. [CrossRef]
10. Jia, P.; Zhou, Y.; Zhang, X.; Zhang, Y.; Dai, R. Cyanobacterium removal and control of algal organic matter (AOM) release by UV/H_2O_2 pre-oxidation enhanced Fe(II) coagulation. *Water Res.* **2018**, *131*, 122–130. [CrossRef]
11. Rodriguez-Narvaez, O.M.; Peralta-Hernandez, J.M.; Goonetilleke, A.; Bandala, E.R. Treatment technologies for emerging contaminants in water: A review. *Chem. Eng. J.* **2017**, *323*, 361–380. [CrossRef]
12. Zhu, Y.; Wu, M.; Gao, N.; Chu, W.; Li, K.; Chen, S. Degradation of phenacetin by the UV/chlorine advanced oxidation process: Kinetics, pathways, and toxicity evaluation. *Chem. Eng. J.* **2018**, *335*, 520–529. [CrossRef]
13. Lee, M.; Merle, T.; Rentsch, D.; Canonica, S.; von Gunten, U. Abatement of Polychoro-1,3-butadienes in Aqueous Solution by Ozone, UV Photolysis, and Advanced Oxidation Processes (O_3/H_2O_2 and UV/H_2O_2). *Environ. Sci. Technol.* **2017**, *51*, 497–505. [CrossRef] [PubMed]
14. Yang, Y.; Pignatello, J.J.; Ma, J.; Mitch, W.A. Effect of matrix components on UV/H_2O_2 and UV/$S_2O_8^{2-}$ advanced oxidation processes for trace organic degradation in reverse osmosis brines from municipal wastewater reuse facilities. *Water Res.* **2016**, *89*, 192–200. [CrossRef] [PubMed]
15. Rozas, O.; Vidal, C.; Baeza, C.; Jardim, W.F.; Rossner, A.; Mansilla, H.D. Organic micropollutants (OMPs) in natural waters: Oxidation by UV/H_2O_2 treatment and toxicity assessment. *Water Res.* **2016**, *98*, 109–118. [CrossRef] [PubMed]
16. Ali, M.A.; Maafa, I.M.; Qudsieh, I.Y. Photodegradation of Methylene Blue Using a UV/H_2O_2 Irradiation System. *Water* **2024**, *16*, 453. [CrossRef]
17. Song, W.; Li, C.; Du, Z.; Yue, J.; Sun, W.; Hou, L.; Liu, J.; Jia, R. Research about Organic Matter Removal and Biofilms Development of Pilot-Scale UV/H_2O_2-BAC Process. *Water* **2021**, *13*, 565. [CrossRef]
18. Zhou, W.; Meng, X.; Gao, J.; Alshawabkeh, A.N. Hydrogen peroxide generation from O_2 electroreduction for environmental remediation: A state-of-the-art review. *Chemosphere* **2019**, *225*, 588–607. [CrossRef] [PubMed]
19. Salmerón, I.; Plakas, K.V.; Oller, I.S.I.; Maldonado, M.I.; Karabelas, A.J.; Malato, S. Optimization of electrocatalytic H_2O_2 production at pilot plant scale for solar-assisted water treatment. *Appl. Catal. B Environ. Energy* **2019**, *242*, 327–336. [CrossRef]
20. Wang, Y.; Yu, G.; Deng, S.; Huang, J.; Wang, B. The electro-peroxone process for the abatement of emerging contaminants: Mechanisms, recent advances, and prospects. *Chemosphere* **2018**, *208*, 640–654. [CrossRef]
21. Zhao, Y.; Deng, N.; Fan, Z.; Hu, Z.; Fan, L.; Zhou, J.; Huang, X. On-site H_2O_2 electro-generation process combined with ultraviolet: A promising approach for odorous compounds purification in drinking water system. *Chem. Eng. J.* **2022**, *430*, 132829. [CrossRef]
22. Li, Y.; Zhang, Y.; Xia, G.; Zhan, J.; Yu, G.; Wang, Y. Evaluation of the technoeconomic feasibility of electrochemical hydrogen peroxide production for decentralized water treatment. *Front. Environ. Sci. Eng.* **2020**, *15*, 122417–122851. [CrossRef]
23. Ma, Y.; Zhao, E.; Xia, G.; Zhan, J.; Yu, G.; Wang, Y. Effects of water constituents on the stability of gas diffusion electrode during electrochemical hydrogen peroxide production for water and wastewater treatment. *Water Res.* **2023**, *229*, 119503. [CrossRef] [PubMed]
24. Liao, P.; Yuan, S.; Chen, M.; Tong, M.; Xie, W.; Zhang, P. Regulation of electrochemically generated ferrous ions from an iron cathode for Pd-catalytic transformation of MTBE in groundwater. *Environ. Sci. Technol.* **2013**, *47*, 7918–7926. [CrossRef] [PubMed]
25. Li, L.; Li, Y.; Liang, J.; Zhao, Z. Elimination method of H_2O_2 interference in determination of COD_{Cr}. *Exp. Technol. Manag.* **2021**, *38*, 47–53+60. (In Chinese) [CrossRef]
26. Liu, Y.; Quan, X.; Fan, X.; Wang, H.; Chen, S. High-yield electrosynthesis of hydrogen peroxide from oxygen reduction by hierarchically porous carbon. *Angew. Chem.* **2015**, *54*, 6837–6841. [CrossRef]
27. Zhang, Q.; Zhou, M.; Ren, G.; Li, Y.; Li, Y.; Du, X. Highly efficient electrosynthesis of hydrogen peroxide on a superhydrophobic three-phase interface by natural air diffusion. *Nat. Commun.* **2020**, *11*, 1731. [CrossRef] [PubMed]
28. Jin, Q.; Wang, H.; Hu, C.; Chen, Z.; Wang, X. Effects of NOM on the degradation of chloramphenicol by UV/H_2O_2 and the characteristics of degradation products. *Sep. Purif. Technol.* **2018**, *191*, 108–115. [CrossRef]
29. Sarathya, S.R.; Stefan, M.I.; Royce, A.; Mohseni, M. Pilot-scale UV/H_2O_2 advanced oxidation process for surface water treatment and downstream biological treatment: Effects on natural organic matter characteristics and DBP formation potential. *Environ. Technol.* **2011**, *33*, 1709–1718. [CrossRef]
30. Wang, Y.; Li, G.; Song, W.; Sun, S.; Zhang, S.; Li, M.; Jia, R. Removal Characteristics of Organic Matter in Effluent from Conventional Sand Filtration by UV/H_2O_2 Process. *China Water Wastewater* **2022**, *38*, 60–66. (In Chinese) [CrossRef]
31. García-Morales, M.A.; Roa-Morales, G.; Barrera-Díaz, C.; Miranda, V.M.; Hernández, P.B.; Silva, T.B.P. Integrated Advanced Oxidation Process (Ozonation) and Electrocoagulation Treatments for Dye Removal in Denim Effluents. *Int. J. Electrochem. Sci.* **2013**, *8*, 8752–8763. [CrossRef]
32. Valencia, S.; Marín, J.; Velásquez, J.; Restrepo, G.; Frimmel, F.H. Study of pH effects on the evolution of properties of brown-water natural organic matter as revealed by size-exclusion chromatography during photocatalytic degradation. *Water Res.* **2011**, *46*, 1198–1206. [CrossRef] [PubMed]
33. Świetlik, J.; Sikorska, E. Application of fluorescence spectroscopy in the studies of natural organic matter fractions reactivity with chlorine dioxide and ozone. *Water Res.* **2004**, *38*, 3791–3799. [CrossRef]
34. Valencia, S.; Marín, J.M.; Restrepo, G.; Frimmel, F.H. Evaluation of natural organic matter changes from Lake Hohloh by three-dimensional excitation–emission matrix fluorescence spectroscopy during TiO_2/UV process. *Water Res.* **2014**, *51*, 124–133. [CrossRef] [PubMed]

35. Guo, K.; Wu, Z.; Yan, S.; Yao, B.; Song, W.; Hua, Z.; Zhang, X.; Kong, X.; Li, X.; Fang, J. Comparison of the UV/chlorine and UV/H_2O_2 processes in the degradation of PPCPs in simulated drinking water and wastewater: Kinetics, radical mechanism and energy requirements. *Water Res.* **2018**, *147*, 184–194. [CrossRef]
36. Miklos, D.B.; Remy, C.; Jekel, M.; Linden, K.G.; Drewes, J.E.; Hübner, U. Evaluation of advanced oxidation processes for water and wastewater treatment—A critical review. *Water Res.* **2018**, *139*, 118–131. [CrossRef]

water

MDPI

Article

Effect of UV-LED Wavelength on Reactive Species Photogeneration from Dissolved Organic Matter

Ze-Chong Guo [1,2,3], Lu Zhang [3], Yong Chen [1,*], Cheng Huang [1,2,3] and Zhi-Min Liao [2]

1 School of Environmental and Engineering, Huazhong University of Science and Technology, Wuhan 430074, China; guozechong@just.edu.cn (Z.-C.G.)
2 Jiangxi Jindalai Environmental Protection Co., Ltd., Nanchang 330100, China
3 School of Environmental and Chemical Engineering, Jiangsu University of Science and Technology, Zhenjiang 212100, China
* Correspondence: ychen@mail.hust.edu.cn

Abstract: The photogeneration of reactive species from dissolved organic matter (DOM) plays a crucial role in the photochemical and photobiochemical processes in natural aquatic systems. However, the impact of the ultraviolet (UV) wavelength on the photogeneration of reactive species by different sources of DOM remains unclear. In this study, UV light at four wavelengths (365 nm, 310 nm, 280 nm, and 260 nm) provided by UV-LEDs were irradiated onto three types of DOM: humic acid (HA), fulvic acid (FA), and effluent organic matter (EfOM). Three reactive species produced by DOM, including excited triplet-state DOM ($^3DOM^*$), singlet oxygen ($^1O_2$), and hydroxyl radicals (•OH), were determined. UV_{365} proved to be the most efficient wavelength for generating $^1O_2$ and •OH, with formation rates of 3.47×10^{-6} M s^{-1} and 1.67×10^{-8} M s^{-1}, respectively, with the addition of FA and EfOM. The highest steady-state concentrations of all three reactive species were also generated under UV_{365}, reaching 3.00×10^{-13} M ($^3DOM^*$) and 1.64×10^{-11} M ($^1O_2$) with the FA addition, and 1.44×10^{-10} M (•OH) with the EfOM. Across the different DOM sources, UV_{365} obtained the maximum quantum yields of reactive species, indicating the stronger effect of UV_{365} on inducing the photosensitization of DOM compared to the other shorter wavelengths. This study expands our understanding of the photochemistry of DOM in aquatic environments.

Keywords: dissolved organic matter; quantum yield; reactive species; steady-state concentration; UV-LED irradiation

1. Introduction

Ultraviolet (UV) radiation has garnered significant attention due to its specific impact on photochemical processes [1,2]. It has been proven to play a key role in remediating contamination in aquatic systems [3]. UV radiation encompasses wavelengths ranging from 100 to 400 nm and is further divided into four bands, namely, UVA (long-wave UV, 320~400 nm), UVB (middle-wave UV, 290~320 nm), UVC (short-wave UV, 200~290 nm), and UVD (vacuum UV, 100~200 nm) [4]. Sunlight serves as the primary source of UV radiation in the environment, with UVA making up approximately 95% of the sunlight radiating to the ground and UVB making up the remaining 5%. Other radiations that are considered harmful to life are largely filtered by the atmosphere [5].

UV irradiation can directly degrade certain classes of contaminants in an aqueous solution, such as iodinated trihalomethanes [6], ametryn [7], fluconazole [8], diclofenac [9], and so on. In addition to direct photodegradation, another significant process in contaminant degradation is the photosensitization that occurs in the presence of dissolved organic matter (DOM). DOM is prevalent in natural aquatic environments and has strong absorption for the sunlight spectrum. When exposed to light irradiation, DOM is excited to the singlet state ($^1DOM^*$) from the ground state (S_0), and the non-steady-state species are prone to transformation through two pathways: (i) the release of energy via fluorescence emissions and a radiationless transition back to the S_0, and (ii) by undergoing intersystem crossings through spin–orbital couplings to the excited triplet state ($^3DOM^*$) [10,11]. $^3DOM^*$ can directly participate in contaminant remediation through energy or electron-transfer interactions and induce the formation of singlet oxygen ($^1O_2$) and hydroxyl radicals (•OH) in the presence of oxygen [12]. •OH are well known for their high efficacy in attenuating a broad range of contaminants, while $^1O_2$ has also been confirmed to enhance the degradation of certain contaminants and pathogenic microorganisms [13]. For instance, the photodegradation of 17 α-ethinylestradiol was notably enhanced by the addition of DOM, with the contributions of direct and photosensitization degradation identified as 27% and 73%, respectively [14]. The degradation of roxithromycin was also confirmed to be promoted by DOM, with the photogeneration of hydroxyl radicals (•OH) identified as the major contributor to the photosensitization effect [15]. Understanding the photoreactivity of DOM is essential for assessing the fate of organic contaminants in aquatic systems.

The photochemical behaviors of different types of DOM are known to be closely correlated with their source and molecular characteristics [16]. Natural DOM, primarily originating from allochthonous sources, is composed of humic acid (HA) and fulvic acid (FA), with an apparent molecular weight ranging from 1.5 to 7 kDa. With the increasing capacity of wastewater treatment, the discharge of effluent into the natural environment is on the rise. In some waterways, effluent from wastewater treatment has become the dominant component, which emphasizes the importance of understanding the photochemical effects of effluent organic matter (EfOM) [12]. EfOM is primarily autochthonous and mainly comprises soluble microbial products and macromolecules with a higher apparent molecular weight exceeding 20 kDa [17]. In comparison to terrestrial-origin DOM, such as HA and FA, EfOM generally exhibits fewer aromatic constituents, specific UV absorbance at 254 nm, and higher apparent quantum yields of reactive species [18].

The photochemical behaviors of various types of DOM induced by UV irradiation have been extensively reported. However, the impact of the UV wavelength on the efficiency of reactive species photogeneration is still unrevealed. In previous investigations, mercury lamps and xenon lamps were frequently utilized as illumination sources, which normally provide broad emission spectra rather than single-wavelength emission spectra [16]. With the advancement of light-emitting diode (LED) technology, UV-LEDs have emerged as a promising alternative to traditional UV-light sources. They offer advantages such as emissions at a single wavelength, higher energy efficiency, a longer lifespan, more constant light intensity, and improved heating control [19].

In this work, four UV-LED lights with wavelengths of 365 nm (UVA), 310 nm (UVB), 280 nm (UVC), and 260 nm (UVC) were employed to investigate their impacts on the generation of photochemically produced reactive intermediates. The photosensitization effects of three DOM sources, namely, HA, FA, and EfOM, were compared across various UV-LED wavelengths. The steady-state concentrations and quantum yields of the reactive species ($^3DOM^*$, $^1O_2$, and •OH) were systematically assessed. The results of this study enhance our understanding of DOM photosensitization and reactive species generation in natural water systems, and they can contribute to the development of UV-LED-based photochemical water treatment technologies.

2. Materials and Methods

2.1. Photochemical Irradiation Experiment

Irradiation experiments were conducted in a customized quartz beaker featuring a diameter of 3.85 cm, an empty volume of 50 mL, and a working volume of 40 mL, as illustrated in Figure 1. The UV-LED light source (Yonglin Optoelectronics Co., Ltd., Shenzhen, China) was affixed to the outside of the bottom of the beaker, as described in previous literature [20]. The photochemical reactor was placed on a magnetic stirrer that provided a stirring speed of 400 rpm to achieve a homogeneous mixture. A hole with a rubber stopper was made in the cover of the quartz lid to facilitate sampling with disposable syringes. In the experimental setup, a DOM solution (5.0 mg C L^{-1}) with a probe molecule at pH 7.0 was added into the reactor. Then, the solution underwent irradiation using an LED light source, and aliquots of 0.5 mL were periodically withdrawn at predetermined intervals from the reactor for analysis.

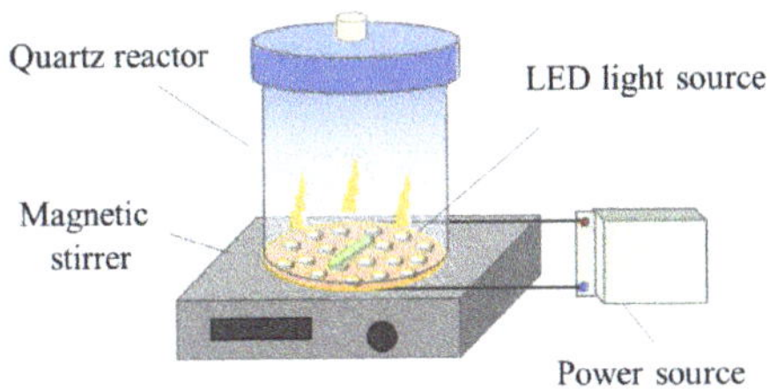

Figure 1. Schematic diagram of reactor configuration.

2.2. Materials

Wastewater effluent (EfOM) was obtained from a local wastewater treatment plant in Wuhan, China. The EfOM sample was isolated by a solid-phase extraction process using PPL cartridges (Agilent Technologies, Santa Clara, CA, USA), as previously reported. Humic acid (HA) and fulvic acid (FA) were purchased from Henan Changsheng Chemical Co., Ltd (Luohe, China) and Sigma-Aldrich (St. Louis, MO, USA), respectively, and they were purified according to the method widely accepted by the International Humic Substances Society (IHSS). All DOM samples were stored in the dark and kept at 4.0 °C before the experiments.

Molecule probes, including 2,4,6-trimethylphenol (TMP) (99%), Furfuryl alcohol (FFA) (99%), and Terephthalic acid (TPA) (99%), were all purchased from Sigma-Aldrich (St. Louis, MO, USA). Mobile phases, including acetonitrile and methanol, were liquid-chromatographic-grade. Others were analytical-reagent-grade. Deionized water was used in all experiments.

2.3. UV-LED Intensity Determination

The intensities of the four UV-LED light sources were determined before the photochemical reaction experiment. The light intensities of the UV-LEDs with wavelengths of 260 nm, 280 nm, and 310 nm were measured using iodide/iodate chemical actinometry [21]. The 365 nm UV-LED intensity was estimated by potassium ferrioxalate actinometry [22]. The light intensities and corresponding irradiances of the UV-LED light sources under different wavelengths are presented in Table 1.

Table 1. Light intensities and irradiances of UV-LEDs under different wavelengths.

Wavelength (nm)	Light Intensity (mW cm^{-2})	Light Irradiance (Einstein L^{-1} s^{-1})
260	0.93	5.87×10^{-7}
280	3.16	2.15×10^{-6}
310	5.49	4.14×10^{-6}
365	1.18	1.05×10^{-6}

2.4. Detection of Reactive Species

TMP, FFA, and TPA were commissioned as molecular probes for the ^{3}DOM*, $^1O_2$, and •OH, respectively. These compounds were chosen because their direct degradation processes were inactive, and their molar absorption coefficients and quantum yields were found to be negligible. The quantification of the TMP, FFA, and photoproduct of TPA (HTPA) was carried out using high-performance liquid chromatography (Essentia LC-15C, Shimadzu, Kyoto, Japan) equipped with a UV–vis detector (Essentia SPD-15C) and an HC-C18 column (5 μm, 250 mm × 4.6 mm, Agilent, USA). TMP was determined at a detection wavelength of 280 nm, utilizing acetonitrile, water, and phosphoric acid (*v*:*v*:*v* = 70:30:0.1) as the mobile phases with a flow rate of 1.0 mL min^{-1}; FFA was determined at 220 nm with a mobile phase of methanol and water (v:v = 35:65) under a flow rate of 1.0 mL min^{-1}; HTPA was detected at 315 nm using acetonitrile and water (v:v = 33:67) with a flow rate of 0.8 mL min^{-1}.

In the determination of the ^{3}DOM*, the solutions were sparged with N_2 to eliminate the dissolved oxygen during the entire reaction period. For the measurements of both the ^{3}DOM* and $^1O_2$, methanol (0.1 M) was introduced to quench the photogeneration of the •OH. The degradation rate constants of the TMP (k_{TMP}) and FFA (k_{FFA}), as well as the formation rate constant of the HTPA (k_{HTPA}), were obtained by fitting the data based on pseudo-first-order kinetics.

In this study, the value of the formation rate of the ^{3}DOM* ($R_{^3DOM^*}$) is approximately equal to the initial transformation rate of the TMP (R_{TMP}) because the scavenging rate of the TMP was much greater than the physical quenching constant of the ^{3}DOM*. The $R_{^3DOM^*}$ and the steady-state concentration of the ^{3}DOM* can be calculated with the following equations:

$$\mathrm{TMP} + {}^3\mathrm{DOM}^* \rightarrow \mathrm{photoproduct} \tag{1}$$

$$\frac{d[\mathrm{TMP}]}{dt} = -k_{TMP,{}^3DOM*}[\mathrm{TMP}][{}^3\mathrm{DOM}^*]_{\mathrm{ss}} \tag{2}$$

$$R_{^3DOM^*} \approx R_{TMP} = k_{TMP}[\mathrm{TMP}]_0 \tag{3}$$

where $[^3\mathrm{DOM}^*]_{\mathrm{ss}}$ is the steady-state concentration of the ^{3}DOM*, M; [TMP] is the concentration of TMP, M; $[\mathrm{TMP}]_0$ is the initial concentration of TMP, 500 μM; $k_{TMP,{}^3DOM*}$ is the reaction rate constant, 3.0×10^9 M^{-1} s^{-1} [23,24].

The steady-state concentration of $^1O_2$ and the formation rate of $^1O_2$ ($R_{^1O_2}$) was quantified by the concentration variation in the FFA according to the following equations:

$$FFA + {}^1O_2 \rightarrow \mathrm{photoproduct} \tag{4}$$

$$\frac{d[\mathrm{FFA}]}{dt} = -k_{FFA,{}^1O_2}[FFA][{}^1\mathrm{O}_2]_{\mathrm{ss}} \tag{5}$$

$$k_{FFA} = k_{FFA,{}^1O_2}\frac{R_{^1O_2}}{k_{FFA,{}^1O_2}[FFA]_0 + k_{^1O_2}} \tag{6}$$

where $[^1O_2]_{\mathrm{ss}}$ is the steady-state concentration of $^1O_2$, M; [FFA] is the concentration of FFA, M; $[\mathrm{FFA}]_0$ is the initial concentration of FFA, 50 μM; $k_{FFA,{}^1O_2}$ is the reaction rate constant, 8.3×10^7 M^{-1} s^{-1} [25,26]; $k_{^1O_2}$ is the physical quenching rate of $^1O_2$, 2.5×10^5 M^{-1} s^{-1} [10].

The steady-state concentration of the •OH and the formation rate of the •OH ($R_{\bullet\mathrm{OH}}$) were quantified by the detection of the HTPA (fluorescent), which is the single-hydroxylation product of TPA (non-fluorescent) [27], according to the following equations:

$$TPA + \bullet OH \rightarrow HTPA \tag{7}$$

$$\frac{\mathrm{d[HTPA]}}{\mathrm{dt}} = 0.35 \times k_{TPA,\cdot OH}[\mathrm{TPA}][\bullet\mathrm{OH}]_{\mathrm{ss}} = 0.35 \times R_{\bullet\mathrm{OH}} \tag{8}$$

where $[\bullet\mathrm{OH}]_{\mathrm{ss}}$ is the steady-state concentration of •OH, M; [HTPA] and [TPA] are the concentrations of HTPA and TPA, M; $k_{TPA,\cdot OH}$ is the reaction rate constant, 3.3×10^9 M^{-1} s^{-1} [26,28].

2.5. Quantum Yield Calculation

The quantum yields of the ^{3}DOM*, $^1O_2$, and •OH were calculated as follows:

$$\Phi_i = \frac{R_i}{I_0\left(1 - 10^{\varepsilon_\lambda bc}\right)} \tag{9}$$

where R_i is the formation rate of the $^3DOM^*$, $^1O_2$, and •OH in the DOM solutions, M·s^{-1}; I_0 is the incident light intensity obtained by the actinometers, Einstein L^{-1} s^{-1}; ε_λ is the absorption coefficient of the DOM at a specific wavelength, L mg^{-1} cm^{-1}; b is the path length, cm; c is the concentration of the DOM, mg L^{-1}.

3. Results and Discussion

3.1. DOM Characteristics

The absorption spectra of the FA, HA, and EfOM utilized in this study exhibited broad and unstructured profiles, displaying an expected near-exponential decrease with the increasing wavelength in the ultraviolet and visible wavelengths (Figure 2). The absorption coefficients of the FA, HA, and EfOM were determined at wavelengths of 260 nm, 280 nm, 310 nm, and 365 nm and are summarized in Table 2. Among the three types of DOM, the FA displayed the highest absorption, followed by the HA, while the EfOM presented the lowest absorption. Based on the absorption spectra, the E2/E3 ratios (the ratio of absorbance at 254 nm and 365 nm), which are believed to be negatively correlated with the humification, aromaticity degree, and molecular weight of the DOM [29], were 2.3, 2.5, and 4.2 for the HA, FA, and EfOM, respectively. The evident differences in the UV–vis absorption spectra among the three DOM samples highlighted significant variations in their structures and compositions, which strongly influence their photochemistry.

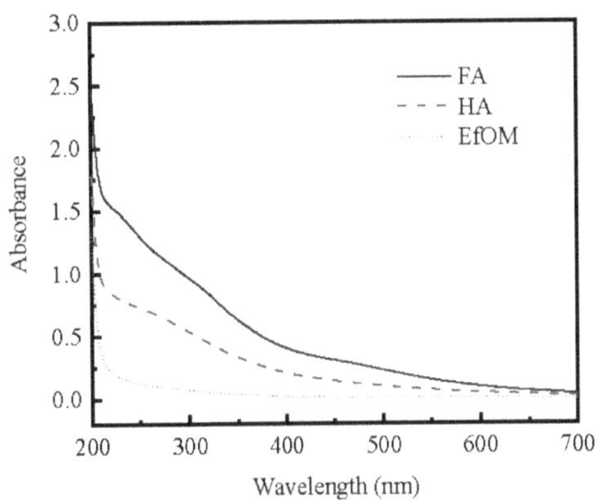

Figure 2. UV–vis absorption spectra of HA, FA, and EfOM ([DOM] = 10 mg C L^{-1}, pH = 7.0).

Table 2. Absorption coefficients of DOM samples at different wavelengths.

DOM	Absorption Coefficient (ε_λ, L mg^{-1} cm^{-1})			
	365 nm	310 nm	280 nm	260 nm
FA	0.0542	0.0902	0.1086	0.1215
HA	0.0285	0.0488	0.0620	0.0698
EfOM	0.0028	0.0063	0.0090	0.0110

3.2. Reactive Species Formation under UV-LED Irradiation

The reactivities of the reactive species with the probe molecules under various wavelengths of UV-LED irradiation are depicted in Figure 3. In the case of the TMP, the molecule probe for the $^3DOM^*$, a linear degradation pattern was observed, with the slopes of UV-LED$_{365}$ noticeably higher than those of the other irradiation wavelengths when HA and FA were added. UV-LED$_{280}$ demonstrated a comparable effect to UV-LED$_{365}$ in the EfOM group. Concerning the FFA, the molecule probe for the $^1O_2$, UV-LED$_{365}$ exhibited the highest degradation effect in all the DOM conditions, indicating the most efficient $^1O_2$ formation compared to the other wavelengths. The rate of HTPA (the molecule probe for

•OH) formation was higher with the HA and FA additions under UV-LED_{280}; however, UV-LED_{365} appeared to be more efficient with the EfOM addition.

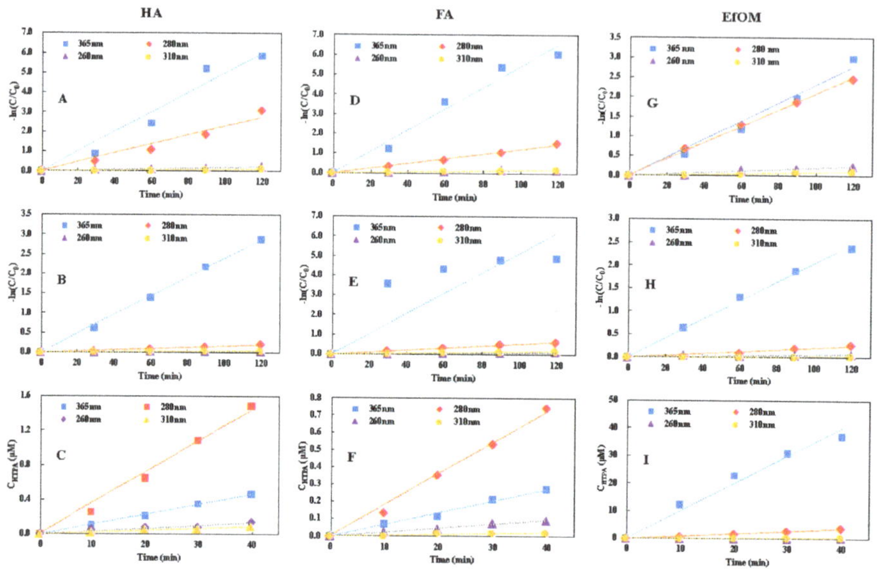

Figure 3. Reactivities of $^3DOM^*$ ((**A**) $^3HA^*$; (**D**) $^3FA^*$; (**G**) $^3EfOM^*$), $^1O_2$ (**B**,**E**,**H**), and •OH (**C**,**F**,**I**) with probe molecules as a function of time under various wavelengths of UV-LED irradiation. (**A**,**D**,**G**) TMP, 500 µM; (**B**,**E**,**H**) FFA, 50 µM; and (**C**,**F**,**I**) TPA, 1 mM ([DOM] = 5 mg C L^{-1}, pH = 7.0).

The formation of reactive species was significantly influenced by the variation in the DOM type, as detailed in Table 3. The HA and FA exhibited similar $^3DOM^*$ reaction rates under all irradiation conditions, with the highest k_{TMP} values recorded at 0.0536 and 0.0541 min^{-1}, respectively, under UV-LED_{365}. The corresponding formation rates ($R_{^3DOM^*}$) were 4.62×10^{-7} and 4.66×10^{-7} M s^{-1}, respectively. For the EfOM, UV-LED_{365} and UV-LED_{280} provided similar k_{TMP} values of 0.0232 and 0.0209 min^{-1}, with $R_{^3DOM^*}$ values of 2.00×10^{-7} and 1.80×10^{-7} M s^{-1}, respectively.

Table 3. Reaction rate constants (k) for probe molecules and formation rates (R) of reactive species at various UV-LED light wavelengths.

DOM	Wavelength (nm)	k_{TMP} (min^{-1})	k_{FFA} (min^{-1})	k_{HTPA} (min^{-1})	$R_{^3DOM^*}$ (10^{-7} M s^{-1})	$R_{^1O_2}$ (10^{-7} M s^{-1})	$R_{\bullet OH}$ (10^{-10} M s^{-1})
HA	365	0.0536	0.0244	0.0118	4.62	9.52	5.48
	310	0.0011	0.0004	0.0022	0.09	0.17	1.10
	280	0.0252	0.0018	0.0381	2.17	0.77	17.10
	260	0.0019	0.0003	0.0031	0.16	0.13	1.57
FA	365	0.0541	0.0817	0.0068	4.66	34.72	1.10
	310	0.0016	0.0015	0.0006	0.14	0.64	0.10
	280	0.0125	0.0053	0.0180	1.08	2.25	3.00
	260	0.0016	0.0010	0.0024	0.14	0.43	0.40
EFOM	365	0.0232	0.0205	0.9997	2.00	8.71	166.60
	310	0.0010	0.0002	0.0138	0.09	0.09	2.30
	280	0.0209	0.0021	0.0934	1.80	0.89	15.60
	260	0.0020	0.0005	0.0059	0.17	0.21	0.90

The peak reaction rates of the FFA appeared at 365 nm, irrespective of the DOM type, surpassing those at the other wavelengths by at least one order of magnitude. The highest k_{FFA} was 0.0817 min^{-1} by the FA, being 2.3- and 3.0-fold higher than those by the HA and EfOM, respectively. Correspondingly, the maximum formation rate of $^1O_2$ ($R_{^1O_2}$) was 34.72 $\times$ 10^{-7} M s^{-1} with the FA addition, and it was considerably lower in the HA and EfOM, recorded at 9.52 and 8.71 $\times$ 10^{-7} M s^{-1} under UV-LED$_{365}$. The k_{HTPA} in the EfOM group was noticeably higher than in the HA and FA groups, reaching 0.9997 min^{-1} at the wavelength of 365 nm, corresponding to a •OH ($R_{\bullet OH}$) formation rate of 166.60 $\times$ 10^{-10} M s^{-1}, two orders of magnitude higher than in the HA and FA. These results indicate that the EfOM, particularly under the irradiation of 365 nm, was much more efficient at •OH formation.

3.3. Steady State Concentration of Reactive Species

The steady-state concentration of reactive species is a key parameter for investigating photochemical processes and their environmental impacts. The steady-state concentrations of the ^{3}DOM*, $^1O_2$, and •OH are summarized in Figure 4. ^{3}DOM* is an important source of reactive oxygen species and can transform into $^1O_2$ and •OH upon UV irradiation. In the group with the HA addition, the steady-state concentrations of the ^{3}HA* were 10.6 $\times$ 10^{-15} M, 140.0 $\times$ 10^{-15} M, 6.1 $\times$ 10^{-15} M, and 297.8 $\times$ 10^{-15} M, respectively, under irradiations of 260 nm, 280 nm, 310 nm, and 365 nm UV-LEDs (see Figure 4A). This indicates that UV-LED$_{280}$ and UV-LED$_{365}$ were more efficient at the photosensitization of ^{3}HA* generation, with the [^{3}HA*]$_{ss}$ of UV-LED$_{365}$ being about 1.1-fold higher than that of UV-LED$_{280}$. UV-LED$_{365}$ was beneficial for the photogeneration of $^1O_2$ and yielded an [$^1O_2$]$_{ss}$ of 4.9 $\times$ 10^{-12} M, which was about 60, 13, and 80 times higher than that under UV-LED$_{260}$, UV-LED$_{280}$, and UV-LED$_{310}$, respectively. In contrast to the ^{3}HA* and $^1O_2$, the peak [•OH]$_{ss}$ of 51.8 $\times$ 10^{-17} M was recorded under UV-LED$_{280}$ irradiation, and it was 2.2-fold higher than that under 365 nm.

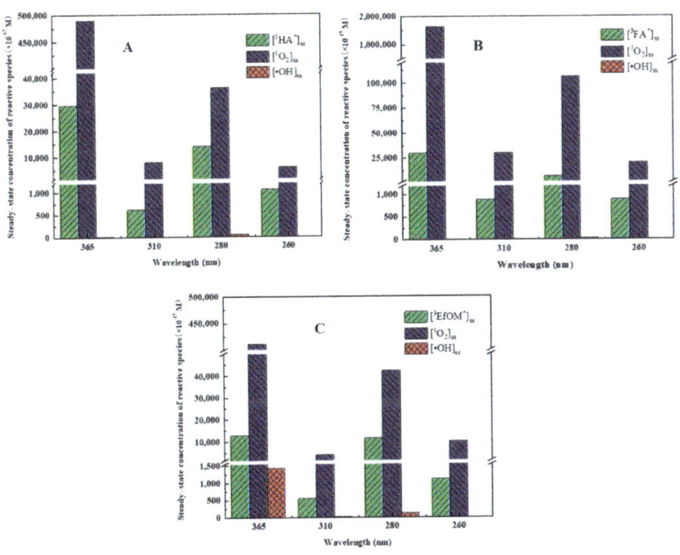

Figure 4. Steady−state concentrations of ^{3}DOM*, $^1O_2$, and •OH at HA (**A**), FA (**B**), and EfOM (**C**), adding conditions at varying wavelengths of 260 nm, 280 nm, 310 nm, and 365 nm.

For the FA, the influences of the UV-LED wavelength on the photogeneration of $^1O_2$ and ^{3}FA* were quite similar (Figure 4B). The maximum [^{3}FA*]$_{ss}$ of 300.6 $\times$ 10^{-15} M and [$^1O_2$]$_{ss}$ of 16.41 $\times$ 10^{-12} M were observed under UV-LED$_{365}$, which were about

2.3-fold and 15-fold higher those of UV-LED$_{280}$. Irradiations of UV-LED$_{310}$ and UV-LED$_{260}$ were inefficient, with steady-state concentrations two orders of magnitude lower. UV-LED$_{280}$ was the most efficient for •OH generation, demonstrating a [•OH]$_{ss}$ of 26.0 × 10^{-17} M, followed by UV-LED$_{365}$ (9.8 × 10^{-17} M) and UV-LED$_{260}$ (3.5 × 10^{-17} M). The photosensitization for •OH formation was inferior under UV-LED$_{310}$ irradiation.

In the experiment with the EfOM addition (Figure 4C), UV-LED$_{365}$ appears to be the most efficient for reactive species generation compared to UV-LED$_{310}$, UV-LED$_{280}$, and UV-LED$_{260}$. The highest steady-state concentrations of [^{3}EfOM*]$_{ss}$, [$^1O_2$]$_{ss}$, and [•OH]$_{ss}$ were 128.9 × 10^{-15} M, 41.2 × 10^{-13} M, and 1.4 × 10^{-14} M, respectively. UV-LED$_{280}$ exhibited a similar performance in ^{3}EfOM* photogeneration to that of UV-LED$_{365}$, with a [^{3}EfOM*]$_{ss}$ of 116.1 × 10^{-15} M. Similarly, the photosensitizations of UV-LED$_{310}$ and UV-LED$_{260}$ were relatively weak, and the production of reactive species was remarkably low.

Among all the reactive species, $^1O_2$ emerged as the predominant species under various wavelength conditions, with the [$^1O_2$]$_{ss}$ consistently surpassing both the ^{3}DOM* and [•OH]$_{ss}$, fluctuating within a wide range. Typically, the steady-state concentration of $^1O_2$ in natural aquatic environments is reported to be from around 10^{-15} to 10^{-13} M L^{-1} [30]. However, in this study, the maximum [$^1O_2$]$_{ss}$ reached 4.1 × 10^{-10} M L^{-1} in the presence of the FA addition under UV$_{365}$, indicating a substantial formation of $^1O_2$ through the photosensitization of FA induced by UV$_{365}$. In addition to $^1O_2$, •OH also play a critical role in the degradation of pollutants and other processes in natural waters. As previous reported, the steady-state concentration of •OH in natural water is typically from around 10^{-18} to 10^{-16} M L^{-1} [31]. The maximum [•OH]$_{ss}$ in this study obtained with the addition of EfOM under UV$_{365}$ was notably higher, reaching a value of 3.6 × 10^{-13} M L^{-1}. With such a considerable level of •OH, the system with the EfOM addition under UV$_{365}$ would be highly beneficial for the removal of organic pollutants.

In general, the steady-state concentrations of the ^{3}DOM*, $^1O_2$, and •OH were remarkably influenced by the types of DOM. Higher [^{3}DOM*]$_{ss}$ values were obtained in the HA and FA addition groups, and the [$^1O_2$]$_{ss}$ was most pronounced in the presence of FA, while EfOM was found to be most beneficial for •OH generation. The UV-LED wavelengths exhibited a consistent impact on the varying trends of the reactive species concentrations across all the DOM addition conditions. Among these, UV-LED$_{365}$ demonstrated the most pronounced benefits for reactive species generation, followed by UV-LED$_{280}$, which was generally from one to two orders of magnitude higher than those observed under UV-LED$_{310}$ and UV-LED$_{260}$.

3.4. Quantum Yields

Quantum yields of reactive species serve as crucial indicators to unveil the formation efficiency and dynamics of reactive species, offering a comprehensive insight into the photochemical behaviors of different types of DOM. The quantum yields were estimated using the reactive species formation rates (refer to Table 3), irradiation intensities (see Table 1), and absorption coefficients (see Table 2), as shown in Table 4.

Table 4. Quantum yields of ^{3}DOM*, $^1O_2$, and •OH at various UV-LED wavelengths ([DOM] = 5 mg C L^{-1}, pH = 7.0).

Wavelength (nm)	HA			FA			EfOM		
	$\Phi_{^3HA^*}$ ($\times10^{-2}$)	$\Phi_{^1O_2}$ ($\times10^{-2}$)	$\Phi_{\bullet OH}$ ($\times10^{-4}$)	$\Phi_{^3FA^*}$ ($\times10^{-2}$)	$\Phi_{^1O_2}$ ($\times10^{-2}$)	$\Phi_{\bullet OH}$ ($\times10^{-4}$)	$\Phi_{^3EfOM^*}$ ($\times10^{-2}$)	$\Phi_{^1O_2}$ ($\times10^{-2}$)	$\Phi_{\bullet OH}$ ($\times10^{-4}$)
365	65.2	134.5	7.7	50.4	375.5	1.2	180.0	785.2	1501.4
310	0.3	0.5	0.3	0.3	1.6	0.02	0.9	0.9	2.5
280	11.0	3.9	8.7	5.1	10.6	1.4	27.8	13.8	24.1
260	2.9	2.3	2.8	2.4	7.3	0.7	8.3	10.2	4.3

The quantum yields of reactive species were significantly influenced by the DOM types. Specifically, the EfOM exhibited higher quantum yields than the other DOM samples across various reactive species. It was efficient in ^{3}DOM* formation, with the $\Phi_{^3DOM^*}$ 2.5–5.5 times higher compared to those of the HA and FA. The trends closely mirrored those of the $\Phi_{^3DOM^*}$, and the $\Phi_{^1O_2}$ of the EfOM surpassed those of the HA and FA, except under UV_{310}. Notably, the $\Phi_{\bullet OH}$ of the EfOM exceeded those of the other DOM samples, with a maximum increase of 1251 times (compared to FA under 365 nm).

Previous reports have indicated higher triplet-state photo quantum yields for EfOM than the reference natural organic matter [32]. Another study observed higher quantum yields of ^{3}DOM*, $^1O_2$, and •OH formation under simulated sunlight irradiation for EfOM compared to HA and FA [10], which is consistent with this study's findings. According to prior literature [16,33], the quantum yield of reactive species is positively correlated with the E2/E3 of the DOM. The higher E2/E3 ratio of EfOM (4.2) compared with those of HA (2.3) and FA (2.5) explains its outstanding performance in reactive species production.

In addition to DOM, the wavelength notably impacts the quantum yield of each reactive species. UV-LED_{365} exhibited the maximum $\Phi_{^3DOM^*}$ across all the DOM samples, being 5.9–9.8 times higher than that of UV-LED_{280}, and remarkably higher than those under 260 nm and 310 nm. This indicates that the DOM samples effectively absorbed incident photons and were excited to ^{3}DOM* under 365 nm UV irradiation. The $\Phi_{^1O_2}$ for all the DOM samples exhibited consistent patterns under different wavelengths, with 34.5–56.8 times higher values than UV-LED_{280}, and two orders of magnitude higher values than the other wavelengths. DOM is a major source of •OH in natural water systems, and •OH generation is related to the ^{3}DOM* oxidation process. In this study, the $\Phi_{\bullet OH}$ was two orders of magnitude lower than the $\Phi_{^3DOM^*}$ and $\Phi_{^1O_2}$. With the addition of HA and FA, the $\Phi_{\bullet OH}$ was similar under 365 nm and 280 nm irradiation and obviously higher than at 310 nm and 260 nm. In the EfOM addition group, UV-LED_{365} substantially enhanced the $\Phi_{\bullet OH}$, being 62 times that under 280 nm. Given •OH's strong oxidizability, the high-level photosensitization of EfOM at UV-LED_{365} will be favorable for the photodegradation of organic contaminants.

The quantum yields of reactive species are influenced by both the formation rates and light intensities. With higher formation rates and a relatively low light intensity (slightly higher than 310 nm but lower than 260 nm and 280 nm), UV_{365} achieved the maximum quantum yields of all the reactive species across the different sources of DOM. This suggests that UV-LED_{365} is more effective at inducing the photosensitization of DOM compared to other shorter wavelengths in the UVB and UVC bands. In water photocatalytic treatment using UV-LEDs, UVC light is typically chosen and believed to facilitate photocatalytic degradation more easily due to its higher energy. However, the results of this study present an alternative perspective. Specifically, the findings of the efficient generation of $^1O_2$ and •OH under UV_{365} with the EfOM addition provides a new technical option based on UV-LED for treating micro-polluted water containing EfOM.

4. Conclusions

In the current study, the impact of the UV-LED wavelength on the reactive species formation from various types of DOM was investigated. The different DOM samples exhibited diverse preferences in reactive species generation: FA was found to be beneficial for the $^1O_2$ and ^{3}DOM* formation, while EfOM facilitated the •OH generation. Among all the wavelengths, UV_{365} displayed the highest steady-state concentrations of reactive species, with 3.00×10^{-13} M [^{3}DOM*]ss and 1.64×10^{-11} M $[^1O_2]_{ss}$ by the FA addition, and 1.44×10^{-10} M $[\bullet OH]_{ss}$ by the EfOM. Moreover, the quantum yields of UV_{365} for all the reactive species were remarkably higher than those of the other wavelengths across all the DOM samples. This result not only broadens our understanding of the photochemistry of DOM in natural water systems, but it also provides guidance for the development of UV-LED-based water treatment technology.

Author Contributions: Z.-C.G.: investigation and writing—original draft. L.Z.: data processing. Y.C.: conceptualization and supervision. C.H.: funding, review, and editing. Z.-M.L.: writing—review and editing. All authors have read and agreed to the published version of the manuscript.

Funding: This research was funded by the National Natural Science Foundation of China, No. 52000090, the Natural Science Foundation of Jiangsu Province, No. BK20201001, and the China Postdoctoral Science Foundation, No. 2021M701511.

Data Availability Statement: Data are contained within the article.

Conflicts of Interest: Authors Ze-Chong Guo, Cheng Huang, Zhi-Min Liao were employed by the company Jiangxi Jindalai Environmental Protection Co., Ltd. The remaining authors declare that the research was conducted in the absence of any commercial or financial relationships that could be construed as a potential conflict of interest.

References

1. Qian, L.; Georgi, A.; Gonzalez-Olmos, R.; Kopinke, F.-D. Degradation of perfluorooctanoic acid adsorbed on Fe-zeolites with molecular oxygen as oxidant under UV-A irradiation. *Appl. Catal. B Environ.* **2020**, *278*, 119283. [CrossRef]
2. Silva, M.P.; Lastre-Acosta, A.M.; Mostafa, S.; McKay, G.; Linden, K.G.; Rosario-Ortiz, F.L.; Teixeira, A.C.S.C. Photochemical generation of reactive intermediates from urban-waste bio-organic substances under UV and solar irradiation. *Environ. Sci. Pollut. Res. Int.* **2017**, *24*, 18470–18478. [CrossRef] [PubMed]
3. Chen, J.; Loeb, S.; Kim, J.H. LED revolution: Fundamentals and prospects for UV disinfection applications. *Environ. Sci. Water Res. Technol.* **2017**, *3*, 188–202. [CrossRef]
4. Hockberger, P.E. A History of Ultraviolet Photobiology for Humans, Animals and Microorganisms. *Photochem. Photobiol.* **2002**, *76*, 561–579. [CrossRef]
5. Widel, M.; Krzywon, A.; Gajda, K.; Skonieczna, M.; Rzeszowska-Wolny, J. Induction of bystander effects by UVA, UVB, and UVC radiation in human fibroblasts and the implication of reactive oxygen species. *Free Radic. Biol. Med.* **2014**, *68*, 278–287. [CrossRef]
6. Xiao, Y.; Fan, R.; Zhang, L.; Yue, J.; Webster, R.D.; Lim, T.-T. Photodegradation of iodinated trihalomethanes in aqueous solution by UV 254 irradiation. *Water Res.* **2014**, *49*, 275–285. [CrossRef] [PubMed]
7. Martins de Oliveira, D.; Pereira Cavalcante, R.; de Melo da Silva, L.; Sans Mazón, C.; Esplugas Vidal, S.; Oliveira, S.C.D.; Machulek, A., Jr. Identification of intermediates, acute toxicity removal, and kinetics investigation to the Ametryn treatment by direct photolysis (UV254), UV254/H_2O_2, Fenton, and photo-Fenton processes. *Environ. Sci. Pollut Res. Int.* **2019**, *26*, 4348–4366. [CrossRef] [PubMed]
8. Chen, Z.-F.; Ying, G.-G.; Jiang, Y.-X.; Yang, B.; Lai, H.-J.; Liu, Y.-S.; Pan, C.-G.; Peng, F.-Q. Photodegradation of the azole fungicide fluconazole in aqueous solution under UV-254: Kinetics, mechanistic investigations and toxicity evaluation. *Water Res.* **2014**, *52*, 83–91. [CrossRef]
9. Kim, I.; Tanaka, H. Photodegradation characteristics of PPCPs in water with UV treatment. *Environ. Int.* **2009**, *35*, 793–802. [CrossRef]
10. Wan, D.; Sharma, V.K.; Liu, L.; Zuo, Y.; Chen, Y. Mechanistic Insight into the Effect of Metal Ions on Photogeneration of Reactive Species from Dissolved Organic Matter. *Environ. Sci. Technol.* **2019**, *53*, 5778–5786. [CrossRef]
11. McNeill, K.; Canonica, S. Triplet state dissolved organic matter in aquatic photochemistry: Reaction mechanisms, substrate scope, and photophysical properties. *Environ. Sci. Process. Impacts* **2016**, *18*, 1381–1399. [CrossRef]
12. O'Connor, M.; Helal, S.R.; Latch, D.E.; Arnold, W.A. Quantifying photo-production of triplet excited states and singlet oxygen from effluent organic matter. *Water Res.* **2019**, *156*, 23–33. [CrossRef]
13. Mostafa, S.; Rosario-Ortiz, F.L. Singlet oxygen formation from wastewater organic matter. *Environ. Sci. Technol.* **2013**, *47*, 8179–8186. [CrossRef]
14. Ren, D.; Huang, B.; Bi, T.; Xiong, D.; Pan, X. Effects of pH and dissolved oxygen on the photodegradation of 17alpha-ethynylestradiol in dissolved humic acid solution. *Environ. Sci. Process. Impacts* **2016**, *18*, 78–86. [CrossRef]
15. Lv, B.L.; Li, W.; Yu, X.L.; Zhang, D.; Zhang, Y.L. Effect of dissolved organic matter on the photodegradation of roxithromycin. *Acta Sci. Circumstantiae* **2019**, *39*, 747–754.
16. Cai, T.; Zhang, X.; Zhang, S.; Ming, Y.; Zhang, Q. Photochemical behaviors of dissolved organic matter in aquatic environment: Generation, characterization, influencing factors and practical application. *Environ. Res.* **2023**, *231*, 116174. [CrossRef] [PubMed]
17. Cai, M.-H.; Wu, Y.-P.; Ji, W.-X.; Han, Y.-Z.; Li, Y.; Wu, J.-C.; Shuang, C.-D.; Korshin, G.V.; Li, A.-M.; Li, W.-T. CCharacterizing property and treatability of dissolved effluent organic matter using size exclusion chromatography with an array of absorbance, fluorescence, organic nitrogen and organic carbon detectors. *Chemosphere* **2020**, *243*, 125321. [CrossRef] [PubMed]
18. Wan, D.; Kong, Y.; Wang, X.; Selvinsimpson, S.; Sharma, V.K.; Zuo, Y.; Chen, Y. Effect of permanganate oxidation on the photoreactivity of dissolved organic matter for photodegradation of typical pharmaceuticals. *Sci. Total Environ.* **2022**, *813*, 152647. [CrossRef] [PubMed]
19. Muramoto, Y.; Kimura, M.; Nouda, S. Development and future of ultraviolet light-emitting diodes: UV-LED will replace the UV lamp. *Semicond. Sci. Technol.* **2014**, *29*, 084004. [CrossRef]

20. Wan, D.; Yang, J.; Wang, X.; Xiang, W.; Selvinsimpson, S.; Chen, Y. Wavelength-Dependent Photoreactivity of Root Exudates from Aquatic Plants under UV-LED Irradiation. *ACS EST Water* **2022**, *2*, 2613–2622. [CrossRef]
21. Wang, W.-L.; Wu, Q.-Y.; Li, Z.-M.; Lu, Y.; Du, Y.; Wang, T.; Huang, N.; Hu, H.-Y. Light-emitting diodes as an emerging UV source for UV/chlorine oxidation: Carbamazepine degradation and toxicity changes. *Chem. Eng. J.* **2016**, *310*, 148–156. [CrossRef]
22. Hatchard, C.G.; Parker, C.A.; Bowen, E.J. A new sensitive chemical actinometer—II. Potassium ferrioxalate as a standard chemical actinometer. *Proc. R. Soc. Lond. Ser. A Math. Phys. Sci.* **1956**, *235*, 518–536.
23. Mccabe, A.J.; Arnold, W.A. Reactivity of triplet excited states of dissolved natural organic matter in stormflow from mixed-use watersheds. *Environ. Sci. Technol.* **2017**, *51*, 9718–9728. [CrossRef] [PubMed]
24. Halladja, S.; ter Halle, A.; Aguer, J.-P.; Boulkamh, A.; Richard, C. Inhibition of Humic Substances Mediated Photooxygenation of Furfuryl Alcohol by 2,4,6-Trimethylphenol. Evidence for Reactivity of the Phenol with Humic Triplet Excited States. *Environ. Sci. Technol.* **2007**, *41*, 6066–6073. [CrossRef] [PubMed]
25. Latch, D.E.; Stender, B.L.; Packer, J.L.; Arnold, W.A.; McNeill, K. Photochemical Fate of Pharmaceuticals in the Environment: Cimetidine and Ranitidine. *Environ. Sci. Technol.* **2003**, *37*, 3342–3350. [CrossRef] [PubMed]
26. Zhang, D.; Yan, S.; Song, W. Photochemically induced formation of reactive oxygen species (ROS) from effluent organic matter. *Environ. Sci. Technol.* **2014**, *48*, 12645–12653. [CrossRef] [PubMed]
27. Page, S.E.; Arnold, W.A.; McNeill, K. Terephthalate as a probe for photochemically generated hydroxyl radical. *J. Environ. Monit.* **2010**, *12*, 1658–1665. [CrossRef] [PubMed]
28. Mark, G.; Tauber, A.; Laupert, R.; Schuchmann, H.-P.; Schulz, D.; Mues, A.; von Sonntag, C. OH-radical formation by ultrasound in aqueous solution—Part II: Terephthalate and Fricke dosimetry and the influence of various conditions on the sonolytic yield. *Ultrason. Sonochem.* **1998**, *5*, 41–52. [CrossRef] [PubMed]
29. Dalrymple, R.M.; Carfagno, A.K.; Sharpless, C.M. Correlations between Dissolved Organic Matter Optical Properties and Quantum Yields of Singlet Oxygen and Hydrogen Peroxide. *Environ. Sci. Technol.* **2010**, *44*, 5824–5829. [CrossRef] [PubMed]
30. Ossola, R.; Jönsson, O.M.; Moor, K.; McNeill, K. Singlet Oxygen Quantum Yields in Environmental Waters. *Chem. Rev.* **2021**, *121*, 4100–4146. [CrossRef]
31. Vione, D.; Minella, M.; Maurino, V.; Minero, C. Indirect Photochemistry in Sunlit Surface Waters: Photoinduced Production of Reactive Transient Species. *Chem. Eur. J.* **2014**, *20*, 10590–10606. [CrossRef] [PubMed]
32. Zhou, H.; Yan, S.; Lian, L.; Song, W. Triplet-State Photochemistry of Dissolved Organic Matter: Triplet-State Energy Distribution and Surface Electric Charge Conditions. *Environ. Sci. Technol.* **2019**, *53*, 2482–2490. [CrossRef] [PubMed]
33. Zhou, H.; Lian, L.; Yan, S.; Song, W. Insights into the photo-induced formation of reactive intermediates from effluent organic matter: The role of chemical constituents. *Water Res.* **2017**, *112*, 120–128. [CrossRef] [PubMed]

water

MDPI

Article

Co-Pyrolysis of Fenton Sludge and Pomelo Peel for Heavy Metal Stabilization: Speciation Mechanism and Risk Evaluation

Cheng Huang [1,2,3], Lixian Wang [1], Lingyi Fan [3] and Yong Chen [1,*]

1 School of Environmental and Engineering, Huazhong University of Science and Technology, Wuhan 430074, China; chenghuang@just.edu.cn (C.H.)
2 Jiangxi Jindalai Environmental Protection Co., Ltd., Nanchang 330100, China
3 School of Environmental and Chemical Engineering, Jiangsu University of Science and Technology, Zhenjiang 212003, China
* Correspondence: ychen@mail.hust.edu.cn; Tel./Fax: +86-511-84400076

Abstract: The safe disposal and resource utilization of Fenton sludge (FS) are challenges due to the presence of heavy metals (HMs). Co-pyrolysis with biomass waste can effectively increase biochar quality and immobilize HMs, but research focusing on heavy metal stabilization from Fenton sludge using the co-pyrolysis approach is scattered. In this study, the co-pyrolysis of FS and pomelo peel (PP) was developed as a strategy to reduce the environmental risk of HMs. The results showed that co-pyrolysis greatly increased the pH and aromaticity of biochars, and the maximum specific surface area was 6.5 times higher than the corresponding FS-based biochar due to the sponge-like structure of PP, which was likely conducive to adsorbing HMs during pyrolysis. Meanwhile, the addition of PP promoted the enrichment of HMs in co-pyrolyzed biochars as well as induced the transformation of bio-available HM fractions to stable forms, especially at high temperatures. Finally, the presence of PP led to the decline in HM leachability in biochars; thus, the potential ecological risks of HMs decreased from considerable pollution levels to moderate and even clean levels. This study demonstrated that co-pyrolysis with PP is a promising approach to reduce the toxicity of HMs and improve the functionality of biochar for industrial sludge management.

Keywords: Fenton sludge; co-pyrolysis; biochar; heavy metals; ecological risk

1. Introduction

The Fenton oxidation process has been widely applied in industrial wastewater treatment for the efficient degradation of refractory organics [1]. However, a large amount of Fenton sludge (FS) is inevitably produced from the neutralization step of acidic effluent, which contains high contents of ferric iron, organic contaminants, sediment impurities, and heavy metals (HMs). Among the above-mentioned contaminants, HMs (Pb, Zn, Cu, Cr, etc.) pose a significant threat to ecosystems and humans due to their non-biodegradability and strong biological toxicity [2]. Therefore, it is necessary to develop a proper sludge management strategy to eliminate or alleviate potential environmental pollution caused via HMs in FS.

In the past decade, pyrolysis has been proven as an alternative technology for sludge disposal, in which a reduction in sludge volume, heavy metal passivation, and resource recovery (e.g., biofuels and biochar) can be achieved simultaneously [3]. It has been well documented that biochar, the main product derived from the partial combustion of biomass waste, exhibits a great potential for application in pollutant removal or soil amendment [4]. Moreover, biochar is a negative carbon emission material that can facilitate achieving carbon neutrality in sludge reutilization systems. Thus, numerous researchers have focused on the optimization of pyrolysis conditions [5,6] and modification strategies of biochar [7] as well as its emerging applications [8]. However, most HMs are concentrated in the biochar matrix after pyrolysis, which may result in the potential risk of HM leaching if the sludge-derived biochar is applied to soil and water environments. Although some studies have shown that pyrolysis contributes to the transformation of HMs from a bioavailable state into stable oxidizable and residual fractions, the immobilization of HMs largely depends on the pyrolysis temperature, species of HMs, and sludge properties. For example, a pyrolysis temperature above 600 °C is commonly necessary to improve the conversion of Mn, Cu, and Zn in sewage-sludge-based char to more stable fractions [9], while in another case, Udayanga et al. reported that Cd, Pb, and Hg in industrial sludge readily migrate at such high temperatures [10]. In addition, the characteristics of sludge (especially industrial sludge) with high ash content and low calorific value may cause more energy consumption and poor pyrolysis performance [11]. Therefore, in order to overcome these limitations, a more efficient method to immobilize HMs is urgently needed.

Recently, the co-pyrolysis of sludge with biomass waste has gained increasing attention from researchers due to its multiple benefits, including the improvement in biochar properties, the stabilization of HMs, and the reduction in energy consumption, which can be achieved simultaneously [12]. In the context of "treating the waste with the waste", various types of biomass, such as rice straw, sawdust, vegetable waste, and microalgae have been co-pyrolyzed with sludge [12–14], which has provided the potential to modify the properties of biochar as well as reduce the leaching toxicity of HMs. For instance, Dong et al. found that the co-pyrolysis of sewage sludge and rice straw exhibited great benefits for fabricating hierarchical porous structures, improving high ash content, and immobilizing heavy metals in char samples [15]. However, previous studies have mostly focused on organic-dominated sludge from municipal wastewater treatment plants; in-depth and comprehensive investigations on the co-pyrolysis of industrial sludge (such as FS) and biomass waste are still scattered. In addition, more attention should be paid to the leaching toxicity and underlying environmental risks of HMs in sludge-derived biochar.

Pomelo (*Citrus grandis*) is one of the most widely consumed fruits around the world, resulting in a large amount of pomelo peel (PP) being discarded as biomass waste. Many researchers have demonstrated that PP is a promising ingredient for the preparation of engineered carbonaceous materials due to its polymeric network structure with cellulose microfibril, hemicellulose, and lignin, which provides large specific surface areas and abundant functional groups after thermal conversion. To date, several PP-based carbon materials have been developed and applied in the adsorption of HMs, catalytic degradation and solar evaporation [16–18]. These studies inspired us to investigate whether the co-pyrolysis of FS and PP may synergistically enhance the stabilization of HMs in biochars. However, until now, the feasibility of the co-pyrolysis of pomelo peel and sludge with high concentrations of HM has not been studied.

In this study, for the first time, HMs, including Pb, Zn, Cr, and Cu, are immobilized in a biochar matrix via conducting the co-pyrolysis of pomelo peel and Fenton sludge. The main objectives of this work are to (1) explore the effect of PP addition on the characteristics of biochar and heavy metal immobilization during co-pyrolysis and (2) evaluate the single and multiple environmental risks of HMs in the biochars. This work aimed to provide a new solution to stabilize HMs in industrial sludge.

2. Materials and Methods

2.1. Feedstocks and Chemicals

Raw FS was collected from the Fenton oxidation process of an electroplating wastewater treatment plant in Wuxi, China. The fresh pomelo peel was obtained from a local fruit shop around the campus. All the feedstocks were oven dried at 105 °C for 24 h, ground finely, screened through 100 mesh and then stored in a desiccator for further use. The main properties of FS and pomelo peel are listed in Table 1.

All the chemicals used in this study, including CH_3COOH, NaOH, $NH_2OH{\cdot}HCl$, HNO_3, H_2O_2, NH_4Ac, HCl and HF were analytical grade and purchased from Shanghai Macklin Biochemical Co., Ltd., Shanghai, China.

Table 1. The main properties of FS and pomelo peel.

Parameter	FS	Pomelo Peel
pH	5.2 ± 0.1	6.9 ± 0.1
Total organic matter (TOC) (%)	4.58 ± 0.47	92.1 ± 1.05
Cu (mg/kg)	818 ± 21	ND [a]
Zn (mg/kg)	16,380 ± 231	ND
Cr (mg/kg)	128 ± 17	ND
Pb (mg/kg)	483 ± 24	ND

Note: [a] ND, Not detected.

2.2. Co-Pyrolysis of FS and PP

Co-pyrolysis experiments were conducted in a tubular furnace (OTF-1200X, Hefei Kejing Material Technology Co., Ltd., Hefei, China) as shown in Figure S1. Firstly, the pretreated FS and PP were thoroughly mixed in the agitator as the mass ratio of 1:1 based on the previous publication [19]. Then, 30.0 g of the mixed feedstock was loaded into the quartz tube and heated at different pyrolysis temperatures (300, 400, 500 and 600 °C) for 2 h. The heating rate for all the experimental groups was set invariably as 10 °C/min. During the entire pyrolysis and cooling stages, the high-pure nitrogen gas was flushed with the constant flow rate of 100 mL/min to ensure oxygen-free atmosphere. The pyrolysis products were taken out and stored after natural cooling to ambient temperature. The as-prepared samples from different temperatures were labeled as SPC300, SPC400, SPC500 and SPC600, respectively. Meanwhile, the pyrolysis of sole FS was conducted as the control test, and the procedure was same as mentioned above. The biochar samples derived from sole FS were labeled as FSC300, FSC400, FSC500 and FSC600, respectively. The specific surface area was examined using N_2 adsorption–desorption isotherm measurements equipped with an automatic specific surface analyzer (ASAP2020, Micromeritics Instruments Co., Ltd., Norcross, GA, USA). The surface functional groups of biochars were characterized using Fourier transform infrared (FTIR) spectroscopy (Nicolet iS50R, Thermo Fisher Scientific Inc., Waltham, USA). The thermogravimetric analysis was conducted using a thermogravimetric analyzer (TGA/SDTA851e MET, Shanghai Mettler Toledo Scientific Co., Ltd., Shanghai, China), following the procedure previously described [14].

2.3. Heavy Metals Analysis

2.3.1. Sequential Extraction of Heavy Metals

A modified European Communities Bureau of Reference (BCR) sequential extraction method was used to investigate the speciation of HMs in biochars, by which HMs can be classified into four fractions, namely the acid-soluble or exchangeable fraction (F1), reducible fraction (F2, bound to iron-manganese oxide), oxidizable fraction (F3, bound to organic matter and sulfide) and residual fraction (F4, primary minerals and secondary silicate lattice metals) [20]. Normally, the mobility and bio-toxicity of HMs decrease with the sequential extraction, and the order is as follows: F1 > F2 > F3 > F4. The extraction steps are provided in Table S1.

2.3.2. Leaching Experiment

The toxic characteristic leaching procedure (TCLP) recommended by the U.S. EPA was implemented to assess the leachability and ecological risk of HMs in biochar [21]. Briefly, 0.5 g of biochar was extracted by the prepared glacial acetic acid solution (pH 2.88) with the liquid-to-solid ratio of 20:1. The mixture was placed in a shaker (120 rpm) at a mesophilic temperature of 25 °C for 18 h. The extracted samples were centrifuged at 5000 rpm for

15 min, and the supernatant was filtered through a 0.22 μm PFS membrane filter prior to analysis [22].

2.3.3. Determination of Heavy Metals

The total HMs content in the samples were determined by a microwave-assisted digestion method according to the previous study [23]. Briefly, 0.200 g of sample was dissolved in the mixture of nitric acid, hydrochloric acid and hydrofluoric acid (HNO_3:HCl:HF = 6:2:1, $v/v/v$) and digested in a microwave digester (GH08Z-Z, Shanghai Metash Instruments, Shanghai, China). The detailed procedures are shown in Table S2. The total concentrations of Pb, Zn and Cr in digestion solution were determined by an inductively coupled plasma optical emission spectrometry (ICP-OES) instrument (OPTIMA 8300DV, Perkin Elmer Instrument Co., Ltd., Waltham, MA, USA).

2.4. Ecological Risk Assessment

The risk assessment code (RAC) and Nemerow pollution index (NPI) were applied to evaluate the potential environmental risk of biochar samples in this study. RAC is expressed by the proportion of F1 fraction, which represents the availability of single HM [24]. According to the RAC value, the environmental risk can be classified five levels: RAC < 1% indicates "no risk", 1% < RAC < 10% means "low risk", 10% < RAC < 30% indicated "medium risk", 30% < RAC < 50% indicates "high risk" and RAC < 50% means "medium risk".

The NPI Is one of the most widely used pollution indexes to evaluate the pollution and potential ecological risk caused by multi-contamination factors. In this study, the NPI was carried out to comprehensively assess the bioavailability and harmfulness of various heavy metals and the formula were shown as follows:

$$P_i = F_i/S_i \quad (1)$$

$$P_N = \sqrt{\left(\overline{P}_i^2 + P_{imax}^2\right)/2} \quad (2)$$

$$\overline{P}_i = \frac{\sum_{i=1}^{n} P_i}{n} \quad (3)$$

where P_i is the single pollution index of HMs in samples, F_i is the extractable concentration of HMs by the TCLP method, S_i is soil permissible limits based on the Soil Environmental Quality Standards for the developed area, in which the contents of Pb, Zn, Cr and Cu are 400, 2000, 30 and 2000 mg/kg, respectively [25]. P_N is the Nemerow integrated pollution index, P_i is the average value of the single pollution index of HMs, and P_{imax} is the maximum value of the single pollution index of HMs.

3. Results and Discussion

3.1. Properties of Sludge and Biochars

3.1.1. General Properties

The properties of biochars are strongly dependent on the elemental compositions and contents of feedstocks as well as pyrolysis conditions [26]. Thus, the properties of the feedstocks and their biochars are shown in Table 2. Both C and H contents of FS were less than 5%, indicating that the main components in the FS were inorganic elements (e.g., Fe, Zn, Cu and other HMs). On the contrary, the organic matter was the dominant component in the PP. These results were in accordance with the thermogravimetric study presented in Figure 1, in which the weight loss rate of FS reached 25%, whereas almost 80% of PP decomposed when the pyrolysis temperature increased to 600 °C.

As shown in Table 2, both the high temperature and co-pyrolysis with PP reduced the yields of biochar. For example, there was a slight decline in the biochar yield from 83.36% to 74.32% at 300 to 600 °C, indicating the high ash contents in FS. Moreover, the yields of blended biochars (47.32–71.83%) were much lower than the biochars derived from FS alone under the same pyrolysis conditions, which can be interpreted by the high organics content in PP. During pyrolysis, the main organic compounds in PP such as hemicellulose, cellulose, lignin and aliphatic compounds readily decomposed via the demethylation, demethoxylation and decarboxylation reactions [27].

Table 2. Yield, pH and ultimate analysis of feedstocks and biochars.

Sample	Yield (%)	pH	C (%)	H (%)	H/C	SSA (m^2/g)
FS	/	5.25 ± 0.05	4.37	1.66	4.56	/
PP	/	6.95 ± 0.05	58.34	13.67	2.81	/
FSC300	83.36 ± 4.15	5.29 ± 0.08	2.29 ± 0.12	0.84 ± 0.03	4.40	2.15
FSC400	82.18 ± 3.18	5.27 ± 0.04	1.79 ± 0.18	0.63 ± 0.05	4.22	3.18
FSC500	79.46 ± 2.87	5.28 ± 0.07	1.77 ± 0.07	0.62 ± 0.08	4.20	3.74
FSC600	74.32 ± 4.43	5.80 ± 0.13	1.67 ± 0.09	0.53 ± 0.02	3.80	4.17
SPC300	71.83 ± 2.79	7.28 ± 0.06	40.67 ± 0.16	2.52 ± 0.10	0.75	18.54
SPC400	60.31 ± 4.63	8.59 ± 0.04	37.14 ± 0.14	3.07 ± 0.08	0.74	22.92
SPC500	52.43 ± 3.98	9.81 ± 0.11	34.52 ± 010	1.73 ± 0.06	0.60	27.25
SPC600	47.32 ± 3.53	10.92 ± 0.08	31.03 ± 0.08	1.14 ± 0.02	0.44	31.26

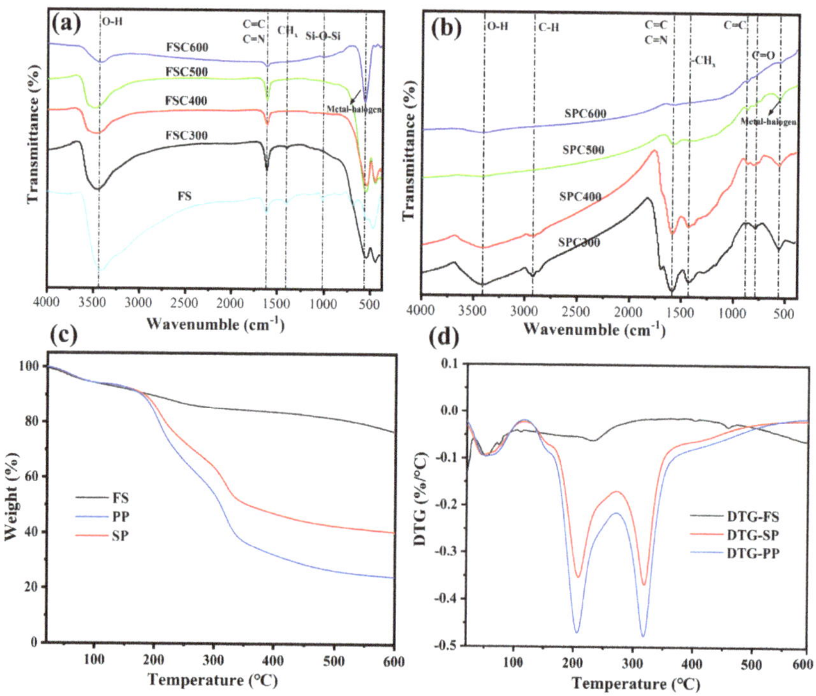

Figure 1. FTIR spectra of FS, FSC (**a**) and SPC (**b**) with different pyrolysis temperatures. TG (**c**) and DTG (**d**) curves of FS, SP and PP.

3.1.2. Elemental Analysis and Specific Surface Area

The elemental analysis of feedstocks and biochars clearly demonstrated that both C and H contents decreased with increasing pyrolysis temperature owing to the rapid decomposition of organic substances. Meanwhile, the H/C molar ratio was used to characterize the organic aromaticity degree of biochars: the H/C ratio ≤0.3 indicated highly condensed aromatic ring structures, whereas H/C ≥0.7 suggested a low condensed structures [20]. As shown in Table 2, the H/C ratios of FSC (3.80–4.40) were much higher than 0.7, showing an obvious non-condensed structure in FSC [28]. On the contrary, the molar H/C ratio of SPC decreased significantly to 0.44–0.75 with the increase in temperature, which represented the increment of aromaticity in SPC. The pH of FSC increased slightly from 300 to 600 °C, which is attributed to the thermal decomposition of acidic functional groups and formation of alkali salts [29]. It should be noted that the pH of FS-based biochar was acidic, which is different from that of organic-dominated sludge in the previous studies [30]. One possible reason for these results is that more metal chlorides were produced during the pyrolysis of FS, contributing to the surface acidity of biochars [31]. Comparing with acidic FSC, the addition of PP resulted in a significant elevation of pH, and all the SPCs were alkaline, which is likely induced by the alkaline species such as metal carbonates and hydroxides in the PP.

Also, it could be observed that the co-pyrolysis with PP addition resulted in the increment of biochars' specific surface area (SSA). For example, the SSA of co-pyrolyzed biochar at 600 °C (31.26 m^2/g) was 6.5 times higher than that of FS-based biochar (4.17 m^2/g). This is possibly because the sponge-like structure of PP was conducive to form well-developed porous biochar during the pyrolysis process.

3.1.3. FTIR Spectra Analysis

The functional groups of FS and biochars were investigated via FTIR analysis. As shown in Figure 1a, most of the peaks, including 3435 cm^{-1} for the −OH stretching vibration, 1655 cm^{-1} for the C=C and C=N stretching vibrations in aromatic annulus and amide bonds, $-CH_x$ groups in aliphatic chains [19], weakened and even completely disappeared with the raising temperature compared with raw FS, suggesting the substantial decomposition of organic compounds. The intensity of the metal–halogen tensile vibration (approximately 570 cm^{-1}) was greatly enhanced with the increase in pyrolysis temperature, which was likely owing to the formation of metal chlorides. This is consistent with the results of acidic characteristics of FSC discussed in Section 3.1.1. After co-pyrolyzing with PP, a new peak of C=O appeared in SPC, and other functional groups were enriched compared with the corresponding peak in FSC, indicating the abundant organics in PP. This was possibly related to the ring opening, depolymerization of hemicellulose and the aromatization of benzene ring in lignin [14,32]. Moreover, the peak of metal–halogen tensile vibration decreased as the temperature rose to 600 °C, which differed from that of FSC. This could be explained by the conversions of metal halide compounds to more stable forms during the co-pyrolysis process. Similar results have also been reported in another publication [33].

3.1.4. Thermogravimetric Analysis

The thermogravimetric results of FS, PP and their mixture are shown in Figure 1c,d. The overall weight loss rates of FS, PP and SP were 23.0%, 59.8% and 76.0%, which are attributed to the higher proportion of volatile matter in PP than that of FS. For FS alone, a small weight loss peak could be observed at about 250 °C, which could be designated to the decomposition of hydrocarbons in the industrial sludge. In contrary, two strong peaks in DTG curves were presented for PP and the mixture, dividing the whole pyrolysis process into two individual stages. The first stage occurred at 180–240 °C and was due to the decomposition of pectin [34]. The second decomposition phase occurred at around 310–360 °C, which was believed to be due to the decomposition of cellulose and hemicellulose in PP [35]. The TG curve became flat after 550 °C with no obvious loss of weight for all the samples, suggesting the good thermal stability of derived biochars.

3.2. The Total Contents of HMs in the Biochars

The total contents of HMs in raw FS and biochars are presented in Table 3. The total amounts of Cu, Cr and Pb of FS were in the range of 128.55–818.68 mg/kg while the concentration of Zn was more than 16,000 mg/kg due to the dominant zinc-plating process in the electroplating factory. Most of the HMs (Pb, Zn and Cr) were enriched in the FSC, and the concentrations increased as the pyrolysis temperature rose, which is likely attributed to the thermal decomposition of organic matters and high stability of HMs [36]. However, the contents of Cu in the FSC were found to be lower compared with that in the FS, indicating that a certain proportion of volatile Cu compounds generated and transferred to bio-oil and bio-gas. This is consistent with the previous study, in which dissatisfactory immobilization rates of Cu (33.62–67.85%) were achieved after the pyrolysis of electroplating sludge [31]. A similar change trend of HM contents could be observed in the co-pyrolyzed biochars, but the concentrations of HM in the SPC were much lower than those in the FSC, which may be attributed to the "dilution effect" caused by the HMs-free PP addition.

Table 3. Total contents and recovery rates of HMs in FS and biochar.

Sample	Pb		Zn		Cr		Cu	
	Total Content (mg/kg)	Recovery Rate (%)	Total Content (mg/kg)	Recovery Rate (%)	Total Content (mg/kg)	Recovery Rate (%)	Total Content (mg/kg)	Recovery Rate (%)
FS	483.45 ± 24.29	/	16,380 ± 231	/	128.55 ± 17.07	/	818.68 ± 21.85	/
FSC300	519.65 ±21.67	89.68 ± 3.74	17,771 ± 373	90.4 ± 1.90	145.76 ± 2.76	94.9 ± 1.80	704.25 ± 45.50	71.77 ± 4.64
FSC400	513.18 ± 26.40	87.31 ± 4.49	18,661 ± 265	93.6 ± 1.33	144.95 ± 6.83	93.06 ± 4.38	765.13 ± 20.38	76.87 ± 2.05
FSC500	530.29 ± 28.66	87.24 ± 4.71	18,939 ± 736	91.9 ± 3.57	144.63 ± 3.58	89.78 ± 2.21	796.50 ± 11.50	77.37 ± 1.12
FSC600	611.43 ± 2.76	94.08 ± 0.42	20,573 ± 891	93.3 ± 4.05	163.80 ± 3.25	95.11 ± 1.89	873.12 ± 17.63	79.33 ± 0.69
SPC300	304.64 ± 3.92	90.61 ± 1.17	10,930 ± 495	95.8 ± 4.34	94.94 ± 4.81	94.94 ± 5.40	450.63 ± 13.13	81.88 ± 2.38
SPC400	356.31 ± 16.65	88.98 ± 4.16	12,854 ± 379	94.3 ± 2.79	116.25 ± 4.25	104.25 ± 4.00	539.50 ± 35.50	79.55 ± 5.23
SPC500	469.38 ± 12.62	101.90 ± 2.74	13,121 ± 208	84.0 ± 1.33	128.94 ± 2.94	103.94 ± 1.69	677.45 ± 27.50	86.85 ± 3.53
SPC600	503.21 ± 7.69	98.60 ± 1.51	15,960 ± 603	95.7 ± 3.49	139.25 ± 0.75	99.25 ± 0.55	748.81 ± 26.19	86.64 ± 3.03

Meanwhile, the recovery rates of HMs were calculated using the following equation:

$$RR = \frac{C_B \times Y}{C_F} \times 100\% \tag{4}$$

where RR is the recovery rate of specific HM (%), C_B and C_F represent the total concentration of HMs in the feedstocks and biochars (mg/kg), and Y is the yield of biochars (%). As shown in Table 3, the RRs of Cu, Zn, Cr and Pb were 71.77–86.85%, 84.0–95.8%, 89.78–104.25% and 87.24–101.90% in FSC and SPC, respectively, indicating that most of the HMs were retained in biochars after pyrolysis. Moreover, it seems that the recovery rates of HMs in the co-pyrolytic biochars (SPC) were higher than those in the sole biochars (FSC) in most cases. For instance, the RRs of Cr in the FSC were in the range of 89.78–95.11%, while the values increased to 94.94–104.25% under co-pyrolysis with PP. These results suggested that the addition of PP was conducive to immobilizing HMs in biochar owing to the high organic carbon content of PP [37].

3.3. *The Speciation of HMs in the Biochars*

It is well known that the toxicity and migration characteristics of HMs in biochar largely depend on the chemical speciation. Therefore, the distribution characteristics of HM fractions in FSC and SPC were analyzed based on a modified BCR extraction procedure, and the results are shown in Figure 2. Generally, the BCR-extractable fractions of HMs can be divided into three categories: (1) both the acid-extractable (F1) and reducible (F2) fractions represent high mobility and direct toxicity, (2) the oxidizable fraction (F3) exhibits a potential toxicity due to its low degradation resistance in the certain environment, and (3) the residual fraction (F4) is non-bioavailable and extremely stable under most conditions [38]. For the raw FS, the bioavailable fractions (F1 + F2) of Pb and Zn were 49.8% and 59.7%, respectively, which indicated high ecological risks from these HMs if the FS was discharged into the aquatic and soil ecosystems. After converting FS into FSC through single pyrolysis, the bioavailable fractions (F1 and F2) of Pb were readily converted into relatively stable forms (F3 and F4) with the increasing temperature. For example, the F1 + F2 fraction of Pb decreased from 40.8% in the FSC300 to 4.42% in the FSC600, suggesting that pyrolysis in higher temperature was more beneficial to the passivation of HMs. Similar change trends were also observed in the speciation distribution of Cu and

Cr. The stable fractions (F3 + F4) of Cu and Cr in FSC were more than 90% and did not change significantly at 300–600 °C, which could be interpreted by their strong affinity to organic ligands or minerals in raw FS [19]. The F1 and F2 portions of Zn remained stable and still more than 35.1% in FSC600, while the portion of F4 increased, indicating that the transformation of oxidizable Zn complexes to more stable Zn oxide or co-crystal compounds [39].

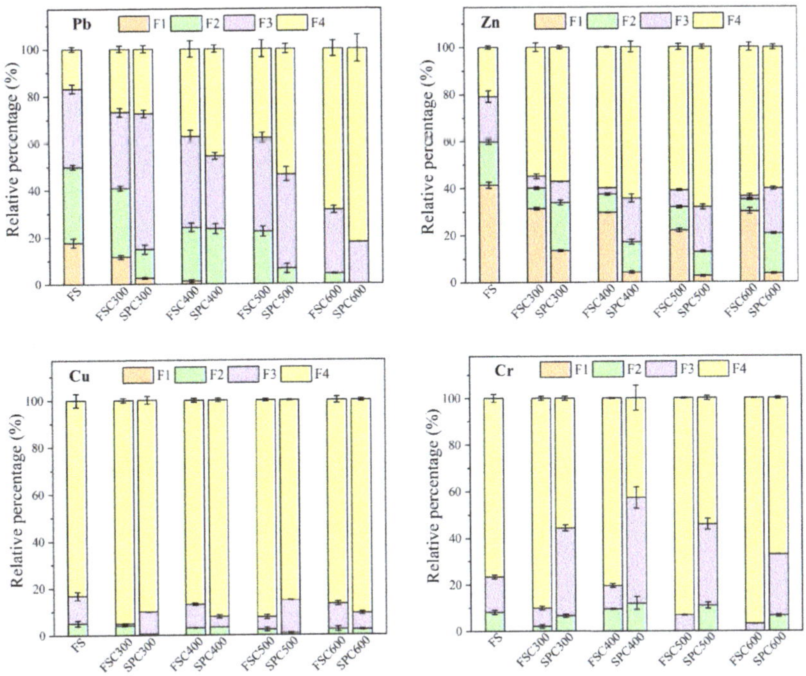

Figure 2. Speciation distribution of HMs in FS and biochars.

Co-pyrolysis with PP exhibited a remarkable effect on the distribution of HM fractions in SPC, and the change trend largely depended on the type of HMs. As to Pb, the fractions of F1 and F2 were reduced by 52.9–100% in SPC at the pyrolysis temperature of 300–600 °C, indicating that co-pyrolysis had a strong passivating effect on the Pb. Compared with the chemical speciation of Zn in FSC, the proportion of F1 and F2 in SPC declined sharply, whereas the F3 fraction increased significantly. Similar changes were observed in Cr. These results were in accordance with a previous study in which the F1, F2 and F4 fractions of all HMs decreased while the content of F3 increased with the mixture ratios of cotton stalks [40]. This phenomenon is likely attributed to the complexation reactions that occurred between the HMs and organic substance in the PP. In summary, co-pyrolysis with PP is conducive to the conversion of unstable HMs to stable species such as oxidizable and residual fractions, and it consequently decreased the bio-toxicity of HMs.

3.4. The Leaching Toxicity of HMs in the Biochars

The TCLP test was conducted to investigate the leachability of HMs in the biochars, and the results are presented in Figure 3. The limit values of HMs were 100 mg/L for Zn and Cu, 15 mg/L for Cr and 5 mg/L for Pb based on the identification standards for hazardous wastes in China [6]. The leachable contents of Cr, Pb and Cu in raw FS did not exceed the threshold values, while the contents of Zn in the TCLP leachate was 642.08 mg/L, which was 5.42 times higher than that of permissible limits, indicating the potential risk of Zn pollution to the environments. After converting FS to FSC, the leaching of HMs was significantly suppressed ($p < 0.05$) and continuously fell off with the rising temperature. For instance, the leached concentrations of Zn, Cr, Cu and Pb in FSC600 exhibited the remarkable declines of 61.4%, 46.0%, 92.8% and 100%, respectively, compared with those in raw FS. These results agreed with the change trend of chemical speciation of HMs and could be attributed to the formation of more stable structures (e.g., metal–silicate and metal–oxide) in the carbon matrix [19].

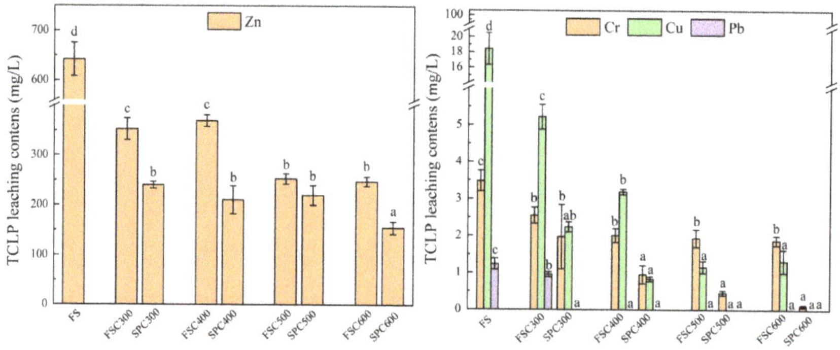

Figure 3. The TCLP leaching contents of HMs in FS and biochars. The lowercase letters above the bars show significant differences of each bar of data ($p < 0.05$, n = 3).

As for the SPC, the toxic leachability of HMs was further decreased after the co-pyrolysis process. Although the leachable content of Zn in SPC remained higher than the limit value, the addition of PP notably decreased the leaching content of Zn from 6.42-fold of threshold values to less than 2.5 times in the temperature range. In addition, no leachable Cu and Pb were detected in SPC500 and SPC600, indicating that the co-pyrolysis of sludge and PP effectively enhanced the immobilization of HMs and reduced the toxic leaching of them to environments. There are some possible reasons for the results: (i) the generation of a more stable structure such as organometallic compounds and metal–oxide in the carbon matrix is stimulated by the presence of PP [39], and (ii) the well-developed pores and abundant functional groups of SPC are capable of trapping or absorbing the metal(loid)s [41]. Furthermore, several mechanisms including a carbon shield, complexation interactions and electrostatic interactions were likely involved in the immobilization of HMs [21]; further study is needed to comprehensively reveal the contribution of the main ingredients in PP during pyrolysis.

3.5. The Environmental Risk Evaluation of HMs in the Biochars

The ecological risks of single and multiple HMs in biochars were evaluated using several indicators including a risk assessment code (RAC) and Nemerow pollution index (NPI). As shown in Figure 4, the risk levels for Pb (17.6) and Zn (41.4) in FS were considered as medium and high pollution levels, respectively (Cr and Cu were not listed due to there being no detectable F1 form in this study). After pyrolysis, the pollution level of Pb in FSC reduced to low risk level (1.2) at 400 °C and further reduced to a clean level at 500–600 °C, while the RAC values of Zn declined by 27.2–46.9% with the increasing temperature, indicating a high or moderate pollution risk for FSC. These results suggested that pyrolysis could improve the transformation of the F1 fraction of HMs to other stable forms, but the passivation was not very satisfactory for Zn. In contrary, co-pyrolysis with PP could further decrease the risk level of HMs in biochars. The RACs of Pb in SPC were significantly decreased to 2.6 at 300 °C, and even Pb is not detected at 400–600 °C, presenting low-risk or no-risk levels in the co-pyrolytic biochars. Apart from SPC300, other SPC samples exhibited a low environmental risk level for Zn, indicating the positive effect of co-pyrolysis on the Zn immobilization.

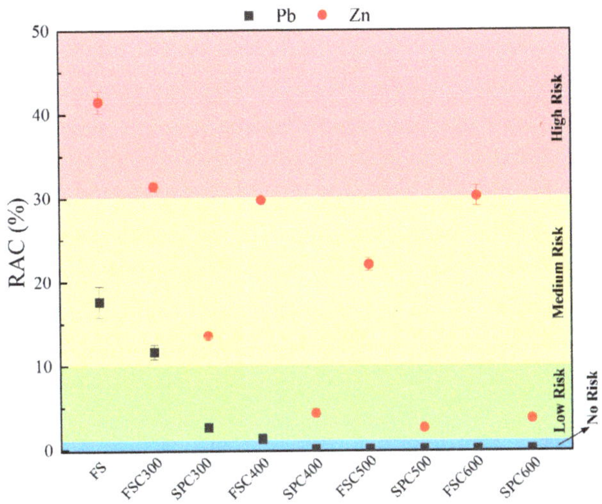

Figure 4. The potential ecological risk (RAC) of Pb and Zn in biochars.

The results of NPI assessment and the classification level of the HMs pollution are presented in Table 4. The P_N value of FS was 9.62, which far exceeded the threshold of the "serious" pollution level. This indicated that it posed an extremely high contamination risk if FS was directly applied into soil. The P_N values of FSC decreased from 9.62 to 5.33, 3.64, 5.53, and 3.13 for FSC300–600, respectively, which were considered at a "severe" risk level. Furthermore, co-pyrolytic biochars showed more positive impacts in decreasing the NPI: especially the comprehensive pollution level of SPC600 was the lowest, belonging to a "moderate" pollution degree. These results suggested that co-pyrolysis with PP is a promising strategy to reduce the ecological risk of HMs in industrial sludge. However, both the RAC and NPI values of co-pyrolyzed biochars were higher than the expected results due to the high contents of HMs in FS, and thereby, further study is still required for a comprehensively technical optimization and potential application prospects of biochar in pilot-scale trials.

Table 4. The Nemerow pollution index (P_i/P_N) of HMs in FS and biochars.

Sample	P_i				P_N	Classification Level
	Pb	Zn	Cr	Cu		
FS	12.84	4.64	0.37	0.12	9.62	Serious
FSC300	7.06	3.40	0.10	0.10	5.33	Severe
FSC400	4.80	2.64	0.04	0	3.64	Severe
FSC500	7.39	2.68	0.06	0	5.53	Severe
FSC600	4.21	1.28	0.02	0	3.13	Severe
SPC300	5.06	2.59	0.02	0	3.82	Severe
SPC400	4.40	0.61	0	0	3.23	Severe
SPC500	4.95	2.51	0.03	0	3.74	Severe
SPC600	3.10	0.15	0	0	2.26	Moderate
Pollution degree	$P_i/P_N < 1$, low contamination; $1 < P_i/P_N < 3$, moderate contamination; $3 < P_i/P_N < 6$, severe contamination; $P_i/P_N > 6$, serious contamination.					

4. Conclusions and Future Perspectives

In this study, the effect of co-pyrolysis of FS and PP on the stabilization of HMs was firstly investigated. The addition of PP decreased the yield rate but improved the properties of co-pyrolytic biochar including pH, aromaticity, specific surface area and functional groups. Furthermore, co-pyrolysis with PP could reduce the total contents of HMs (i.e., Pb,

Zn, Cr and Cu) in biochar, and the transformation from bio-available (F1 + F2) to stable fractions (F3 + F4) was achieved simultaneously. The TCLP leaching test demonstrated that the leachable concentrations of HMs in co-pyrolyzed biochar were greatly decreased, thereby reducing the potential ecological risks of single (RAC) and multiple (NPI) HMs from "high" to "moderate" and even "low" risk degrees. Therefore, co-pyrolysis with PP was considered as a feasible strategy to immobilize HMs in industrial sludge. However, the immobilization performance of HMs in FS should be further improved via technical parameter optimization and other carbonization technologies application due to high HM contents in industrial sludge. In addition, the co-pyrolysis of PP and different type of industrial sludge in the long-term application should be comprehensively evaluated before its practical application to soil.

Supplementary Materials: The following supporting information can be downloaded at: https://www.mdpi.com/article/10.3390/w15213733/s1, Figure S1: Fixed bed reactor system for co-pyrolysis; Table S1: Details of BCR extraction procedure and extraction fluids; Table S2: Microwave digestion procedure used in this study.

Author Contributions: Conceptualization, Y.C. and C.H.; methodology, L.W. and L.F.; validation, C.H., L.W. and L.F.; data curation, L.W.; writing—original draft preparation, C.H.; writing—review and editing, C.H.; visualization, C.H.; supervision, Y.C.; project administration, Y.C.; funding acquisition, C.H. All authors have read and agreed to the published version of the manuscript.

Funding: This study was funded by the Natural Science Foundation of Jiangsu Province (Grant No. BK20201001), the Postdoctoral Science Foundation of China (Grant No. 2021M701512), the Post-doctoral Preferred Funding Project of Jiangxi (Grant No. 2020KY36) and the Open Research Fund of Changjiang Survey Planning Design and Research Co., Ltd. (CX2020K09).

Institutional Review Board Statement: Not applicable.

Informed Consent Statement: Not applicable.

Data Availability Statement: Not applicable.

Conflicts of Interest: The authors declare no conflict of interest.

References

1. Gao, L.; Cao, Y.; Wang, L.; Li, S. A review on sustainable reuse applications of Fenton sludge during wastewater treatment. *Front. Environ. Sci. Eng.* **2021**, *16*, 77. [CrossRef]
2. Yaqoob, A.A.; Ahmad, H.; Parveen, T.; Ahmad, A.; Oves, M.; Ismail, I.M.I.; Qari, H.A.; Umar, K.; Mohamad Ibrahim, M.N. Recent Advances in Metal Decorated Nanomaterials and Their Various Biological Applications: A Review. *Front. Chem.* **2020**, *8*, 341. [CrossRef] [PubMed]
3. Chai, Y.; Huang, C.; Sui, M.; Yin, Y.; Sun, N.; Chen, Y.; Liao, Z.; Sun, X.; Shen, W.; Tang, S. Fe-loaded alginate hydrogel beads activating peroxymonosulfate for enhancing anaerobic fermentation of waste activated sludge: Performance and potential mechanism. *J. Environ. Manag.* **2023**, *341*, 118079. [CrossRef]
4. He, M.; Xu, Z.; Hou, D.; Gao, B.; Cao, X.; Ok, Y.S.; Rinklebe, J.; Bolan, N.S.; Tsang, D.C.W. Waste-derived biochar for water pollution control and sustainable development. *Nat. Rev. Earth Environ.* **2022**, *3*, 444–460. [CrossRef]
5. Hossain, M.K.; Strezov, V.; Chan, K.Y.; Ziolkowski, A.; Nelson, P.F. Influence of pyrolysis temperature on production and nutrient properties of wastewater sludge biochar. *J. Environ. Manag.* **2011**, *92*, 223–228. [CrossRef] [PubMed]
6. Jin, J.; Li, Y.; Zhang, J.; Wu, S.; Cao, Y.; Liang, P.; Zhang, J.; Wong, M.H.; Wang, M.; Shan, S.; et al. Influence of pyrolysis tem-perature on properties and environmental safety of heavy metals in biochars derived from municipal sewage sludge. *J. Hazard. Mater.* **2016**, *320*, 417–426. [CrossRef]
7. Zhang, J.; Shao, J.; Jin, Q.; Li, Z.; Zhang, X.; Chen, Y.; Zhang, S.; Chen, H. Sludge-based biochar activation to enhance Pb(II) adsorption. *Fuel* **2019**, *252*, 101–108. [CrossRef]
8. Mian, M.M.; Alam, N.; Ahommed, M.S.; He, Z.; Ni, Y. Emerging applications of sludge biochar-based catalysts for environ-mental remediation and energy storage: A review. *J. Clean. Prod.* **2022**, *360*, 132131. [CrossRef]
9. Li, D.; Shan, R.; Jiang, L.; Gu, J.; Zhang, Y.; Yuan, H.; Chen, Y. A review on the migration and transformation of heavy metals in the process of sludge pyrolysis. *Resour. Conserv. Recy.* **2022**, *185*, 106452. [CrossRef]
10. Udayanga, W.D.C.; Veksha, A.; Giannis, A.; Liang, Y.N.; Lisak, G.; Hu, X.; Lim, T.-T. Insights into the speciation of heavy metals during pyrolysis of industrial sludge. *Sci. Total Environ.* **2019**, *691*, 232–242.

11. Feng, S.; Zhang, G.; Yuan, D.; Li, Y.; Zhou, Y.; Lin, F. Co-pyrolysis of paper mill sludge and textile dyeing sludge with high calorific value solid waste: Pyrolysis kinetics, products distribution, and pollutants transformation. *Fuel* **2022**, *329*, 125433. [CrossRef]
12. Ma, M.; Xu, D.; Zhi, Y.; Yang, W.; Duan, P.; Wu, Z. Co-pyrolysis re-use of sludge and biomass waste: Development, kinetics, synergistic mechanism and industrialization. *J. Anal. Appl. Pyrol.* **2022**, *168*, 105746. [CrossRef]
13. Li, Y.; Yu, H.; Liu, L.; Yu, H. Application of co-pyrolysis biochar for the adsorption and immobilization of heavy metals in contaminated environmental substrates. *J. Hazard. Mater.* **2021**, *420*, 126655. [CrossRef] [PubMed]
14. Peng, B.; Liu, Q.; Li, X.; Zhou, Z.; Wu, C.; Zhang, H. Co-pyrolysis of industrial sludge and rice straw: Synergistic effects of biomass on reaction characteristics, biochar properties and heavy metals solidification. *Fuel Process. Technol.* **2022**, *230*, 107211. [CrossRef]
15. Dong, Q.; Zhang, S.; Wu, B.; Pi, M.; Xiong, Y.; Zhang, H. Co-pyrolysis of Sewage Sludge and Rice Straw: Thermal Behavior and Char Characteristic Evaluations. *Energy Fuels* **2020**, *34*, 607–615. [CrossRef]
16. Lu, J.; Kumar Mishra, P.; Hunter, T.N.; Yang, F.; Lu, Z.; Harbottle, D.; Xu, Z. Functionalization of mesoporous carbons de-rived from pomelo peel as capacitive electrodes for preferential removal/recovery of copper and lead from contaminated water. *Chem. Eng. J.* **2022**, *433*, 134508. [CrossRef]
17. Wang, W.; Chen, M. Catalytic degradation of sulfamethoxazole by peroxymonosulfate activation system composed of nitrogen-doped biochar from pomelo peel: Important roles of defects and nitrogen, and detoxification of intermediates. *J. Colloid Interf. Sci.* **2022**, *613*, 57–70. [CrossRef]
18. Geng, Y.; Sun, W.; Ying, P.; Zheng, Y.; Ding, J.; Sun, K.; Li, L.; Li, M. Bioinspired Fractal Design of Waste Biomass-Derived Solar–Thermal Materials for Highly Efficient Solar Evaporation. *Adv. Funct. Mater.* **2021**, *31*, 2007648. [CrossRef]
19. Wang, X.; Wei-Chung Chang, V.; Li, Z.; Song, Y.; Li, C.; Wang, Y. Co-pyrolysis of sewage sludge and food waste digestate to synergistically improve biochar characteristics and heavy metals immobilization. *Waste Manag.* **2022**, *141*, 231–239. [CrossRef]
20. Wang, X.; Chang, V.W.-C.; Li, Z.; Chen, Z.; Wang, Y. Co-pyrolysis of sewage sludge and organic fractions of municipal solid waste: Synergistic effects on biochar properties and the environmental risk of heavy metals. *J. Hazard. Mater.* **2021**, *412*, 125200. [CrossRef]
21. Cui, Z.; Xu, G.; Ormeci, B.; Liu, H.; Zhang, Z. Transformation and stabilization of heavy metals during pyrolysis of organic and inorganic-dominated sewage sludges and their mechanisms. *Waste Manag.* **2022**, *150*, 57–65. [CrossRef] [PubMed]
22. Huang, C.; Wang, W.; Sun, X.; Shen, J.; Wang, L. A novel acetogenic bacteria isolated from waste activated sludge and its potential application for enhancing anaerobic digestion performance. *J. Environ. Manag.* **2020**, *255*, 109842. [CrossRef]
23. Zhang, P.; Zhang, X.; Li, Y.; Han, L. Influence of pyrolysis temperature on chemical speciation, leaching ability, and environmental risk of heavy metals in biochar derived from cow manure. *Bioresour. Technol.* **2020**, *302*, 1228500. [CrossRef]
24. Gu, W.; Guo, J.; Bai, J.; Dong, B.; Hu, J.; Zhuang, X.; Zhang, C.; Shih, K. Co-pyrolysis of sewage sludge and Ca(H2PO4)2: Heavy metal stabilization, mechanism, and toxic leaching. *J. Environ. Manag.* **2022**, *305*, 114292. [CrossRef]
25. Qi, Q.; Hu, C.; Lin, J.; Wang, X.; Tang, C.; Dai, Z.; Xu, J. Contamination with multiple heavy metals decreases microbial diver-sity and favors generalists as the keystones in microbial occurrence networks. *Environ. Pollut.* **2022**, *306*, 119406. [CrossRef]
26. Fan, Z.; Zhou, X.; Peng, Z.; Wan, S.; Gao, Z.F.; Deng, S.; Tong, L.; Han, W.; Chen, X. Co-pyrolysis technology for enhancing the functionality of sewage sludge biochar and immobilizing heavy metals. *Chemosphere* **2023**, *317*, 137929. [CrossRef]
27. Chen, J.; Liu, J.; He, Y.; Huang, L.; Sun, S.; Sun, J.; Chang, K.; Kuo, J.; Huang, S.; Ning, X. Investigation of co-combustion characteristics of sewage sludge and coffee grounds mixtures using thermogravimetric analysis coupled to artificial neural net-works modeling. *Bioresour. Technol.* **2017**, *225*, 234–245. [CrossRef]
28. Shen, X.; Zeng, J.; Zhang, D.; Wang, F.; Li, Y.; Yi, W. Effect of pyrolysis temperature on characteristics, chemical speciation and environmental risk of Cr, Mn, Cu, and Zn in biochars derived from pig manure. *Sci. Total Environ.* **2020**, *704*, 135283. [CrossRef]
29. Zhang, X.; Zhao, B.; Liu, H.; Zhao, Y.; Li, L. Effects of pyrolysis temperature on biochar's characteristics and speciation and environmental risks of heavy metals in sewage sludge biochars. *Environ. Technol. Inno.* **2022**, *26*, 102288. [CrossRef]
30. Liu, Z.; Liu, H.; Zhang, Y.; Lichtfouse, E. Efficient phosphate recycling by adsorption on alkaline sludge biochar. *Environ. Chem. Lett.* **2023**, *21*, 21–30. [CrossRef]
31. Zhang, S.; Gu, W.; Geng, Z.; Bai, J.; Dong, B.; Zhao, J.; Zhuang, X.; Shih, K. Immobilization of heavy metals in biochar by co-pyrolysis of sludge and $CaSiO_3$. *J. Environ. Manag.* **2023**, *326*, 116635. [CrossRef] [PubMed]
32. Xiaorui, L.; Longji, Y.; Xudong, Y. Evolution of chemical functional groups during torrefaction of rice straw. *Bioresour. Technol.* **2021**, *320*, 124328. [CrossRef] [PubMed]
33. Chen, X.; Ma, R.; Luo, J.; Huang, W.; Fang, L.; Sun, S.; Lin, J. Co-microwave pyrolysis of electroplating sludge and municipal sewage sludge to synergistically improve the immobilization of high-concentration heavy metals and an analysis of the mechanism. *J. Hazard. Mater.* **2021**, *417*, 126099. [CrossRef]
34. Tang, F.; Yu, H.; Yassin Hussain Abdalkarim, S.; Sun, J.; Fan, X.; Li, Y.; Zhou, Y.; Chiu Tam, K. Green acid-free hydrolysis of wasted pomelo peel to produce carboxylated cellulose nanofibers with super absorption/flocculation ability for environmental remediation materials. *Chem. Eng. J.* **2020**, *395*, 125070. [CrossRef]
35. Fang, S.; Yu, Z.; Lin, Y.; Hu, S.; Liao, Y.; Ma, X. Thermogravimetric analysis of the co-pyrolysis of paper sludge and municipal solid waste. *Energ. Convers. Manag.* **2015**, *101*, 626–631. [CrossRef]
36. Yang, Y.-Q.; Cui, M.-H.; Guo, J.-C.; Du, J.-J.; Zheng, Z.-Y.; Liu, H. Effects of co-pyrolysis of rice husk and sewage sludge on the bioavailability and environmental risks of Pb and Cd. *Environ. Technol.* **2021**, *42*, 2304–2312. [CrossRef]

37. Huang, H.; Yao, W.; Li, R.; Ali, A.; Du, J.; Guo, D.; Xiao, R.; Guo, Z.; Zhang, Z.; Awasthi, M.K. Effect of pyrolysis temperature on chemical form, behavior and environmental risk of Zn, Pb and Cd in biochar produced from phytoremediation residue. *Bioresour. Technol.* **2018**, *249*, 487–493. [CrossRef]
38. Chen, G.; Tian, S.; Liu, B.; Hu, M.; Ma, W.; Li, X. Stabilization of heavy metals during co-pyrolysis of sewage sludge and ex-cavated waste. *Waste Manag.* **2020**, *103*, 268–275. [CrossRef]
39. Xu, Y.; Qi, F.; Bai, T.; Yan, Y.; Wu, C.; An, Z.; Luo, S.; Huang, Z.; Xie, P. A further inquiry into co-pyrolysis of straws with manures for heavy metal immobilization in manure-derived biochars. *J. Hazard. Mater.* **2019**, *380*, 120870. [CrossRef]
40. Wang, Z.; Xie, L.; Liu, K.; Wang, J.; Zhu, H.; Song, Q.; Shu, X. Co-pyrolysis of sewage sludge and cotton stalks. *Waste Manag.* **2019**, *89*, 430–438. [CrossRef] [PubMed]
41. Li, Z.; Huang, Y.; Zhu, Z.; Yu, M.; Cheng, H.; Shi, H.; Zuo, W.; Zhou, H.; Wang, S. Co-pyrolysis of industrial sludge with phytoremediation residue: Improving immobilization of heavy metals at high temperature. *Fuel* **2023**, *351*, 128942. [CrossRef]

water

MDPI

Article

Study on Preparation of Rabbit Manure Biochar and Activation of Peroxymonosulfate for Rhodamine B Degradation

Feng Ouyang [1], Yujie Liu [1], Jiao Chen [2], Chenghan Tang [1], Aojie Wang [1], Yixin Lu [1,2,3,*] and Yiping Yuan [2]

1 Faculty of Geosciences and Environmental Engineering, Southwest Jiaotong University, Chengdu 611756, China; fengouyang@163.com (F.O.); lyjenter@163.com (Y.L.); 15201822120@163.com (C.T.); waj565943952@126.com (A.W.)

2 School of Materials and Environmental Engineering, Chengdu Technological University, Chengdu 611730, China; chjicn@foxmail.com (J.C.); yyping0517@163.com (Y.Y.)

3 Sichuan Development Environmental Science and Technology Research Institute Co., Ltd., Chengdu 610094, China

* Correspondence: yxlu61@163.com; Tel.: +86-159-2862-8507

Abstract: Using rabbit manure as raw material, three distinct types of rabbit manure biochar (RBC400, RBC500, and RBC600) were prepared via pyrolysis at 400 °C, 500 °C, and 600 °C, respectively. The effects of pyrolysis temperature on the physicochemical properties of biochar were examined by scanning electron microscopy, Brunauer–Emmett–Teller analysis, Fourier transform infrared spectroscopy, X-ray photoelectron spectroscopy, and X-ray diffraction techniques. Rabbit manure biochar was used to activate permonosulfate (PMS) to degrade rhodamine B (RhB). The findings revealed that RBC600 prepared at 600 °C showed the strongest catalytic activity due to its abundant pores and pore structure, high graphitization, and high carbonization degree. Under optimal reaction conditions (0.4 g/L PMS and 0.6 g/L RBC600), the degradation rate of 50 mg/L RhB reached 93.38% within 60 min. RBC600 can be effectively recovered, and effective catalytic activity can be sustained after five cycles. The quenching and EPR experiments showed that both free-radical (SO_4^- and $\cdot OH$) and non-free-radical ($^1O_2$) pathways contributed to the degradation of RhB, in which $^1O_2$ performed a dominant role. In conclusion, the new PMS activator prepared in this study not only realizes the "waste to waste" use of manure waste but also provides technical support for the efficient resource-based treatment of organic wastewater.

Keywords: rabbit manure; biochar; permonosulfate; advanced oxidation; rhodamine B

1. Introduction

Printing and dyeing wastewater has caused severe water pollution due to its content of highly saturated colors, high toxicity, complex composition, and difficulty of degradation [1,2]. Owing to the considerable progress in synthetic dye technology, dye antiphotolysis, antibiodegradation, and anti-oxidation abilities, a substantial amount of refractory dye wastewater has emerged. Rhodamine B (RhB), a type of synthetic dye, has emerged as one of the dyes that are being widely implemented in industrial production, and its impact on the environment cannot be underestimated. Traditional wastewater treatment methods, such as physical adsorption, electrochemical methods, and biodegradation [3–5], pose considerable challenges in terms of their applicability to printing and dyeing wastewater. In this context, how printing and dyeing wastewater can be efficiently degraded has become an urgent problem that necessitates immediate resolution.

The Opinions on Implementing the Strategy of Rural Revitalization issued by the CPC Central Committee and the State Council in 2018 clearly emphasized the importance of realizing waste resources, replacing organic fertilizers with chemical fertilizers, treating livestock and poultry manure waste, and promoting comprehensive utilization of crop straw. As the largest rabbit-raising province in China, Sichuan Province has ranked first in the country in recent years. According to statistics, since 2020, the annual output of rabbits in Sichuan Province has stabilized at approximately 170 million, accounting for approximately 50% of the national output during the same period. Although a relatively mature treatment mode has been developed for the resulting large amount of rabbit manure, several challenges persist. Under the backdrop of this new era, it is essential to explore a novel resource-based treatment approach for rabbit manure.

Biochar is a type of carbon-rich, porous solid material (similar to charcoal); its production can be realized through solid waste transformation under anaerobic or anoxic conditions. It features inexpensive prices, an elevated carbon content, a large cation exchange capacity, a stable structure, a large specific surface area, and a strong adsorption capacity, among other desirable properties [6–8]. When evaluating the application potential of biochar, most researchers focus on its adsorption of organic pollutants [9]. However, many recent studies have shown that in the presence of biochar and oxidants, some organic contaminants may decompose [10–12]. Simultaneously, due to the electronic conductivity of the carbon structure, biochar can also obtain electrons from electron donors or provide electrons to acceptors and act as electron shuttles or electron conductors to degrade organic pollutants [13,14].

PMS can be activated in a variety of ways, such as ultraviolet activation, microwave irradiation activation [15], ultrasonic activation, alkali activation [16], metal activation [17], carbon material activation, etc. These methods possess their own advantages and disadvantages. RhB degradation via UV/PMS, US/PMS, and other systems has been investigated before, and some scholars have even studied the degradation of RhB through persulfate catalyzed by carbonaceous materials [18,19]. However, studies exploring the reaction process and mechanism of RhB degradation using rabbit manure biochar/PMS are scarce. In this paper, the reaction mechanism was preliminarily explored through the study of the system. It was found that the rabbit manure biochar/PMS system possesses a good degradation effect on RhB, and the rabbit manure biochar exhibits excellent material stability and recyclability. A novel RhB degradation system was proposed in this paper. Carbon activation is easier to control and more economical than UV activation, ultrasonic activation, etc. This important conclusion is a breakthrough and contribution to previous studies and lays the foundation for subsequent studies. Therefore, in this study, rabbit manure served as a raw material in the preparation of biochar material, which served as a catalyst and activator in the activation of permonosulfate (PMS) to degrade RhB in printing and dyeing wastewater. At the same time, the degradation efficiency of RhB in water via single biochar systems, single PMS systems, and rabbit manure biochar and PMS systems under different conditions was investigated. Combined with free-radical quenching and EPR experiments, the possible pathway of RhB degradation was analyzed. Finally, the application potential of the catalyst was evaluated via a stability experiment in an effort to provide a theoretical basis for an efficient, novel, and cheap dye wastewater treatment method.

2. Materials and Methods

2.1. Reagents and Instruments

Reagents: PMS ($KHSO_5 \cdot 0.5KHSO_4 \cdot 0.5K_2SO_4$, $KHSO_5 \geq 42\%$) was purchased from Shanghai Aladdin Biochemical Technology Co., Ltd. (Shanghai, China), Methanol (MeOH), ethanol (ETA), tert-butanol (TBA), *p*-benzoquinone (BQ), L-histidine (L-His), sodium hydroxide, sodium bicarbonate, sodium chloride, sodium nitrate, hydrochloric acid, humic acid (HA), and RhB were all purchased from Chengdu Colon Chemicals Co., Ltd. (Chengdu, China). The chemical reagents used in the experiment were all analytically pure, and the laboratory water was ultrapure (resistivity of $\geq$18.25 MΩ/cm).

Instruments: Laboratory pH meter (PHSJ-3F, Shanghai Lei Magnetic (Shanghai, China)); numerical control ultrasonic cleaner (KH5200DB type, Kunshan Ultrasonic Instruments Co. Ltd. (Kunshan, China)); UPHW-IV-90T, Sichuan Upu Water Instrument (Chengdu, China); ultraviolet visible spectrophotometer (WFZ UV-4802H, Unique, Shanghai, China); electric blast drying oven (101-3AB, Tianjin Test (Tianjin, China)); constant-temperature oscillating box (BS-2E, Changzhou Jintan Liangyou (Changzhou, China)); table low-speed centrifuge (TD-420, Sichuan Shuke (Chengdu, China)); analytical balance, FA2004B, Shanghai Youke (Shanghai, China); muffle furnace (SX2-3-10, Shenyang Energy-saving Electric Furnace Factory (Shenyang, China)); and high-speed multifunctional crusher (SS-1022, Wuyi Haina (Wuyi, China)). Scanning electron microscope energy spectrometer (ZEISS Gemini SEM 300, ZEISS, Jena, Germany); Brunauer–Emmett–Teller (BET) ratio surface analyzer (Micromeritics ASAP 2460, Micromeritics Instruments Co., Norcross, GA, USA); Fourier transform infrared spectrometer (Thermo Scientific Nicolet 6700, Thermo Fisher Scientific, Waltham, MA, USA); X-ray photoelectron spectrometer (Thermo Scientific K-Alpha, USA); X-ray diffractometer (Rigaku Ultima IV, Rigaku, Tokyo, Japan); paramagnetic resonance spectrometer ESR/EPR (Bruker ELEXSYS-II E500, Bruker GmbH, Mannheim, Germany); and a temperament combination instrument (Agilent 7890A-5975C, Agilent Technologies, Santa Clara, CA, USA).

2.2. Preparation of Biochar

Rabbit manure from a rabbit breeding base in Chengdu City was selected as the raw material for biochar. Large impurities, including grass and leaves, were removed from the manure using a mesh sieve and tweezers. The manure was then broken up using a crusher. The crushed rabbit excrement was dried in an oven at 105 °C for 24 h. An appropriate amount of the material was collected for weighing, placed in a crucible, compacted, and covered with a tin foil-wrapped crucible in a muffle furnace. The material was heated in the muffle furnace at a speed of 10 °C/min at set temperatures (400 °C, 500 °C, and 600 °C). After constant-temperature heating for 2 h, the material was taken out and cooled to room temperature. After the cooling pyrolysis products were mixed with 1 mol/L HCl at a 1:25 (m/V, g/mL) solid–liquid ratio, ultrasonic cleaning was performed for 10 min, the supernatant was skimmed off, and the same volume of HCl was added. The process was repeated thrice. Finally, the pyrolysis products obtained after HCl cleaning were centrifuged and repeatedly rinsed with deionized water until the clear liquid was neutral. The collected solids were dried at 105 °C to a constant weight, then ground with a mortar and screened with 100 mesh (0.15 mm) to obtain RBC400, RBC500, and RBC600. The prepared rabbit manure biochar was sealed in a ziplock bag, marked, and kept in a brown vacuum dryer. Figure 1 depicts the precise preparation procedure.

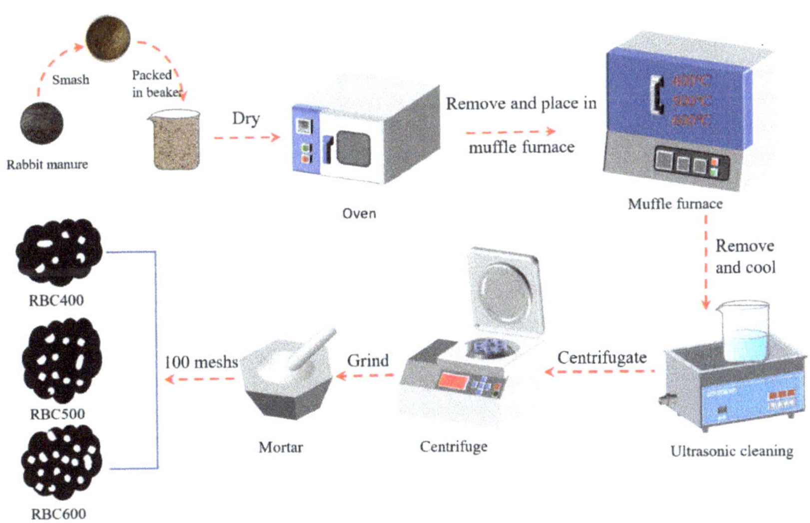

Figure 1. Preparation process of RBCs.

2.3. Experimental Scheme

Certain amounts of PMS (0.2–1.0 g/L) and biochar (0.2–1.0 g/L) were added to 150 mL of a 50 mg/L RhB solution. Furthermore, the conical flask was placed in a constant-temperature oscillation chamber and oscillated at 200 r/min at 15–45 °C. A total of 2.0 mL of the reaction liquid was collected at the set time interval point. After the sample passed through the 0.45-μm filter membrane, 2 mL of MeOH solution was quickly added for quenching, and the absorbance of the mixture was determined by a spectrophotometer after dilution with pure water. To learn more about how diverse pH levels affect the activation properties of the solution, 1 mol/L of NaOH or HCl was used to adjust the pH of the solution. NaCl, $NaHCO_3$, $NaNO_3$, and HA were used to regulate the concentrations of different background components in the solution to study the effects of background components (Cl^-, HCO_3^-, NO_3^-, and HA) on the degradation of RhB. MeOH, TBA, BQ, and L-His were added to the solution as quenchers to investigate the formation of different active substances.

2.4. Analysis Method

The absorbance of RhB was determined by a UV-visible spectrophotometer (WFZ UV-4802H, Unic, Shanghai, China). The measurement wavelength range was 190–1100 nm, and the wavelength accuracy was ±0.3 nm. At the maximum visible wavelength of 554 nm, the absorbance of RhB was measured, and its concentration was calculated based on the standard curve of RhB. In this research, all experiments were performed thrice in parallel, and the degradation rate η (%) was calculated using the average data. The calculation formulas of RhB degradation rate η and first-order reaction kinetics are shown in Equations (1) and (2), respectively.

$$\eta = \frac{C_0 - C_t}{C_0} \times 100\% \quad (1)$$

$$\ln(C/C_0) = -k_{obs}t \quad (2)$$

where C_0 and C_t are the mass concentrations of RhB in solution (mg/L), respectively, with 0 denoting the initial concentration and t denoting the concentration of RhB at time t; C is the concentration of RhB in the solution at any given time (mg/L); V is the volume of RhB solution (mL); and k_{obs} is the pseudo-first-order reaction rate constant (min^{-1}).

3. Results and Discussion

3.1. Structural Characteristics of RBC

The prepared rabbit manure biochar was placed under a scanning electron microscope, and its microstructure was observed (Figure 2). The acceleration voltage of the scanning electron microscope was 2 kV, and its magnification was 2–10 k. The sample preparation process is as follows: The powder sample was dispersed on the clean, smooth surface of the silicon wafer (0.2 cm × 0.2 cm). After natural drying, the other side of the silicon wafer is pasted on the conductive tape of the sample table. After gold spraying, the powder sample was placed in the sample room for testing. As depicted in Figure 2, after pyrolysis, the surfaces of the three types of biochar showed irregular concave–convex shapes. When the pyrolysis temperature was 400 °C, almost no pore structure was observed on the surface of the biochar, and the surface exhibited a rough and caked state. Some tiny particles were deposited on the surface (Figure 2a). This finding indicates that at 400 °C, the biomass pyrolysis was insufficient, and the pores were not fully opened. Figure 2b shows that when the pyrolysis temperature rose to 500 °C, the surface roughness of the biochar increased, and some pores appeared. However, the pore size was generally large. When the pyrolysis temperature was 600 °C, more pores and channels appeared on the biochar surface, and a certain amount of fold structure appeared, indicating that biochar generated at 600 °C formed abundant pore and channel structures after full pyrolysis at high temperature. According to the scanning electron microscopy (SEM) analysis, with increasing pyrolysis temperatures during rabbit feces biochar preparation, biochar pores, channels, and fold structures increased.

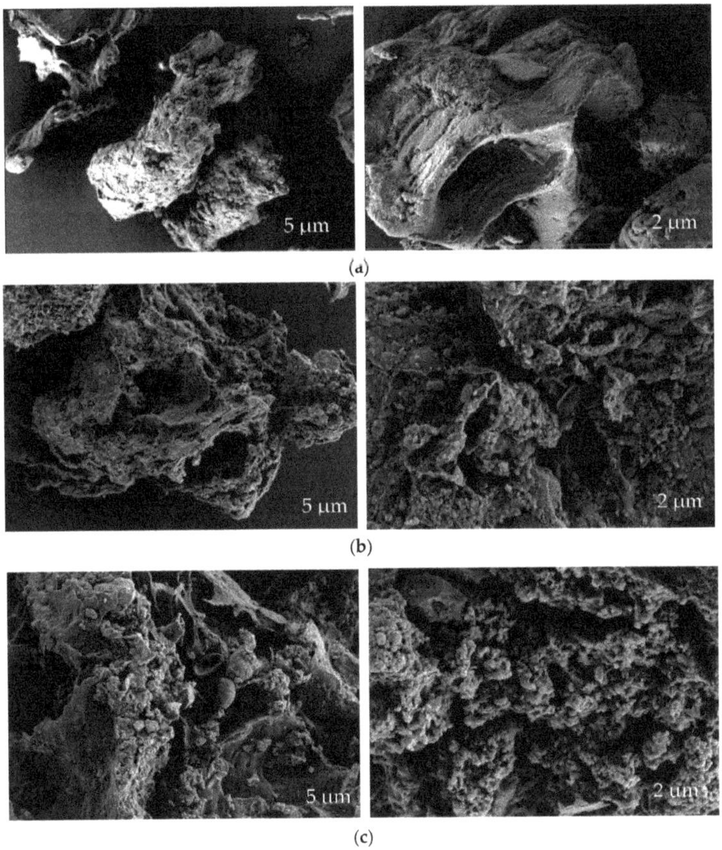

Figure 2. SEM images of different activators. (**a**) RBC400. (**b**) RBC500. (**c**) RBC600.

The specific surface area, total pore volume, and average pore size of the biochar were determined using the static volumetric method. N_2 (99.999%) was used as an adsorbent, and the degassing time was 8 h. The BET results of biochar in Table 1 and Figure 3 showed that the pore sizes of the three types of biochar ranged from 2 nm to 50 nm, and they all belonged to mesoporous materials. Table 1 shows that the specific surface area, total pore volume, and micropore volume of RBC600 were significantly improved compared with those of RBC400 and RBC500. Combined with Figure 3, the surface characteristics and microstructure diagrams can be further comprehensively analyzed. The surface of RBC400 was rare with medium pores and folds, and the surface roughness was low, which resulted in a small specific surface area and total pore volume. Given the increase in the pyrolysis temperature, the pores of RBC500 were opened to a certain extent, and the surface roughness increased, displaying a specific number of folds and pores. When the pyrolysis temperature was further increased, RBC600 formed a more favorable surface structure, and compared with RBC500, the micropores, surface folds, and roughness further improved. The surface characteristic test results also showed that the specific surface area and total pore volume were larger.

Table 1. Test results of biochar surface characteristics.

Biochar	Specific Surface Area (m^2/g)	Total Pore Volume (cm^3/g)	Micropore Volume (cm^3/g)	Mean Aperture (nm)
RBC400	16.887	0.023	0.0001	13.196
RBC500	57.095	0.084	0.0008	6.493
RBC600	89.027	0.136	0.0057	6.855

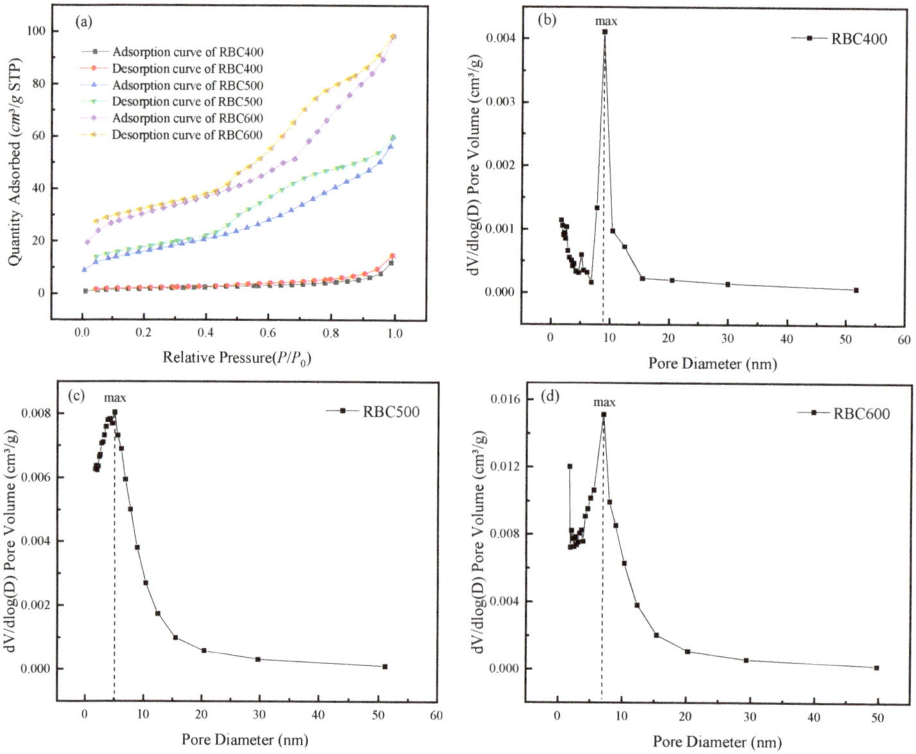

Figure 3. Effect of pyrolysis temperature on N_2 physisorption isotherms and pore characteristics.

The pore structural properties of biochar can also be obtained from the N_2 adsorption–desorption curve and pore size analysis during BET characterization. Figure 3 shows the N_2 adsorption–desorption curve and pore size distribution curves of three biochar types. Figure 3a displays that the three biochar species had a high adsorption pressure ($P/P_0 > 0.8$). The gas adsorption line rose sharply, the adsorption rate of N_2 accelerated significantly, and capillary condensation occurred in mesoporous pores, showing a VI N_2 adsorption–desorption isotherm [20]. The hysteresis loops of RBC500 and RBC600 appeared when the relative pressure was between 0.4 and 1.0, and the width of the hysteresis loops was wider than that of RBC400; thus, the pore structure size distributions of RBC500 and RBC600 were wider, and their pore structure was mainly mesoporous [21].

Combined with the BET characterization results of the three types of rabbit manure biochar described above, the specific surface area of rabbit manure biochar showed a gradual increase with the increase in the pyrolysis temperature of rabbit feces biochar. Moreover, the pore size and pore structure of rabbit manure biochar gradually became complex. In general, with the increase in the pyrolysis temperature, the specific surface area, pore size, and pore volume of rabbit manure biochar changed in a direction conducive to the reaction.

The surface functional groups of three types of rabbit manure biochar were characterized by Fourier transform infrared spectroscopy (FT-IR). Before the test, the samples were first pressed and measured in the wavelength range of 400–4000 cm^{-1} after deducting the background value, and the FT-IR characterization results are shown in Figure 4. As shown in Figure 4, the types of functional groups on the surface of RBC400, RBC500, and RBC600 biochar were similar, and the main characteristic peaks appeared at 3438, 2958, 1630, 1388, and 1093 cm^{-1}. The stretching peak near 3438 cm^{-1} was the characteristic peak of –OH, and the stretching peak near 2958 cm^{-1} was the C–H of alkane. The strength of the stretching peak of RBC500 and RBC600 was weak here, indicating that the carbonization degrees of both were high at high pyrolysis temperatures. The C=C or C=O bond of aromatic ring transformation in biomass wood fiber appeared at the stretching peak near 1630 cm^{-1} [22]; near 1388 cm^{-1}, a small but sharp stretching peak appeared, which was the absorption peak formed by the deformation and vibration between aromatic C–H [23]. The stretching peak near 1093 cm^{-1} is an ether-type C–O stretching vibration absorption peak [24].

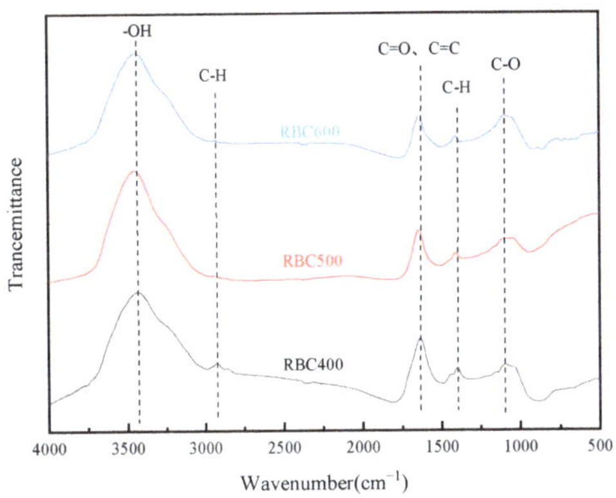

Figure 4. FT-IR diagrams of different activators.

The surface element compositions of RBC400, RBC500, and RBC600 were characterized and analyzed using X-ray photoelectron spectroscopy (XPS) spectra. The test target was a copper target with a scanning range of 10–80° and a step size of 2°/min. Figure 5 depicts the XPS spectra of the three materials, and Table 2 shows the elemental composition of the materials. The XPS also showed that three main characteristic peaks were detected on the three biochar samples, corresponding to O 1 s (532.08 eV), C 1 s (400.08 eV), and N 1 s (248.08 eV), respectively. The elemental composition analysis showed that the main element components of the three biochar surfaces were O, C, and N. As shown in Table 2, with the increase in the pyrolysis temperature of rabbit biochar, the C content on the surface of the generated biochar also increased to a certain extent, which indicates that the increase in the pyrolysis temperature can improve the carbonization degree of rabbit coprobiochar to a certain extent, which matched the FT-IR characterization findings.

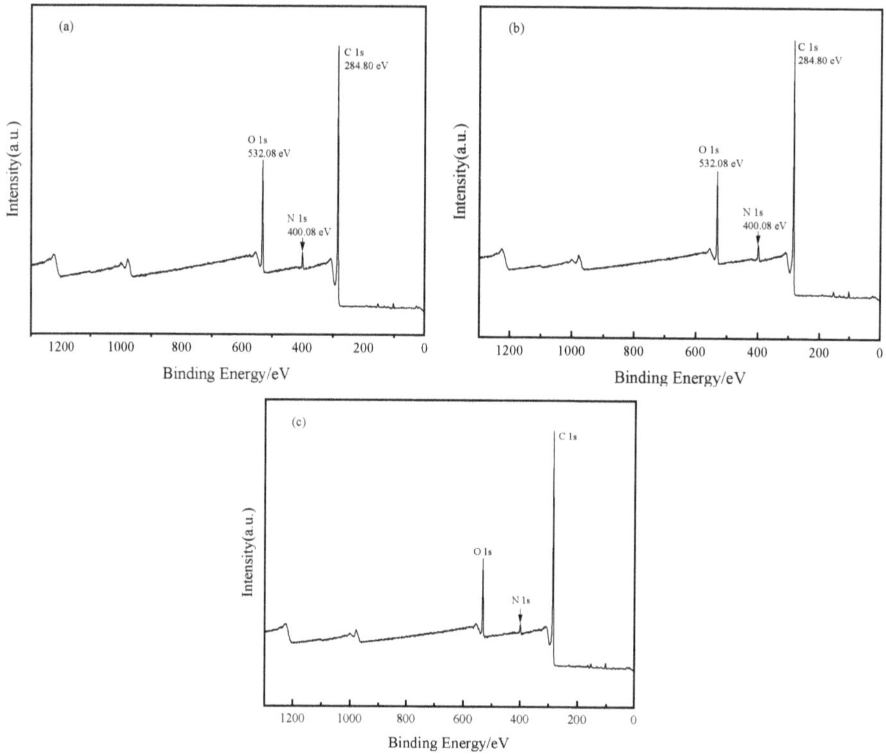

Figure 5. XPS diagrams of different biochar: (**a**) RBC400, (**b**) RBC500, and (**c**) RBC600.

Table 2. Biochar surface element composition.

Biochar	Surface Element Percentage/%			
	O 1 s	C 1 s	N 1 s	Other
RBC400	15.78	79.24	4.55	0.43
RBC500	14.38	79.72	5.32	0.58
RBC600	13.61	81.44	4.44	0.51

The crystal structure information inside rabbit manure biochar can be obtained using X-ray diffraction (XRD). Before the test, appropriate amounts of samples were taken, pressed, and affixed to the sample tray. Furthermore, the samples were placed in the sample chamber of a Thermo Scientific K-Alpha XPS instrument. When the pressure in the sample chamber was less than 2.0×10^{-7} mbar, the samples were sent to the analysis chamber. The spot size, working voltage, and filament current were 400 μm, 12 kV, and 6 mA, respectively. The full-spectrum scanning energy and step size were 150 eV and 1 eV, while the narrow-spectrum scanning energy and step size were 50 eV and 0.1 eV, respectively. The XRD characterization results are shown in Figure 6. After the phase index of three types of rabbit manure biochar was determined, two types of crystal structures were mainly found on their surface: SiO_2 (PDF card number 46-1045) and graphitic carbon (PDF card number 26-1076). Thus, the three types of biochar produced graphitized crystal structures after high-temperature pyrolysis. Further comparison revealed that, unlike those of RBC400 and RBC500, the peaks of RBC600 were significant and sharp at 26.5°, indicating that its particle size and crystallinity were likely greater.

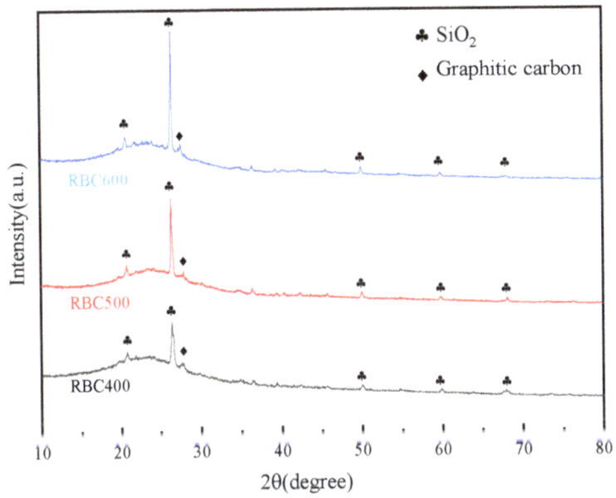

Figure 6. Effect of pyrolysis temperature on the crystal structure of rabbit manure biochar.

3.2. Evaluation of Catalyst Activity

For investigating the differences among the hB removal performance of different biochars and their activated PMS, this part of the experiment explored the RhB removal effect of different systems under the same reaction conditions, and the experimental results are depicted in Figure 7. The removal efficiencies of RhB in different systems from low to high were as follows: blank (1.7%), RBC400 (6.94%), PMS (24.61%), RBC400 + PMS (29.31%), RBC500 (32.39%), RBC600 (35.28%), RBC500 + PMS (60.15%), and RBC600 + PMS (93.98%). According to the analysis of the experimental results, the degradation rate of RhB in the blank control group was 1.7%. Given that the experiment was conducted under normal indoor light, a small amount of RhB was possibly photolyzed in this environment. However, the degree of photolysis was low and negligible. The addition of biochar alone can adsorb RhB, and RBC600 showed the best adsorption performance. RhB in water can be oxidized by a single PMS system because a small amount of $SO_4^{\cdot-}$ is produced through the hydrolysis of PMS in the solution, and RhB can be oxidized by $SO_4^{\cdot-}$ [25]. The RBC/PMS systems can degrade RhB in water more effectively than the RBC or PMS systems and the degradation rate of RhB in the RBC600/PMS system was significantly higher than that in the RBC400/PMS and RBC500/PMS systems. According to the analysis of Table 1 and Figure 3, this is due to the larger specific surface area and more abundant surface functional groups (C=O) of RBC600, which can provide more activation sites and facilitate the catalytic degradation reaction.

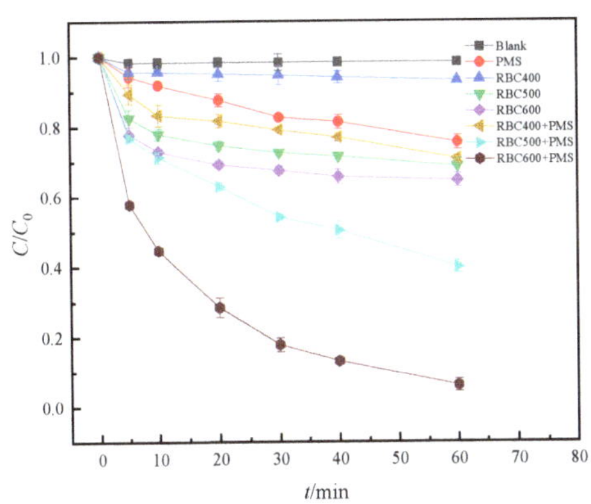

Figure 7. RhB removal effect of different systems (Reaction condition: PMS = 0.4 g/L, RBC = 0.6 g/L, RhB = 50 mg/L, and T = 25 °C).

To better study the reaction kinetics of RhB degradation in the RBC/PMS system, a pseudo-first-order kinetic model was introduced to perform kinetic fitting on the reaction results of RhB catalytic degradation in the RBC/PMS system (Figure 8).

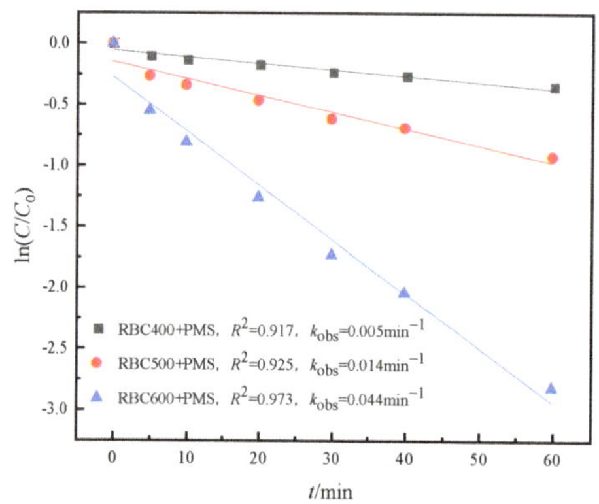

Figure 8. Pseudo-first-order kinetic model of RhB degradation in different systems.

The fitting results revealed that the degradation process of RhB in RBC400/PMS, RBC500/PMS, and RBC600/PMS can follow well the pseudo-first-order reaction kinetics, and the corresponding k_{obs} of the three systems were 0.005, 0.014, and 0.044 min^{-1}, respectively. These findings show that the RBC600/PMS system had a higher apparent rate constant than the other two systems. To better explore the performance, we used the RBC600/PMS system under optimal reaction conditions and the reaction mechanism of RhB degradation by rabbit manure biochar-activated PMS for follow-up experiments.

3.3. Analysis of Adsorption Influencing Factors

3.3.1. Effect of Biochar Dosage

Biochar dosage is an essential factor affecting the process of biochar-activated PMS in degrading pollutants. It determines the cost input required by the system to degrade pollutants to a certain extent. Therefore, it is crucial to select the most appropriate biochar dosage. The experimental results of the influence of biochar dosage are shown in Figure 9.

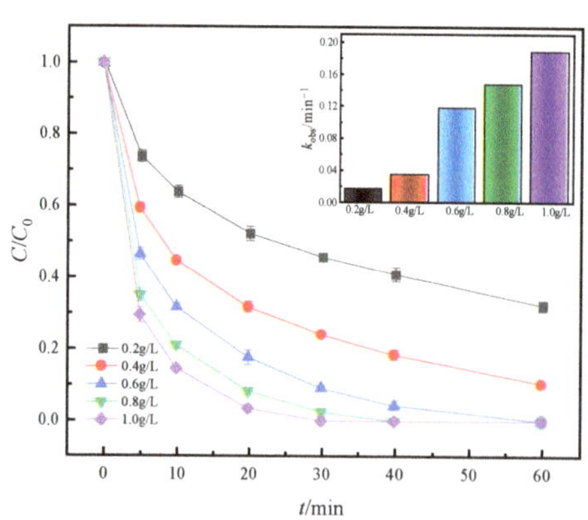

Figure 9. Degradation performance under different biochar dosages. (Reaction condition: PMS = 0.4 g/L, RBC = 0.2, 0.4, 0.6, 0.8, 1.0 g/L, RhB = 50 mg/L, and T = 25 °C).

As depicted in Figure 9, when the biochar dosage increased from 0.2 g/L to 0.6 g/L, the degradation rate of RhB increased from 67.83% to 99.97%, showing an evident upward trend, and the degradation rate of k_{obs} increased from 0.017 min^{-1} to 0.118 min^{-1}. On the one hand, with the increase in biochar dosage, the biochar surface area and the number of adsorption sites involved in pollutant adsorption increased, and the adsorption performance was gradually enhanced. On the other hand, increasing biochar dosage can provide more activation sites for PMS and improve the catalytic degradation efficiency of RhB in the RBC600/PMS system. When the dosage of RBC600 increased from 0.6 g/L to 1.0 g/L, the reaction k_{obs} increased from 0.118 min^{-1} to 0.189 min^{-1}, but the catalytic degradation efficiency did not improve significantly. This result may be due to the limited number of free radicals generated by biochar-catalyzed PMS at certain dosages of PMS. With the increase in the dosage of RBC600, the excess active sites provided by biochar cannot be effectively utilized, and the degradation efficiency of RhB cannot be significantly increased by increasing the dosage of biochar; this condition resulted in a "plateau" in the degradation rate of RhB. Huong et al. [26] also reached a similar conclusion in their study on the degradation of organic contaminants in wastewater by activating PMS with rice husk biochar.

3.3.2. Effect of PMS Dosage

The dosage of PMS is one of the key parameters of persulfate advanced oxidation technology, and its dosage largely determines the amount of free radicals produced in the system, which has an important influence on the degradation efficiency of RhB. Figure 10 shows the effect of PMS dosage on the degradation efficiency of RhB.

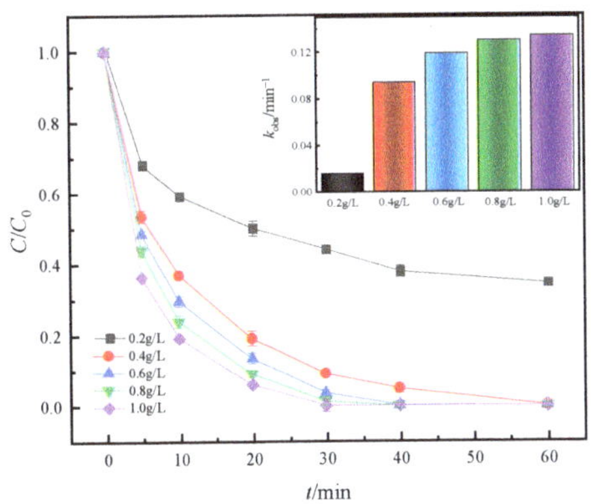

Figure 10. Degradation performance under different PMS dosages. (Reaction condition: PMS = 0.2, 0.4, 0.6, 0.8, 1.0 g/L, RBC = 0.6 g/L, RhB = 50 mg/L, and T = 25 °C).

As presented in Figure 10, with the increase in PMS dosage, the degradation rate of RhB and k_{obs} also improved. When the PMS dosage increased from 0.2 g/L to 1.0 g/L, k_{obs} increased from 0.016 min^{-1} to 0.134 min^{-1}. This result was due to the continuous increase in the rate and amount of free-radical production with the increase in PMS, which increased the chance of free-radical reaction with biochar surface active site contact. However, when the dosage of PMS increased from 0.2 g/L to 0.4 g/L, the degradation efficiency of RhB increased by 34.26%. The apparent rate constant also increased by 4.875 times, from 0.016 min^{-1} to 0.094 min^{-1}. However, when the dosage of PMS increased from 0.4 g/L to 1.0 g/L, the degradation efficiency and apparent rate constant of RhB were not significantly improved, and those of k_{obs} increased by 42.6%, which was relatively slow compared with the 4.875 times mentioned above. This experimental result is consistent with the experimental finding of Chen et al. [27], who used straw-derived biochar to activate PMS to degrade the antibiotic ofloxacin. On the one hand, when the dosage of biochar is certain, the number of biochar surface active sites is limited. When these active sites are occupied, excessive free radicals cannot perform desirable functions in pollutant degradation. On the other hand, the excess free radicals generated by excess PMS may undergo self-quenching (Equations (3)–(5)).

$$SO_4^{\cdot -} + HSO_5^- \rightarrow SO_5^{\cdot -} + HSO_4^- \quad (3)$$

$$OH + HSO_5^- \rightarrow SO_5^{\cdot -} + H_2O \quad (4)$$

$$SO_4^{\cdot -} + H_2O + HSO_5^- \rightarrow HO_2 + 2HSO_4^- \quad (5)$$

3.3.3. Effect of the Initial pH Value of the Solution

The initial pH of the solution is also an important factor affecting the degradation of pollutants in the persulfate system. Within the range of pH set in the experiment, the degradation efficiency of RhB in the system varied (Figure 11).

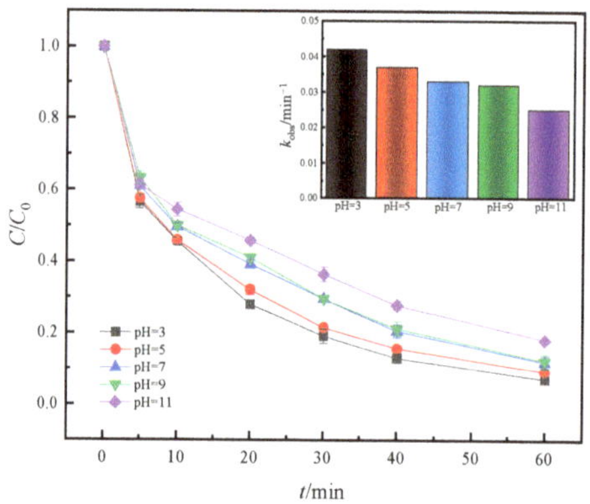

Figure 11. Adsorption performance at different pHs. (Reaction condition: PMS = 0.4 g/L, RBC = 0.6 g/L, RhB = 50 mg/L, T = 25 °C, and pH = 3, 5, 7, 9, and 11).

When the initial pH of the reaction increased from 3 to 11, the degradation rate of RhB decreased by 10.91%, and the corresponding k_{obs} were 0.042, 0.037, 0.033, 0.032, and 0.025 min^{-1}. The experimental results indicated that acidic conditions were more favorable for the reaction because the free radicals involved in the reaction in the solution changed with the pH. When the pH of the solution is lower than 9, more H+ in the system can be activated efficiently to produce more $SO_4^{\cdot -}$. With the increase in pH, $SO_4^{\cdot -}$ gradually transformed into ·OH (Equations (6) and (7)). When the pH of the solution was greater than 9, ·OH quenched $SO_4^{\cdot -}$. Thus, the $SO_4^{\cdot -}$ with a stronger oxidation capacity in the system became ·OH with a weaker oxidation capacity and shorter half-life. Therefore, in RhB degradation by RBC600/PMS, the degradation rate of RhB decreased with the increase in pH [28,29]. Wang Yan et al. [30] also came to a similar conclusion in their study on the degradation of golden oranges by persulfate. On the other hand, biochar can better adsorb RhB under acidic conditions, but when the solution pH is low, the biochar surface functions through group protonation. RhB exists in the form of a cation, which is conducive to the ion exchange between biochar and RhB and improves the RhB degradation rate. However, when the pH of the solution is higher, RhB exists as a zwitterion and forms a macromolecular dimer, which is difficult to be adsorbed by biochar [31]. In general, the RBC600/PMS system has various pH applications and a good degradation effect on RhB in the range of pH 3–11 (the degradation rate is over 80%).

$$SO_4^{\cdot -} + H_2O \rightarrow HSO_4^- + \cdot OH \quad (6)$$

$$SO_4^{\cdot -} + OH^- \rightarrow SO_4^{2-} + \cdot OH \quad (7)$$

3.3.4. Effect of Reaction Temperature

The results showed that PMS could be activated by heating, and the increase in temperature may further improve the degradation effect of the RBC600/PMS system on RhB. This part of the experiment explored the effect of reaction temperature on RhB degradation efficiency, and the experimental results are shown in Figure 12.

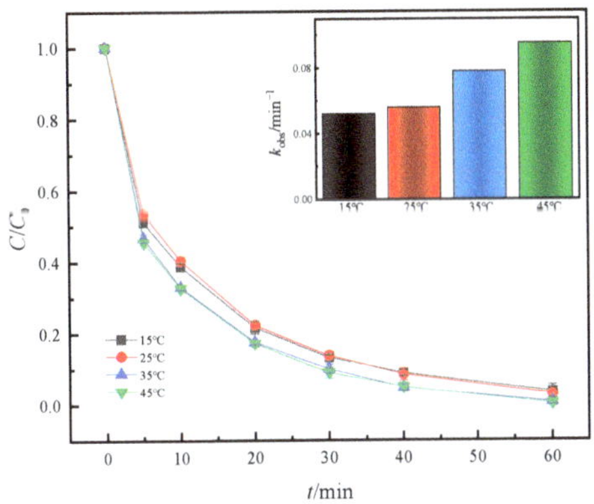

Figure 12. Adsorption performance under different reaction temperatures. (Reaction conditions: PMS = 0.4 g/L, RBC = 0.6 g/L, RhB = 50 mg/L, and T = 15 °C, 5 °C, 35 °C, 45 °C).

As shown in Figure 12, the effect of the reaction temperature change (15–45 °C) on the degradation of RhB in the RBC600/PMS system was very limited, and the degradation rate of RhB in the RBC600/PMS system can reach more than 90% at four reaction temperatures. Although the increase in reaction temperature had no evident effect on the final degradation efficiency of RhB, the apparent rate constant increased with the increase in reaction temperature. The k_{obs} at 15 °C, 25 °C, 35 °C, and 45 °C were 0.052, 0.056, 0.078, and 0.095 min^{-1}, respectively, which indicated that the increase in the reaction temperature could accelerate the reaction. Shi Chenfei et al. [32] showed that compared with rice straw biochar activation of PS, the efficiency of thermal activation was lower. Therefore, the activation efficiency of rabbit manure biochar on PMS was significantly higher than that of thermal activation. In this study, RhB degradation was effectively achieved by activating PMS with rabbit manure biochar. Therefore, significant improvement of the RhB degradation efficiency by thermal activation of PMS was difficult compared with PMS activation by biochar.

3.3.5. Effect of the Water Background Component

Various background components in water may have different degrees of influence on the degradation of RhB by the system. In this experiment, four common background components (Cl^-, HCO_3^-, NO_3^-, and HA) in water were selected to explore the degradation performance of RhB by the system under different concentrations of background components. Figure 13 shows the influence of water background components on RhB degradation efficiency.

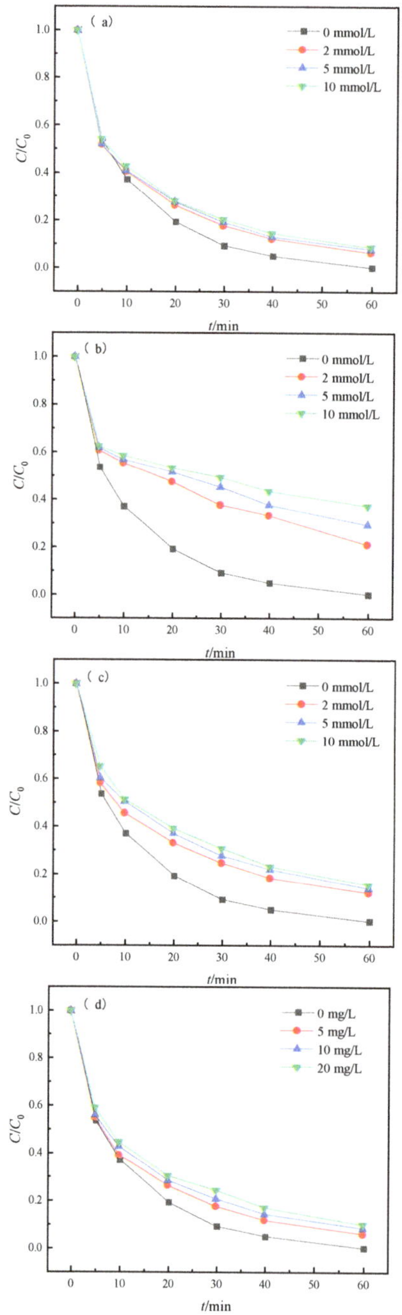

Figure 13. Adsorption performance under different water background components. (**a**) Cl^-, (**b**) HCO_3^-, (**c**) NO_3^-, and (**d**) HA. (Reaction conditions: PMS = 0.4 g/L, RBC = 0.6 g/L, RhB = 50 mg/L, T = 25 °C, Cl^-, HCO_3^-, NO_3^- = 2, 5, 10 mmol/L, and HA = 5, 10, and 20 mg/L).

As shown in Figure 13, the inhibitory effects of Cl^-, HCO_3^-, NO_3^-, and HA on the degradation of RhB in the system were HCO_3^- (37.23%), NO_3^- (15.52%), HA (9.96%), and Cl^- (8.63%) in descending order. As shown in Figure 13a, when Cl^- concentration increased from 0 mmol/L to 10 mmol/L, the degradation rate of RhB decreased from 99.79% to 91.16%, showing a slight inhibition. k_{obs} decreased from 0.094 min^{-1} to 0.042, 0.039, and 0.038 min^{-1}, respectively. This finding was due to the consumption of $\cdot OH$, $SO_4^{\cdot -}$, or HSO_5^- by Cl^- to form $Cl\cdot$ and $Cl_2^{\cdot -}$ with lower activities (Equations (8)–(11)) [33–36]. Guan et al. [37] also reached a similar conclusion in their experiment on the influence of chloride ions on the UV/PMS system:

$$SO_4^{\cdot -} + Cl^- \rightarrow SO_4^{2-} + Cl\cdot \tag{8}$$

$$HSO_5^- + Cl^- \rightarrow SO_4^{2-} + HOCl \tag{9}$$

$$HSO_5^- + 2Cl^- + H^+ \rightarrow SO_4^{2-} + Cl_2 + H_2O \tag{10}$$

$$\cdot OH + Cl^- + H^+ \rightarrow HOCl^{\cdot -} \tag{11}$$

As shown in Figure 13b, when the concentration of HCO_3^- increased from 0 mmol/L to 10 mmol/L, the degradation rate of RhB decreased from 99.79% to 62.56%, indicating a decrease of approximately 37.31%. k_{obs} decreased from 0.094 min^{-1} to 0.022, 0.016, and 0.013 min^{-1}, showing an evident downward trend; this result indicated that the presence of HCO_3^- significantly inhibited the degradation of RhB [38]. This finding can be explained by the rate constant from the work of $\cdot OH$. Kang et al. [39] observed that the rate constant of HCO_3^- was greater than that of OH and that it had a certain scavenging effect on $\cdot OH$ (Equation (12)). In addition, HCO_3^- reacted with $SO_4^{\cdot -}$ to produce $CO_3^{\cdot -}$ (Equation (13)) with a lower redox potential, which inhibited the degradation of RhB to a certain extent.

$$SO_4^{\cdot -} + HCO_3^- \rightarrow SO_4^{2-} + CO_3^{\cdot -} + H^+ \tag{12}$$

$$\cdot OH + HCO_3^- \rightarrow CO_3^{\cdot -} + H_2O \tag{13}$$

Figure 13c shows that when the NO_3^- concentration increased from 0 mmol/L to 10 mmol/L, the degradation rate of RhB decreased from 99.79% to 82.38%. This result indicated that the presence of NO_3^- inhibited the degradation of RhB to a certain extent. Such a result was observed because the NO_3^- in the solution reacted with $SO_4^{\cdot -}$ and $\cdot OH$ to form NO_3 (Equations (14) and (15), respectively), and the redox potential of NO_3 is lower than that of $\cdot OH$ and $SO_4^{\cdot -}$ [40], thus reducing the degradation rate of RhB.

$$NO_3^- + SO_4^{\cdot -} \rightarrow NO_3^{\cdot} + SO_4^{2-} \tag{14}$$

$$NO_3^- + \cdot OH \rightarrow NO_3^{\cdot} + HO^- \tag{15}$$

Figure 13d shows that when the HA concentration increased from 0 mg/L to 20 mg/L, the degradation rate of RhB decreased from 99.79% to 89.83%. k_{obs} decreased from 0.094 min^{-1} to 0.043, 0.038, and 0.035 min^{-1}, respectively, indicating that the presence of HA slightly inhibited the degradation of RhB. The inhibitory effect of HA is mainly reflected in competitive adsorption [41]. As a typical natural organic matter, HA contains groups that are easily degraded by oxidation (such as carboxylic and phenolic hydroxyl groups). These active groups may react with free radicals and compete with RhB to form $\cdot OH$ and $SO_4^{\cdot -}$ [42]. The competitive adsorption of HA on biochar surface active sites and the spatial and electrostatic effects limited the contact between PMS and the biochar surface [43]. However, as the reactive groups of HA react more weakly with free radicals than $\cdot OH$ and $SO_4^{\cdot -}$, the addition of HA can only slightly inhibit the degradation of RhB in the system.

3.4. Exploration of Living Species and Possible Degradation Mechanisms

In general, in PMS catalytic reaction systems, researchers have shown that the degradation of pollutants can be based on free and non-free-radical pathways. To further explore the reaction path of the RBC600/PMS system in the RhB degradation process and the participation of various active substances in the reaction, we selected four different quenchers for the free-radical quenching experiment, and the experimental results are shown in Figure 14.

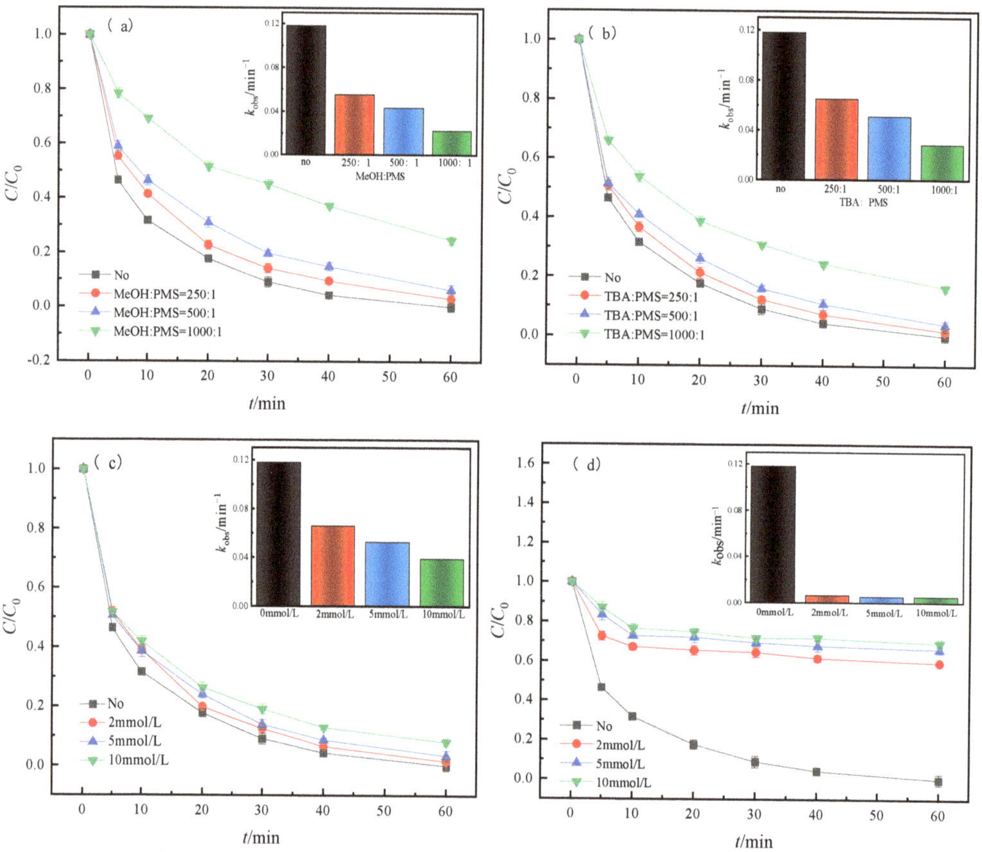

Figure 14. Effect of (**a**) MeOH, (**b**) TBA, (**c**) BQ, and (**d**) L-His on RhB degradation.

MeOH and TBA are usually used to assess the production of $SO_4^{\cdot -}$ and $\cdot OH$ in a reaction system. MeOH can effectively quench the two free radicals ($SO_4^{\cdot -}$ and $\cdot OH$) with different reaction rates ($k_{SO4\cdot -}$ = 1.1 × 10^7 $M^{-1}s^{-1}$, $k_{\cdot OH}$ = 9.7 × 10^8 $M^{-1}s^{-1}$). The reaction rate of TBA with $\cdot OH$ was 6 × 108 $M^{-1}s^{-1}$, but that with $SO_4^{\cdot -}$ was significantly lower (4 × 10^5 $M^{-1}s^{-1}$); thus, TBA can be used as a quencher of $\cdot OH$ [44]. Figure 14a,b shows the experimental results of the quenching agent and PMS at different molar ratios, respectively. When the molar ratio of MeOH and PMS increased from 0 to 1000:1, the degradation rate of RhB decreased from 99.97% to 75.35%, denoting a decrease of 24.61%. k_{obs} decreased from 0.118 min^{-1} to 0.022 min^{-1} (81.36%). When the molar ratio of TBA to PMS increased from 0 to 1000:1, the degradation rates of RhB were 99.97%, 98.47%, 96.19%, and 83.75%, and k_{obs} decreased from 0.118 min^{-1} to 0.065, 0.051, and 0.028 min^{-1}, respectively. These results indicated that both types of free radicals could inhibit the degradation of RhB, and the degree of inhibition of the reaction increased with the increase in the concentration of the quenching agent. Free-radical quenching experiments showed that both SO_4^- and

$\cdot OH$ participated in the degradation of RhB, but $\cdot OH$ performed a more important role than $SO_4^{\cdot -}$.

BQ is usually used as an inhibitor of $O_2^{\cdot -}$. Therefore, BQ was used in this experiment to verify the presence of radical $O_2^{\cdot -}$. Figure 14c shows the presence of $O_2^{\cdot -}$ in the reaction. The results showed that the degradation rate of RhB decreased by 7.99% when the concentration of BQ increased from 0 mmol/L to 20 mmol/L, and the corresponding k_{obs} were 0.118, 0.065, 0.053, and 0.039 min^{-1}. Free-radical quenching experiments showed that $O_2^{\cdot -}$ participated in the degradation of RhB, but the proportion was very low.

L-His is usually used as an inhibitor of $^1O_2$ non-free radicals. In this study, L-His was used to verify the presence of $^1O_2$. The participation of $^1O_2$ in the reaction is shown in Figure 14d. The experimental results revealed that when the L-His concentration increased from 0 mmol/L to 20 mmol/L, the degradation rate of RhB decreased by 69.11% and 69.13%, and the corresponding k_{obs} were 0.118, 0.006, 0.005, and 0.005 min^{-1}. K_{obs} had a very significant decrease of 95.76%. Free-radical quenching experiments showed that $^1O_2$ participated in the degradation of RhB and performed a relatively important role.

To further verify the reactive oxygen species generated during RhB degradation by the RBC600/PMS system, DMPO was used as the spin catcher of SO_4^-, $\cdot OH$, and $O_2^{\cdot -}$ free radicals. TEMP was used as the spin catcher of $^1O_2$ non-free radical. The free and non-free radicals generated during RhB degradation were captured by EPR. The EPR results are shown in Figure 15.

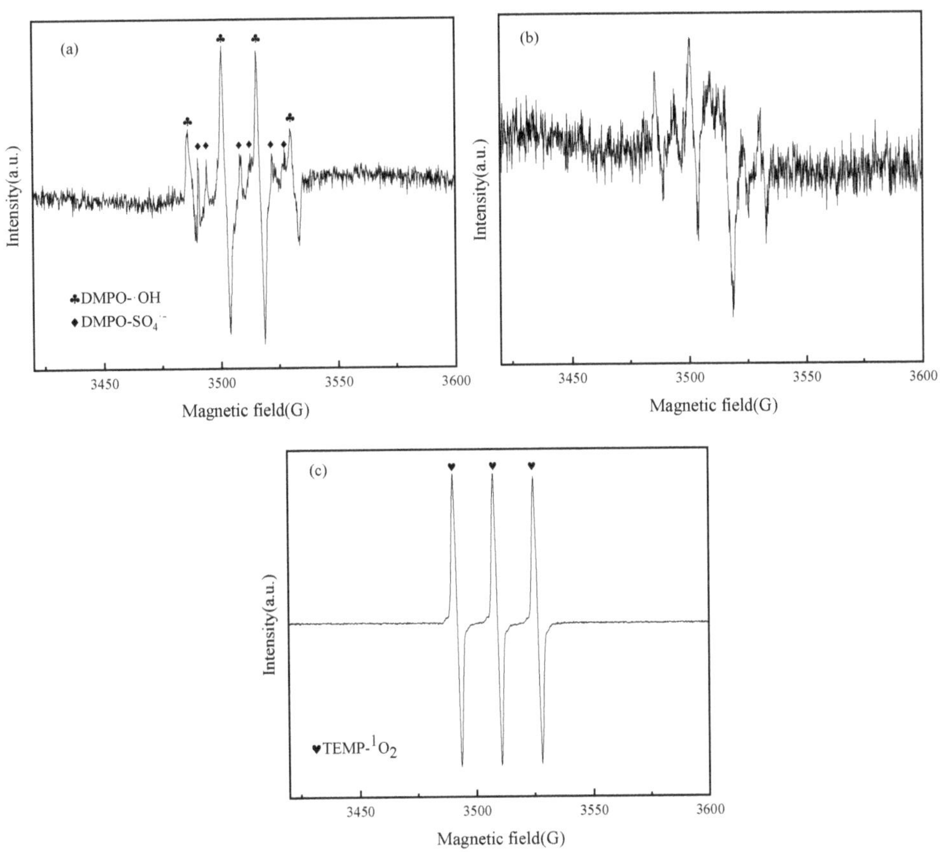

Figure 15. (**a**) SO_4^- and $\cdot OH$, (**b**) O_2^-, and (**c**) $^1O_2$ EPR spectrum.

As shown in Figure 15a, four distinct characteristic peaks with an intensity ratio of 1:2:2:1 can be observed, which proves the existence of DMPO–$SO_4^{\cdot-}$. Meanwhile, a six-row peak with an intensity ratio of 1:1:1:1:1:1 and partially overlapping with the characteristic peaks of DMPO–OH can also be observed, which proves the existence of DMPO–$SO_4^{\cdot-}$. The radical quenching experiment revealed possible traces of $O_2^{\cdot-}$ in the system, but as shown in Figure 15b, EPR did not detect the evident characteristic peak of DMPO–$O_2^{\cdot-}$ possibly because $O_2^{\cdot-}$ wasprotonated, dismutated, and reduced to ·OH. It may also be caused by the extremely strong DMPO–OH and DMPO–$SO_4^{\cdot-}$ signals; similar results were found in the study of Huang Wen et al. [22]. As depicted in Figure 15c, when TEMP was used as the trapping agent, a triple peak with an intensity ratio of 1:1:1 appeared, demonstrating the presence of $^1O_2$.

EPR experiments showed that RBC600 could activate PMS to produce $SO_4^{\cdot-}$, ·OH free radicals, and $^1O_2$ non-free radicals, which is consistent with the experimental results of radical quenching. Two free and non-free radical ways were used to degrade RhB in the RBC6000/PMS system, where the $^1O_2$ non-free radical performed an important role in the degradation process of RhB.

The defective structure of biochar can be used as an electron donor to provide electrons to PMS, thus activating PMS to produce various active substances. In addition, as a good electron acceptor, C=O on biochar can stimulate PMS to produce active substances. Figure 16 shows the mechanism of RBC600 activates PMS to degrade RhB. The activation mechanism of RBC600 to PMS can be described as follows: PMS was first adsorbed on the surface of RBC600. Further, π electron on the RBC600 surface sp^2-C activated PMS to produce SO_4^- and ·OH free radicals as oxidizing substances (Equations (16) and (17), respectively). The C=O functional group on RBC600 can promote the generation of non-free ($^1O_2$) radicals. In the presence of OH^-, HSO_5^- attacked the C=O located at the edge of carbon material to form peroxide adduction and then formed dioxane intermediates through intramolecular nucleophilic substitution. The SO_5^{2-} molecule reacted with the dioxycyclothane intermediate to form $^1O_2$. Finally, SO_4^-, ·OH, and $^1O_2$ attacked the central carbon of RhB to decolorize it. RhB was further degraded by *N*-deethylation, *N*-demethylation, and carboxylation, and the intermediate product was decomposed into smaller monocyclic aromatic compound intermediates. Finally, RhB was completely degraded into CO_2 and H_2O by ring opening and mineralization.

$$HSO_5^- + \pi\text{-electrons} \rightarrow OH^- + SO_4^- \quad (16)$$

$$HSO_5^- + \pi\text{-electrons} \rightarrow \cdot OH + SO_4^{2-} \quad (17)$$

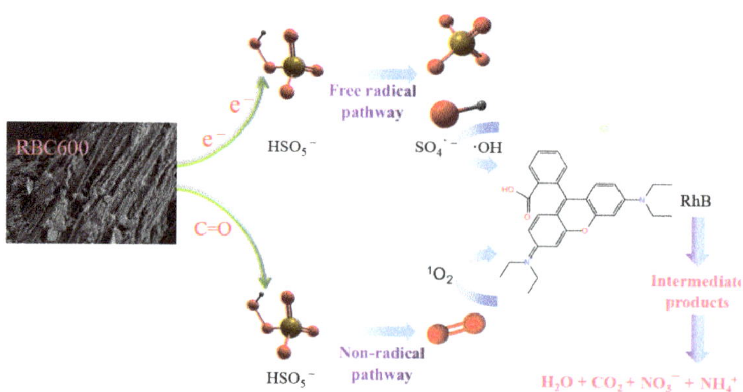

Figure 16. Mechanism diagram of RhB degradation in the RBC600/PMS system.

3.5. Investigation of Catalyst Stability

In addition to studying the degradation conditions of pollutants in the system, this paper explored the stability of materials. Strong stability of materials can not only reduce the environmental impact that may be caused by pollutant treatment but also save the cost of such process, thus increasing the possibility of its application in the actual pollutant degradation process. The experimental results of material recycling five times are shown in Figure 17.

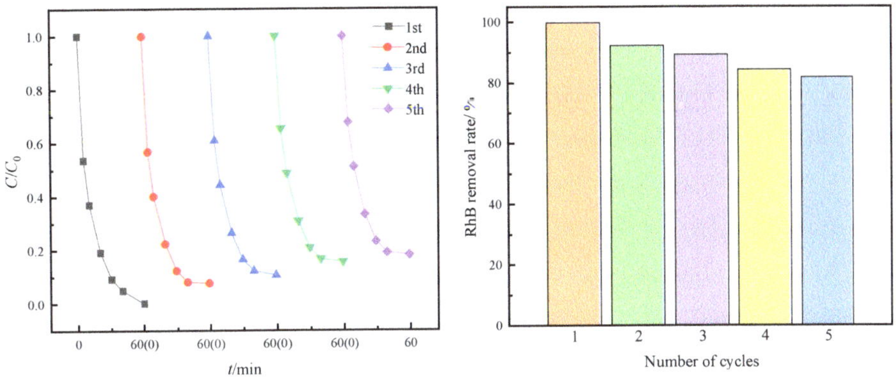

Figure 17. Repeated catalytic activity test of RBC600.

As it depicted in the figure, the reactivity of RBC600 as a catalyst gradually decreased with the increase in the number of cycles. After recycling, the degradation rates of RhB in the RBC600/PMS system were 99.79%, 92.21%, 89.25%, 84.26%, and 81.73%. After five cycles, the degradation rate of RhB decreased by 18.06%. In the process of catalytic degradation, the coverage of byproducts and the collapse of the catalyst structure reduced the catalytic performance of RBC600 to varying degrees. However, in general, the degradation rate of pollutants of RBC600 can remain above 80% after five cycles, which indicates that RBC600 has a good recycling performance.

4. Conclusions

(1) Using rabbit manure as raw material, RBC500 and RBC600 were prepared using pyrolysis and characterized via SEM, BET, FT-IR, XPS, and XRD. The results of characterization and catalytic degradation experiments showed that RBC600 had better catalytic activity than RBC400 and RBC500, which is closely related to its larger specific surface area, highly graphitized structure, and a higher degree of carbonization.

(2) When the RhB concentration was 50 mg/L, the optimum biochar dosage of RBC600-activated PMS for RhB degradation was 0.6 g/L; the optimal dosage of PMS was 0.4 g/L, and the optimum pH was 3. Background components Cl^-, HCO_3^-, NO_3^-, and HA all exhibited different degrees of inhibition on RhB degradation, with HCO_3^- exhibiting the strongest inhibitory effect, followed by NO_3^-, HA, and Cl^- having a weak inhibitory effect on RhB degradation.

(3) Free-radical quenching experiment results showed that the degradation rate of RhB decreased by 24.61%, 16.18%, 7.99%, and 69.11% after MeOH, TBA, BQ, and L-His were added, respectively. EPR analysis revealed that there are two free radicals (SO_4^- and $\cdot OH$) and one non-free ($^1O_2$) radical pathways for RhB degradation by activating PMS with RBC600, and $^1O_2$ performs a leading role in various active substances.

(4) The experimental results of RBC600 material stability analysis showed that RBC600 still had a good activation effect after five cycles, and the degradation rate of RhB by the RBC600/PMS system could still reach 81.73% after five cycles.

The composition of the actual dye wastewater is more complex than that of the simulated dye wastewater studied in this experiment. In the practical application process, it is necessary to consider the control of reaction conditions, operating costs, and other realistic factors. These factors may have a certain restriction effect on the degradation of pollutants in the RBC600/PMS system, which requires further discussion and research. In this study, although RBC600 exhibited good material stability and could still achieve a good pollutant degradation effect after repeated use, it encountered problems, such as difficult material collection and large losses in the recycling process, in the actual process. In the follow-up research, a major research direction is to develop a more convenient and efficient method for material collection and solve the loss problem in material recycling collection.

Author Contributions: Conceptualization, F.O.; Data curation, A.W. and Y.L. (Yixin Lu); Formal analysis, J.C. and C.T.; Investigation, Y.Y. and F.O.; Methodology, Y.L. (Yujie Liu) and J.C.; Writing—original draft, Y.L. (Yujie Liu); Writing—review and editing, F.O. and Y.L. (Yujie Liu). All authors have read and agreed to the published version of the manuscript.

Funding: This work was financially supported by the Sichuan Science and Technology Program (2022YFG0307), Natural Science Foundation of Sichuan Province (2022NSFSC0393), National College Students' Innovation Training Program (202211116025), and Young Seedling Program of CDTU (QM2021064).

Data Availability Statement: Data available on request from the authors.

Conflicts of Interest: The authors declare no conflict of interest.

References

1. Lan, D.; Zhu, H.; Zhang, J. Adsorptive removal of organic dyes via porous materials for wastewater treatment in recent decades: A review on species, mechanisms and perspectives. *Chemosphere* **2022**, *293*, 133464. [CrossRef] [PubMed]
2. Sarkar, P.; Dey, A. Phycoremediation-An emerging technique for dye abatement: An overview. *Process. Saf. Environ.* **2021**, *147*, 214–225. [CrossRef]
3. Natarajan, S.; Bajaj, H.C.; Tayade, R.J. Recent advances based on the synergetic effect of adsorption for removal of dyes from waste water using photocatalytic process. *J. Environ. Sci.* **2018**, *65*, 201–222. [CrossRef] [PubMed]
4. Pattnaik, P.; Dangayach, G.S. Sustainability of textile waste-water management by using an integrated fuzzy AHP-TOPSIS method: A case study. *Int. J. Environ. Sustain. Dev.* **2021**, *20*, 105–128. [CrossRef]
5. Zhang, Y.; Chen, S. Research on the Advanced Oxidation Technology Based on Dyeing Waste-Water Treatment. *Earth Environ. Sci.* **2020**, *495*, 012080. [CrossRef]
6. Ahmad, M.; Rajapaksha, A.U.; Lim, J.E.; Zhang, M.; Bolan, N.; Mohan, D.; Vithanage, M.; Lee, S.S.; OK, Y.S. Biochar as a sorbent for contaminant management in soil and water: A review. *Chemosphere* **2014**, *99*, 19–33. [CrossRef]
7. Pan, X.Q.; Gu, Z.P.; Chen, W.M.; Li, Q.B. Preparation of biochar and biochar composites and their application in a Fenton-like process for wastewater decontamination: A review. *Sci. Total Environ.* **2021**, *754*, 142104. [CrossRef]
8. Wang, J.L.; Wang, S.Z. Preparation, modification and environmental application of biochar: A review. *J. Clean. Prod.* **2019**, *227*, 1002–1022. [CrossRef]
9. Wu, D.P.; Chen, Q.; Wu, M.; Zhang, P.; He, L.P.; Chen, Y.H.; Pan, B. Heterogeneous compositions of oxygen-containing functional groups on biochars and their different roles in rhodamine B degradation. *Chemosphere* **2022**, *292*, 133518. [CrossRef]
10. Gonçalves, M.G.; Veiga, P.A.D.S.; Fornar, M.R.; Peralta-Zamora, P.; Mangrich, A.S.; Silvestri, S. Relationship of the physicochemical properties of novel ZnO/biochar composites to their efficiencies in the degradation of sulfamethoxazole and methyl orange. *Sci. Total Environ.* **2020**, *748*, 141381. [CrossRef]
11. Yu, J.F.; Feng, H.P.; Tang, L.; Pang, Y.; Wang, J.J.; Zou, J.J.; Xie, Q.Q.; Liu, Y.N.; Feng, C.Y.; Wang, J.J. Insight into the key factors in fast adsorption of organic pollutants by hierarchical porous biochar. *J. Hazard. Mater.* **2021**, *403*, 123610. [CrossRef] [PubMed]
12. Zhang, K.; Chen, B.L.; Mao, J.F.; Zhu, L.Z.; Xing, B.S. Water clusters contributed to molecular interactions of ionizable organic pollutants with aromatized biochar via π-PAHB: Sorption experiments and DFT calculations. *Environ. Pollut.* **2018**, *240*, 342–352. [CrossRef]
13. Chen, G.H.; Zhang, Z.R.; Zhang, Z.Y.; Zhang, R.D. Redox-active reactions in denitrification provided by biochars pyrolyzed at different temperatures. *Sci. Total Environ.* **2018**, *615*, 1547–1556. [CrossRef]
14. Xie, J.L.; Li, L.Y.; Sun, L.X.; Pei, Z.G.; Wen, B.; Xing, B.S. Reduced graphene oxide-catalyzed oxidative coupling reaction of 4-methoxyphenol in aerobic aqueous solution. *Carbon* **2017**, *121*, 418–425. [CrossRef]
15. Qi, C.D.; Liu, X.T.; Lin, C.Y.; Zhang, H.J.; Li, X.W.; Ma, J. Activation of peroxymonosulfate by microwave irradiation for degradation of organic contaminants. *Chem. Eng. J.* **2017**, *315*, 201–209. [CrossRef]

16. Qi, C.D.; Liu, X.T.; Ma, J.; Lin, C.Y.; Li, X.W.; Zhang, H.J. Activation of peroxymonosulfate by base: Implications for the degradation of organic pollutants. *Chemosphere* **2016**, *151*, 280–288. [CrossRef]
17. Qi, C.D.; Wen, Y.N.; Zhao, Y.J.; Dai, Y.H.; Li, J.P.; Xu, C.M.; Yang, S.G.; He, H. Enhanced degradation of organic contaminants by Fe(III)/peroxymonosulfate process with l-cysteine. *Chin. Chem. Lett.* **2022**, *33*, 2125–2128. [CrossRef]
18. Zang, T.C.; Wang, H.; Liu, Y.H.; Dai, L.; Zhou, S.; Ai, S.Y. Fe-doped biochar derived from waste sludge for degradation of rhodamine B via enhancing activation of peroxymonosulfate. *Chemosphere* **2020**, *261*, 127616. [CrossRef]
19. Zhu, B.X.; Yu, Y.; Ding, Y.D.; Ge, S.F. Iron-modified granular sludge biochar-based catalysts for improved Rhodamine B degradation by activating peroxymonosulfate. *Biomass Convers. Biorefinery* **2022**, *2022*, 03340. [CrossRef]
20. Liang, C.J.; Lin, Y.T.; Shin, W.H. Persulfate regeneration of trichloroethylene spent activated carbon. *J. Hazard. Mater.* **2009**, *168*, 187–192. [CrossRef]
21. Sang, Q.Q.; Wang, F.J.; Zhao, Y.T.; Zhou, Q.; Cai, Y.Q.; Deng, Y.; Tian, W.Q.; Chen, Y.Z.; Ma, J. Application of Iron and Sulfate-Modified Biochar in Phosphorus Removal from Water. *Huanjing Kexue* **2021**, *42*, 2313–2323. [PubMed]
22. Huang, W.; Zhang, X.P.; Zhang, J.Q.; Tang, Y.X.; Luo, Z. Preparation of ZVI-biochar derived from magnetically modified sheep manure and its activation of peroxymonosulfate to degrade AO7. *Acta Sci. Circumstantiae* **2022**, *42*, 196–208.
23. Cui, Z.-W.; Ren, Y.-F.; Wang, W.; Zhang, L.-M.; Zhang, L.-Y.; Wang, X.-Y.; He, J.-Y. Adsorption Characteristics and Mechanism of Cadmium in Water by Alkali and Magnetic Composite Modified Wheat Straw Biochar. *Huanjing Kexue* **2020**, *41*, 3315–3325.
24. Song, Y.Y.; Hu, J.W.; Liu, J.Y.; Evrendilek, F.; Buyukada, M. Catalytic effects of CaO, Al_2O_3, Fe_2O_3, and red mud on *Pteris vittata* combustion: Emission, kinetic and ash conversion patterns. *J. Clean. Prod.* **2020**, *252*, 119646. [CrossRef]
25. Li, S.Z.; Zhang, Z.H.; Yi, X.H.; Huang, M.Z. Removal of Rhodamine B by modified pig manure made biochar-activated persulfate (PS). *Environ. Chem.* **2022**, *41*, 929–939.
26. Huong, P.T.; Jitae, K.; Al Tahtamouni, T.M.; Tri, N.L.M.; Kim, H.H.; Cho, K.H.; Lee, C. Novel activation of peroxymonosulfate by biochar derived from rice husk toward oxidation of organic contaminants in wastewater. *J. Water Process. Eng.* **2020**, *33*, 101037. [CrossRef]
27. Chen, L.W.; Yang, S.J.; Zuo, X.; Huang, Y.; Cai, T.M.; Ding, D.H. Biochar modification significantly promotes the activity of Co_3O_4 towards heterogeneous activation of peroxymonosulfate. *Chem. Eng. J.* **2018**, *354*, 856–865. [CrossRef]
28. Neta, P.; Grodkowski, J.; Ross, A.B. Rate constants for reactions of aliphatic carbon-centered radicals in aqueous solution. *J. Phys. Chem. Ref. Data* **1996**, *25*, 709–1050. [CrossRef]
29. Rani, S.K.; Easwaramoorthy, D.; Bilal, I.M.; Palanichamy, M. Studies on Mn(II)-catalyzed oxidation of alpha-amino acids by peroxomonosulphate in alkaline medium-deamination and decarboxylation: A kinetic approach. *Appl. Catal. A-Gen.* **2009**, *369*, 1–7. [CrossRef]
30. Wang, Y.; Yang, S.; Zhang, M.X.; Li, C.; Yao, Z.Z.; Zhou, L.L. Degradation of Orange II by ZnFe /BC catalyzed persulfate. *Environ. Chem.* **2018**, *37*, 2630–2637.
31. Wang, Z.H.; Guo, H.Y.; Shen, F.; Duan, D.Q. Production of biochar by vermicompost carbonization and its adsorption to Rhodamine-B. *Acta Sci. Circumstantiae* **2015**, *35*, 3170–3177.
32. Shi, C.F.; Jia, S.M.; Li, Y.M.; Xue, R.J.; Song, Q.Q.; Wang, G.X. Degradation of p-nitrophenol using rice straw-based biochar and persulfate. *Environ. Prot. Chem. Ind.* **2017**, *37*, 632–637.
33. Chen, T.M.; Chen, H.M.; Ma, H.Y.; Tang, K.X.; Zhao, Y.W. Biochar-Co_3O_4 composite activates peroxymonosulfate to degrade atrazine. *China Environ. Sci.* **2020**, *40*, 4786–4794.
34. Shao, Z.C.; Lu, J.; Gao, S.Q.; Xu, J.; Zhang, Z.E. Covalent triazine-based framework activates peroxymonosulfate to degrade sulfamethoxazole. *China Environ. Sci.* **2020**, *40*, 135–142.
35. Wang, Y.; Wei, C.; Huang, T.; Wu, W.; Chen, J. Activation of peroxymonosulfate by nitrogen-doped carbon nanotubes to decolorize acid orange 7. *China Environ. Sci.* **2017**, *37*, 2583–2590.
36. Yuan, R.X.; Ramjaun, S.N.; Wang, Z.H.; Liu, J.S. Effects of chloride ion on degradation of Acid Orange 7 by sulfate radical-based advanced oxidation process: Implications for formation of chlorinated aromatic compounds. *J. Hazard. Mater.* **2011**, *196*, 173–179. [CrossRef] [PubMed]
37. Guan, Y.H.; Ma, J.; Liu, D.K.; Ou, Z.F.; Zhang, W.Q.; Gong, X.L.; Fu, Q.; Crittenden, J.C. Insight into chloride effect on the UV/peroxymonosulfate process. *Chem. Eng. J.* **2018**, *352*, 477–489. [CrossRef]
38. Ghauch, A.; Tuqan, A. Oxidation of bisoprolol in heated persulfate/H_2O systems: Kinetics and products. *Chem. Eng. J.* **2012**, *183*, 162–171. [CrossRef]
39. Kang, Y.M.; Kim, M.K.; Zoh, K.D. Effect of nitrate, carbonate/bicarbonate, humic acid, and H_2O_2 on the kinetics and degradation mechanism of Bisphenol-A during UV photolysis. *Chemosphere* **2018**, *204*, 148–155. [CrossRef]
40. Fu, H.C.; Ma, S.L.; Zhao, P.; Xu, S.J.; Zhan, S.H. Activation of peroxymonosulfate by graphitized hierarchical porous biochar and MnFe2O4 magnetic nanoarchitecture for organic pollutants degradation: Structure dependence and mechanism. *Chem. Eng. J.* **2019**, *360*, 157–170. [CrossRef]
41. Song, H.R.; Yan, L.X.; Ma, J.; Jiang, J.; Cai, G.Q.; Zhang, W.J.; Zhang, Z.X.; Zhang, J.M.; Yang, T. Nonradical oxidation from electrochemical activation of peroxydisulfate at Ti/Pt anode: Efficiency, mechanism and influencing factors. *Water Res.* **2017**, *116*, 182–193. [CrossRef] [PubMed]

42. Guan, Y.H.; Ma, J.; Ren, Y.M.; Liu, Y.L.; Xiao, J.Y.; Lin, L.Q.; Zhang, C. Efficient degradation of atrazine by magnetic porous copper ferrite catalyzed peroxymonosulfate oxidation via the formation of hydroxyl and sulfate radicals. *Water Res.* **2013**, *47*, 5431–5438. [CrossRef] [PubMed]
43. Guan, C.T.; Jiang, J.; Luo, C.W.; Pang, S.Y.; Yang, Y.; Wang, Z.; Ma, J.; Yu, J.; Zhao, X. Oxidation of bromophenols by carbon nanotube activated peroxymonosulfate (PMS) and formation of brominated products: Comparison to peroxydisulfate (PDS). *Chem. Eng. J.* **2018**, *337*, 40–50. [CrossRef]
44. Buxton, G.V.; Greenstock, C.L.; Helman, W.P.; Ross, A.B. Critical review of rate constants for reactions of hydrated electrons, hydrogen atoms and hydroxyl radicals. *Phys. Chem. Ref. Data* **1998**, *17*, 513–886. [CrossRef]

water

MDPI

Article

Methane Promotion of Waste Sludge Anaerobic Digestion: Effect of Typical Metal Meshes on Community Evolution and Electron Transfer

Ling Wang [1,2], Chang Liu [1], Xing Fan [1], Chunxue Yang [3,4], Xiaolin Zhou [1] and Zechong Guo [5,*]

1 School of Environmental and Municipal Engineering, Qingdao University of Technology, Qingdao 266520, China
2 State Key Laboratory of Urban Water Resource and Environment, Harbin Institute of Technology, Harbin 150090, China
3 Heilongjiang Province Key Laboratory of Cold Region Wetland Ecology and Environment Research, School of Geography and Tourism, Harbin University, Harbin 150086, China
4 National Technology Innovation Center of Synthetic Biology, Tianjin Institute of Industrial Biotechnology, Chinese Academy of Sciences, Tianjin 300308, China
5 School of Environmental and Chemical Engineering, Jiangsu University of Science and Technology, Zhenjiang 212100, China
* Correspondence: guozechong@just.edu.cn

Abstract: Anaerobic digestion of waste activated sludge (WAS) to produce methane is a promising pathway for biomass energy recovery. However, a slow organic biodegradation rate and weak microbial cooperation between fermentation bacteria and methanogens lead to low methane production from WAS. Considering the reuse of conductive materials for the regulation of microbial communities, this study chose three kinds of high-mesh metal materials (nickel, copper, and stainless steel) to promote the anaerobic digestion process. All three kinds of metal mesh could effectively increase methane production, and the highest methane production was increased by 61%, reaching 77.52 mL $gVSS^{-1}$. The poor biocompatibility of the stainless steel mesh was the least effective in promoting methane production compared to the biocompatible copper mesh and nickel mesh. The microbiological analysis found that the metal mesh with good biocompatibility can effectively induce and promote the enrichment of key microorganisms in the process of synergistic methane production, and the direct electron transfer process (DIET) of microorganisms on the metal surface contributes to the further improvement of the methane production efficiency. Therefore, the application of metal conductive materials in sludge anaerobic fermentation is feasible to achieve the retention of syntrophic bacteria and methanogens in the system.

Keywords: waste activated sludge; metal meshes; direct interspecies electron transfer; methane

1. Introduction

As a sustainable and promising technology, anaerobic digestion (AD) is widely used in wastewater treatment and solid waste disposal [1,2]. Through the AD process, microorganisms can convert organic wastes such as waste activated sludge (WAS) to biomethane (CH_4). This helps wastewater treatment plants (WWTPs) reduce their carbon footprint if their energy consumption can be compensated for through CH_4 cogeneration [3]. At present, most WWTPs have built sludge anaerobic fermentation facilities, but their practical operation is far from satisfactory due to poor operational stability and low production efficiency [4].

Introducing conductive materials (CMs) into AD has been proven to be an effective way to enhance methane production and improve system stability [5]. The addition of CMs into the anaerobic digester helps to cope with extreme conditions such as digestion deterioration due to temperature changes, over-acidification, and increased H_2 partial pressure [6]. Generally, CMs can thermodynamically and kinetically facilitate AD to form a more efficient synergistic metabolic process and improve the electron capture capacity of methanogens by replacing diffusive electron mediators (such as acetate, formate, and hydrogen), ultimately accelerating the rate of methanogenesis [7,8]. The electrons generated by organic oxidation can be directly transferred to methanogens through CMs, which is called direct interspecies electron transfer (DIET). DIET has always been considered to be the main reason for the promotion of anaerobic methane production by CMs, although it also has other effects, such as the directional enrichment of functional microorganisms, the regulation of pH stability, and adsorption and immobilization of CO_2 by the surface groups of CMs. These reports imply that the mechanism of methanogenesis via CMs in practical reactors is not yet well understood.

It was found that not only typical electrochemically active bacteria (EAB), such as *Geobacter* and *Pseudomonas*, can mediate the DIET process of microorganisms but also some syntrophic microorganisms have the ability to transfer electrons to CMs [5]. These syntrophic microorganisms are critical in anaerobic methanogenesis, as they can convert volatile fatty acids (VFAs) produced by acid-producing bacteria into acetate and hydrogen that are required by methanogens. This "synergistic methanogenesis process" established by methanogens and syntrophic microorganisms is also one of the key factors affecting the rate of methane production. However, these syntrophic microorganisms are relatively rare in conventional AD systems. The effect of CMs on the synergistic methanogenesis process is also uncertain.

To date, a variety of particle CMs, including carbon-based materials [4] and metal nanoparticles [9–13], are reported to significantly improve the performance of AD. In practice, however, the application of particle CMs often encounters material loss and recovery problems [14]. Metal mesh, with good electrical conductivity and high mechanical strength, can be easily fixed in the anaerobic reactors and easily separated from AD residues. Enhancing the AD of WAS by introducing low-cost transition metal meshes (such as nickel (Ni), iron (Fe), and copper (Cu)) are operationally and economically feasible.

In this study, three kinds of metal conductive materials (nickel mesh, copper mesh, and steel stainless mesh) were used to strengthen the methane production process of sludge. In order to promote the hydrolysis of the sludge and prevent the influence of metal corrosion, the sludge was pretreated with alkali, so that the whole process was carried out under weak alkaline conditions. The effects of different metal materials on the substrate metabolism and methanogenesis of the sludge anaerobic digestion process were investigated. Possible DIET processes and the growth of syntrophic microorganisms during methanogenesis were explored. The application and future development of using metal mesh as a method to intensify the AD process is proposed.

2. Materials and Methods

2.1. WAS Characteristics and Pretreatment

All WAS used in this study came from the secondary sedimentation tank of a municipal sewage treatment plant in Harbin, Heilongjiang Province. The ratio of volatile suspended solids (VSS) to total suspended solids (TSS) was approximately 0.62. After being retrieved, the WAS was naturally settled in a refrigerator at 4 °C for 24 h and the supernatant was removed by the siphon method. The concentrated sludge was passed through a 200-mesh sieve to remove large particles such as sand and gravel. The physicochemical indexes of the concentrated sludge after treatment were as follows: VSS of 37.5 g/L, total chemical oxygen demand (COD) of 11,880 mg/L, and pH of 6.90. For the convenience of comparison and calculation, the WAS described above was adjusted to 14.0 gVSS/L by diluting with purified water, and then it was pretreated by alkali to promote the release of intracellular substances. The specific operation was as follows: the original pH of the WAS was adjusted to 10 with 6 mol/L NaOH solution, stirring for 10 min, adjusting the pH of the WAS to 10 again, and repeating the above operation three times [15].

2.2. Reactor Construction and Operation

Eight lab-scale anaerobic fermentation devices were constructed using 550 mL glass bottles with 350 mL of pretreated WAS and 200 mL of headspace (Figure 1A). Each device contained two glass tubes, a short tube for gas collection connected by an air bag and a long tube for liquid sample collection. Nickel mesh (Ni), copper mesh (Cu), and stainless-steel mesh (SS) with a size of 6.25 × 6.25 mm (total area of 200 mm^2, 0.57 $m^2_{surface}/m^3_{WAS}$) were put into six of the anaerobic devices. Two devices were set as controls (C) without any metal meshes. The size and specific surface area of the three metal meshes were the same, and their hole diameter was 0.075 mm. The metal meshes were first soaked in 6 mol/L HCl for 24 h before use to remove the surface oxide film. Before operation, all the devices were aerated with N_2 for 10 min to maintain an anaerobic environment and then placed in a constant temperature water bath shaker at 35 ± 2 °C with a shaking frequency of 105 rpm/min. During operation, the reactor does not discharge sludge. A measure of 4 mL of the sludge samples was taken each time for subsequent analysis and detection.

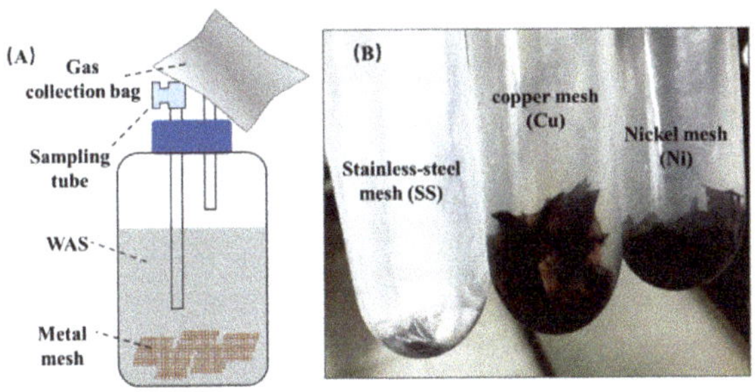

Figure 1. Schematic diagram of fermentation devices (**A**) and metal meshes after experiment (**B**).

2.3. Analysis and Calculation Methods

2.3.1. Chemical Analysis Method

The VFAs (acetic acid, propionic acid, butyric acid, and valeric acid) [16,17] and gas components (H_2, CO_2, and CH_4) were determined by gas chromatography (7890 A, Agilent Technologies (China) Co.,Ltd., Shanghai, China) equipped with a flame ionization detector and a thermal conductivity detector as described in a previous study [18]. The TSS and VSS were measured by the constant weight method at 105 °C and 600 °C, respectively. The TCOD and COD were determined by HACH spectrophotometry (HACH (Shanghai) Co.,Ltd., Shanghai, China) with a reagent model of 2038315. The soluble carbohydrates were determined by the phenol-sulfuric acid method. Absorbance values were measured at a wavelength of 490 nm using a UV–Vis spectrophotometer (UV-1800, Shanghai MAPADA Instruments Co.Ltd., Shanghai, China). The standard curve was drawn using glucose as the standard substance. The slope of the carbohydrate standard curve was 78.72, and the R^2 was 0.99. The soluble proteins were determined by a modified BCA protein kit method (Sangon Biotech, Shanghai, China), and the absorbance value of the sample was measured using a UV–Vis spectrophotometer at a wavelength of 562 nm [19]. The standard curve was drawn with bovine serum albumin as the standard substance. The slope of the protein standard curve was 2573, with an R^2 of 0.99. All the samples for soluble component measurement were centrifuged at 10,000 rpm and filtered through a 0.45 μm filter.

2.3.2. Next-Generation Sequencing and Analysis

At the end of the experiment, metal mesh was taken out from the reactor (Figure 1B), the biofilm formed on metal mesh and the sludge samples were both collected and analyzed. DNA was extracted using a DNA extraction kit (Omega Bio-Tek D55625, Norcross, GA, USA) [20]. The 16S rRNA structures of bacteria and archaea were sequenced using the specific primers 515F (5′-GTGCCAGCMGCCGCGG-3′) and 806R (5′-GGACTACHVGGGTWTCTAAT-3′) [21]. The microbial diversity index, including the Chao1 value, relative abundance (number of

OTUs determined in each sample), principal component analysis (PCoA), hierarchical clustering analysis, and function prediction of PICRUSt2, was calculated by I-Sanger online data processing developed by Majorbio (http://www.majorbio.com/ (accessed on 7 September 2022)). The sampling method of biofilm on the surface of the metal mesh was as follows: First, all the metal meshes were removed with tweezers after the experiment, and the floating sludge on the metal surface was removed gently by washing with sterilized phosphate buffered saline (PBS, 10 mM). Second, the washed metal meshes were fully shaken in a vortex shaker for 5 min and centrifuged at 10,000 rpm for 10 min, and the supernatant was removed to obtain a biofilm on the surface of the metal mesh [22].

3. Results and Discussion

3.1. Methane Production Performance

A sequencing batch anaerobic fermentation experiment was used to compare the effects of adding different metal meshes on an AD methane production process. During the entire operation process, the total methane production (Figure 2A), the total carbon dioxide production (Figure 2B), the daily methane production (Figure 2C), and the methane content (Figure 2D) of the experimental groups with the addition of metal materials were significantly higher than those of the control group, indicating that the addition of metal meshes had an effect on methane production in the sludge fermentation process. The experimental group with Ni mesh and Cu mesh obtained the highest cumulative methane production, reaching 380 ± 18 mL (equivalent to 77.52 mL gVSS^{-1}) and 374 ± 16 mL (76.39 mL gVSS^{-1}), respectively, which increased by 61% and 59% compared with the control group. The SS mesh had a less significant effect on methane production, which was only 24% higher than that of the control group. This may be due to the poor biocompatibility of the SS mesh, which was not helpful for microbes attached to participate in DIET or the synergistic methanogenesis process. After several cycles of experiments, the surface of the SS mesh was still bright, and basically no microbial growth was found on it. Furthermore, the cumulative carbon dioxide production in the SS, Ni, and Cu groups was significantly higher than that of the control group, indicating that the addition of metal meshes had a certain promotion effect on the decomposition and mineralization of complex organic matter.

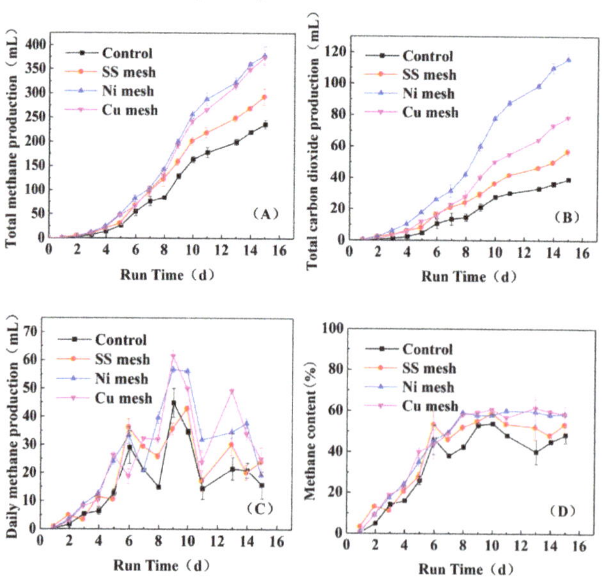

Figure 2. Total methane production (**A**), carbon dioxide production (**B**), daily methane production (**C**), and methane content (**D**) with different metal meshes.

Methane production rate and biogas methane concentration of the experimental group with added metal mesh were significantly higher than those of the control group. With the growth and proliferation of methanogens, the daily methane production increased gradually and reached maximum level during day 6–11. After that, due to the consumption of volatile fatty acids, the daily methane production decreased. The maximum daily methane production of the experimental group with Cu mesh reached 0.123 $m^3/(m^3 \cdot d)$, which was 37% higher than that of the control group. The average methane content of the added metal mesh reached 54%, which was 16% higher than that of the control group. In previous studies, the nanoparticles of three metals were added to the AD system of WAS, and different promoting effects were obtained. For instance, Su et al. [23] obtained a CH_4 increase by >40% by adding 20 nm Fe at a high concentration (1000 mg/L). Panagiotis et al. [24] added Ni-nanoparticles into WAS and slightly (~8%) enhanced the methane yield. Tareq et al. [13] reported a maximum three times increase in methane production by adding 1500 mg/L copper–iron bimetallic nanoparticles. There were also some studies reporting the negative effect of metal nanoparticles on AD [11,25]. Compared with metal nanoparticles, the low-cost metal meshes can also effectively promote methane production from WAS, and according to the results in this study, the promotion effect was Ni mesh ≈ Cu mesh > SS mesh.

3.2. Soluble Organic Matter Variation

The main dissolved organic matter in the sludge mainly includes carbohydrates, proteins, and VFAs [26]. It showed that the addition of metal meshes had little effect on the concentrations of carbohydrates and proteins. During the whole fermentation process, the total carbohydrates concentration was maintained at approximately 200 mg/L (Figure 3A), and the protein concentration was maintained between 450 mg/L and 550 mg/L (Figure 3B).

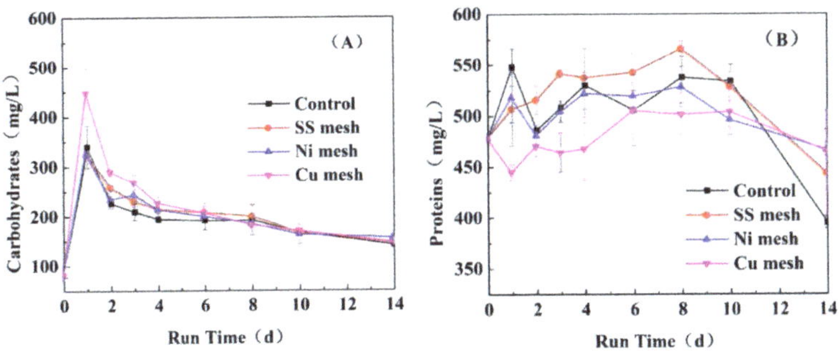

Figure 3. Changes of total (**A**) soluble carbohydrates and (**B**) soluble proteins in sludge.

Various VFAs, mainly including acetic acid (Figure 4A), propionic acid (Figure 4B), iso-butyric acid (Figure 4C), n-butyric acid (Figure 4D), iso-valeric acid (Figure 4E), and n-valeric acid (Figure 4F) were produced from the fermentation of carbohydrate and protein. The concentration of VFAs was affected by both the fermentative acid-producing process and the acid-utilizing process. In early stage, the acid production rate was higher than the consumption rate, thus an accumulation of VFAs was observed, while in the later stage, the situation was just the opposite. Among these VFAs component, only acetic acid can be directly utilized for methanogens, hence its consumption was first observed, whose concentration rapidly decreased after Day 2. From Day 4 to Day 6, it showed that the acetic acid concentration in the control group decreased from 747 mg/L to 512 mg/L, while the acetic acid concentration in the Ni mesh decreased from 582 mg/L to 146 mg/L. This suggested that the degradation rate of acetic acid was significantly accelerated by metal mesh addition.

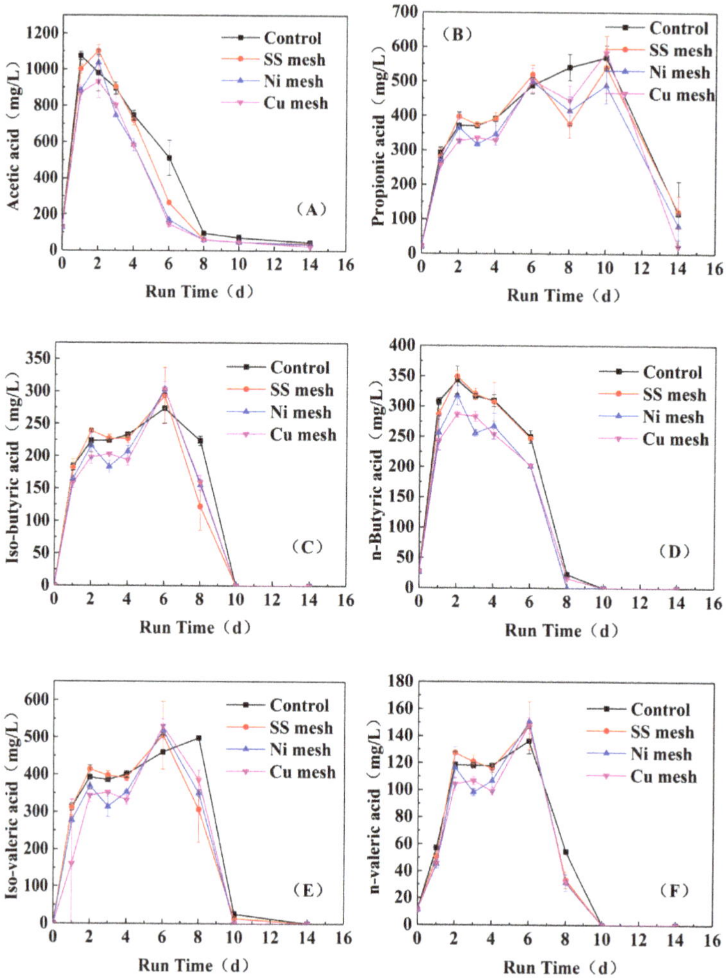

Figure 4. The changes of (**A**) acetic acid, (**B**) propionic acid, (**C**) iso-butyric acid, (**D**) n-butyric acid, (**E**) iso-valeric acid, and (**F**) n-valeric acid concentration during AD reactor operation with different metal meshes.

Except for acetic acid, other VFAs need to be converted into acetic acid before they are used for methane production. Among them, the acetogenesis of butyric acid is relatively easy, followed by valeric acid, and propionic acid is thermodynamically most unfavorable. Therefore, other VFAs were degraded in preference to propionic acid. The VFAs concentrations in the Ni and Cu group were lower than those of the control group most of the time, and after approximately 10 days, all organic acids were degraded and converted to methane, suggesting that metal meshes accelerated the conversion rates compared to the control groups. Little residual propionic acid was detected, and its concentration in the Cu and Ni mesh groups was 20–80 mg/L, while it was 121 mg/L in the control group. These results showed that the addition of Ni mesh and Cu mesh could accelerate the synergistic methanogenesis process in sludge, which was consistent with the change in methane production rate. When acetic acid and butyric acid were exhausted (at day 8), the Ni and Cu groups still maintained a high methane production rate. This proved that the SS mesh showed a weak promotion effect on VFAs conversion compared to the Ni and Cu meshes. This may be caused by the poor biocompatibility of SS, as biomass retention is crucial for the synergistic metabolic process of microorganisms [27].

3.3. Effects of Metal Meshes on the Microbial Community

In PCoA analysis, the relative distance of biofilms formed on the surfaces of Ni mesh (Ni-biofilm) and SS mesh (SS-biofilm) differed from their sludge community structure (Figure 5A). Therefore, a large surface of meshes would extend the promotion effect in the system compared to granular carriers. Furthermore, the relative abundance of some functional bacteria in sludge changed significantly by metal mesh additions. As shown in Figure 5B, with metal mesh additions, the relative abundance of Clostridiales decreased significantly, while Anaerolineales and Synergistales increased significantly, both of which were proven to be key syntrophic microorganisms involved in methanogenesis during AD. The relative abundance of Anaerolineales increased from 1.79% to 8.27% in Ni-sludge. Many genera in Anaerolineales have been confirmed to utilize a variety of substrates, such as sugars, peptides, and organic acids [28], and Synergistales have been confirmed to interact with hydrogenotrophic methanogens for methane production [29]. Hence, their existence was beneficial to promote the synergistic methanogenesis process.

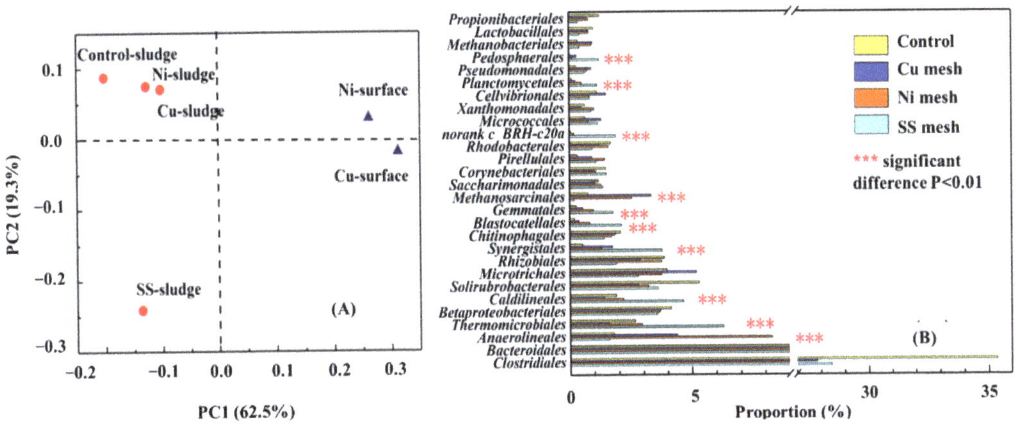

Figure 5. Principal coordinates analysis (PCoA) based on OTUs (**A**) and the differences in relative abundance of sludge microbial community at order level (**B**).

It was Interesting to show that SS mesh had a great influence on sludge communities, while Ni-sludge, Cu-sludge, and Control-sludge were quite similar. Analysis of archaeal community structure found that highly enriched communities of Methanobacteriales (0.91–0.96%) and Methanosarcinales (2.53–3.33%) were found in Ni-sludge and Cu-sludge compared to the Control-sludge (0.35–0.75%), while no significant enrichment of methanogens was found in SS-sludge. Some different enrichment of several fermentative microorganisms was observed in SS sludge, including Thermomicrobiales, Caldilineales, Synergitales, and Blastocatellales, rather than in the sludge with Ni or Cu mesh addition.

The clustering analysis of genera showed that the addition of metal materials played a positive role in the enrichment of some fermentative bacteria and methanogens in the sludge (Figure 6). The addition of Ni mesh and Cu mesh significantly increased the abundance of norank_f_Anaerolineaceae [30] (1.95–4.29%) and Longilinea (1.43–2.42%) compared to the control (0.50–0.97%). Longilinea has been shown to release intracellular electrons through the metabolism of substrates or intermediates [31,32]. The fermentive bacteria Unclassified_f_Blastocatellaceae (1.50%) [33] and Lactivibrio (1.71%) were more dominant in SS sludge than others (<0.14%). The ecological function of Lactivibrio has been confirmed to participate in the metabolism of carbohydrates and VFAs to provide substrates for methanogens [34]. Therefore, the enrichment of these genera can provide abundant substrates for methanogens, thereby promoting methanogenesis. The methanogens detected in the sludge mainly included the following five species: Methanosaeta, Methanosarcina, Methanobacterium, Methanobrevibacter, and Methanospirillum. The most dominant Methanosaeta, as a kind of acetotrophic methanogen, was significantly enriched by metal mesh additions, which was increased from 0.63% to 1.96% and 2.82% in Ni-sludge and Cu-sludge compared to the control. These results suggested that the addition of metal mesh could promote the growth of syntrophic microorganisms and Methanosaeta, which may be responsible for the promotion of methanogenesis.

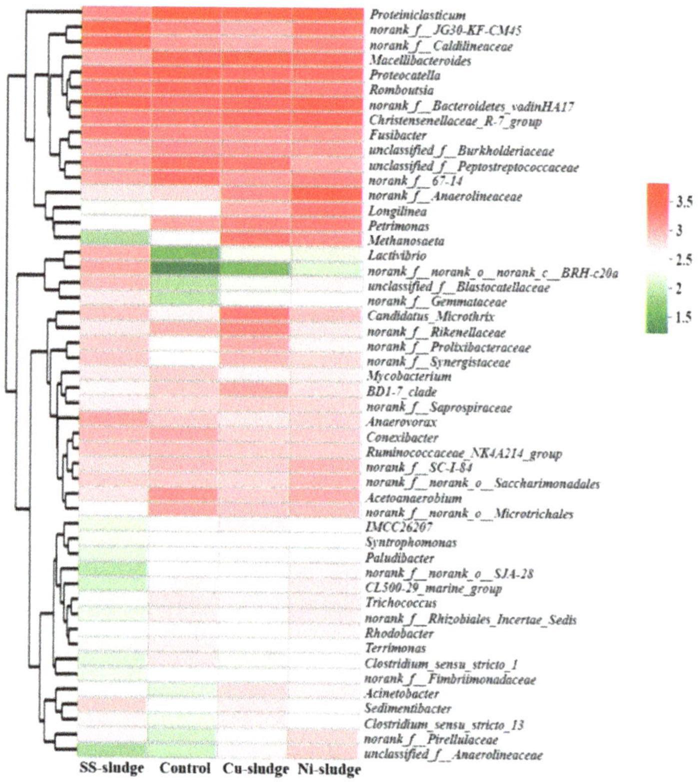

Figure 6. Heatmap of sludge community structure at genus level.

3.4. *Analysis of the Potential DIET and Synergistic Methanogenesis Process Mediated by Metal Mesh*

The metal mesh surface finally formed a different community structure from the sludge community, which meant that there might be different methanogenesis pathways on the metal surface. The top 20 families with significant differences between the metal surface and the corresponding sludge community in Ni mesh group (Figure 7A) and Cu mesh group (Figure 7B) were listed. A relatively higher level of Bacteroidetes_vadinHA17, Methanosaetaceae and Methanobacteraceae was found on both Ni and Cu surfaces compared to the corresponding sludge. Bacteroidetes_vadinHA17 is known as an electrogenic class for direct electricity production, and Methanosaetaceae and Methanobacteraceae were confirmed to be involved in the DIET process of methanogenesis, both of which have been extensively detected in other CM-enhanced systems [35]. Correspondingly, typical fermentative bacteria on the metal surfaces, such as Peptrostreptococcaceae and Clostridiaceae_1, were decreased significantly, while the proportion of some typical methanogenic syntrophic microorganisms, such as Anaerolineaceae, Spirochaetaceae, and Syntrophomonadaceae on the Cu-surface and Rhodocyclaceae on the Ni-surface, increased. Rhodocyclaceae were potentially electrically syntrophic with Methanosaeta via DIET and were selectively enriched in a magnetite-mediated AD system [36]. Hence, it was possible that the high electrical conductivity of metals selectively enriched these genera. The major genera identified were Longilinea (1.93% and 1.41%), Syntrophomonas (0.85% and 0.72%), and Smithella (0.70% and 0.67%), which were identified as potentially involved in the DIET process [32,37,38].

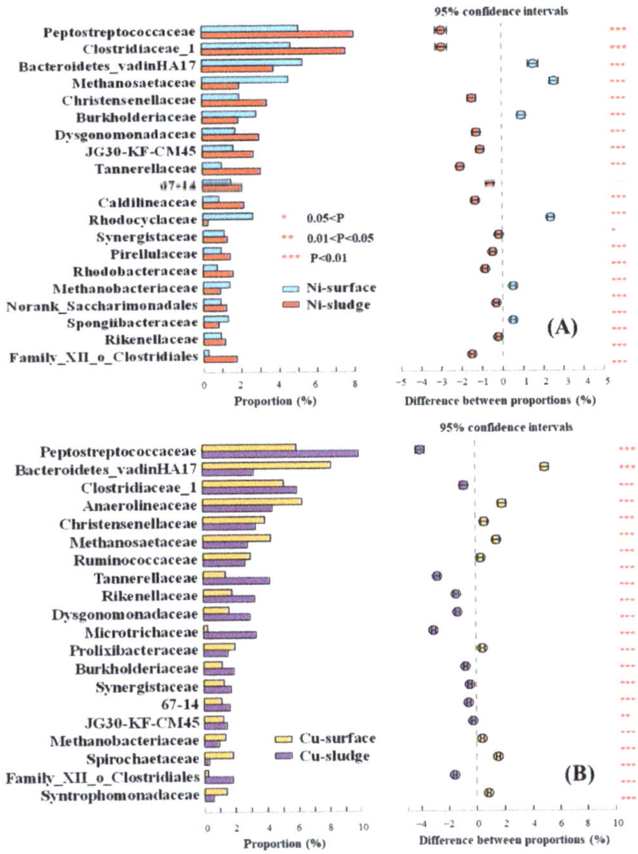

Figure 7. Community differences between (**A**) Ni surface or (**B**) Cu surface and corresponding sludge.

To accurately predict the variation in functional modules of the methanogenesis process associated with DIET, COG functional classification was analyzed using the PICRUSt2 model based on 16S rRNA gene sequencing of sludge and metal surface communities (Figure 8). The results showed that the addition of metal meshes enhanced metabolic activities, including translation, ribosomal structure and biogenesis (J), posttranslational modification, protein turnover, chaperones (O) and intracellular trafficking, secretion, and vesicular transport (U). The activity of Ni-surface and Cu-surface was the highest, followed by the corresponding sludge group and control group. Furthermore, functions responsible for hydrogenotrophic and electron transfer activities, cell wall/membrane/envelope biogenesis (M) and coenzyme transport and metabolism (H) were enhanced, suggesting that the synergistic metabolism and DIET processes were improved. These results showed that the surface of metal materials could promote the growth of methanogens and other syntrophic microorganisms with extracellular electron transfer ability, which was conducive to the formation of DIET and the synergistic methanogenesis process between fermentative microorganisms and methanogens.

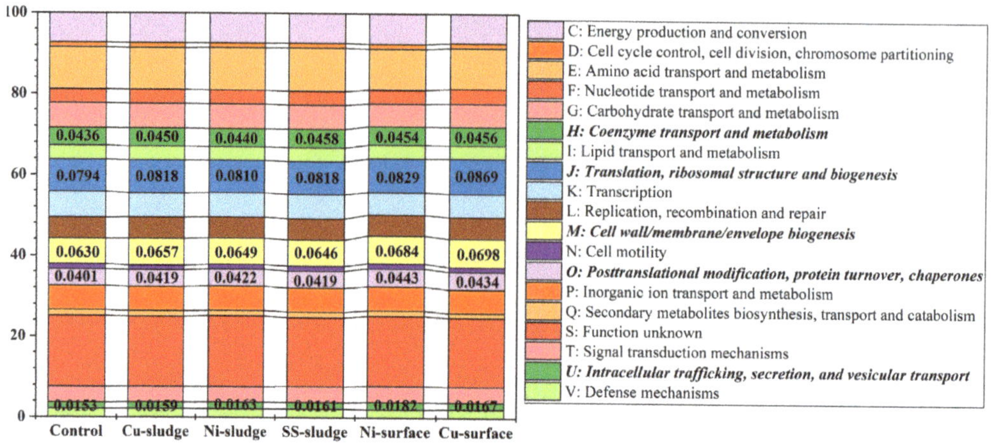

Figure 8. COG function classification based on the 16S rRNA gene sequencing of sludge and metal surface communities.

3.5. Mechanism Analysis of Metal Mesh Promoting Methane Production

The results of this study demonstrated that using metal meshes could promote the growth of methanogens and syntrophic bacteria not only on the surface but also in sludge communities. The relative abundance of the functional microorganisms for methanogenesis on the surface of the metal mesh was higher than that in the corresponding sludge. This result indicated that the SS mesh with poor biocompatibility for methanogens and syntrophic bacteria was significantly lower than that of the Ni mesh and Cu mesh with good biocompatibility, and a much lower methane yield was obtained by SS. The possible microorganisms that participated in the DIET process on the metal surface are shown in Table 1. The relative abundance of Methanosaetaceae, which could directly participate in the DIET process, increased by 6–7 times on the metal surface, and the relative abundance of other methanogens, such as Methanosaetaceae, Methanobacteriaceae, and Methanospirillaceae, all increased. Syntrophomonadaceae and Syntrophaceae are key bacteria in syntrophic methanogenesis and contain many genera that can metabolize long-chain fatty acids (LCFAs) into acetate and hydrogen. They were significantly enriched on the surface of the material (1.27–1.43% and 0.90–0.93%) but were rarely present in the sludge (0.41–0.74% and <0.04%). Spirochaetaceae are apparently involved in the synergistic acetic acid oxidative (SAO) methanogenesis process of hydrogenotrophic methanogens. Pseudomonadaceae had the ability of extracellular electron transport and was enriched on metal surfaces with a relative abundance of ~1% but was not found in sludge. These microorganisms are the dominant species for other dosing of CMs, such as magnetite and granular activated carbon, to induce the DIET process [39,40]. These microorganisms can utilize the high conductivity of CMs to complete the DIET methanogenesis process and induce the sludge to form a better synergistic methanogenesis process, thereby significantly increasing the methanogenesis rate (Figure 9). The enrichment of methanogens and related syntrophic microorganisms observed in this study through the addition of metal meshes is an important reason for improving the methane production rate of sludge, which shows its potential for accelerating the methane production of sludge. The metal mesh can be applied to the AD system in the form of suspended filler, which can be easily recovered and reused by means of sieve filtration. In this way, it is expected to realize the retention of methanogens and syntrophic bacteria in the system and finally realize the rapid production of methane.

Table 1. Possible microorganisms that participate in the DIET process on metal surface.

Family	Relative Abundance (%)					
	Control	SS-Sludge	Ni-Sludge	Cu-Sludge	Ni-Surface	Cu-Surface
Methanosaetaceae	0.63	0.14	1.96	2.82	4.53	4.29
Methanobacteriaceae	0.35	0.43	0.91	0.96	1.41	1.37
Methanosarcinaceae	0.12	0.05	0.57	0.51	0.17	0.21
Methanospirillaceae	0.06	0.02	0.05	0.21	0.70	0.82
Syntrophomonadaceae	0.74	0.41	0.48	0.58	1.27	1.43
Syntrophaceae	0.04	0.01	0.04	0.03	0.90	0.93
Pseudomonadaceae	0.04	0.02	0.10	0.03	1.01	0.74
Anaerolineaceae	1.79	1.60	8.27	4.39	8.18	6.28
Synergistaceae	0.53	3.78	1.31	1.74	1.13	1.29
Rhodocyclaceae	0.30	0.42	0.27	0.45	2.64	0.72
Spirochaetaceae	0.23	0.09	0.20	0.32	1.43	1.85

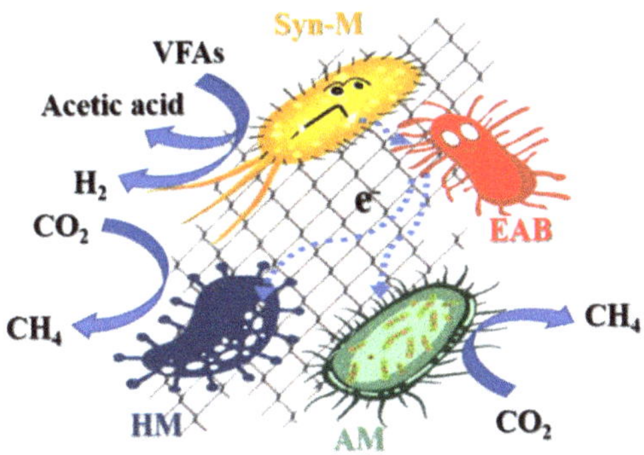

Figure 9. Possible methanogenic process in metal meshes mediated AD (DIET: direct interspecies electron transfer, EAB: electrochemical active bacteria, AM: acetotrophic methanogens, HM: hydrogenotrophic methanogens, Syn-M, syntrophic microorganisms).

4. Conclusions

The results of the study showed that the application of metal mesh could significantly promote the methanogenesis process of sludge. The methane yield of sludge with the addition of Ni mesh and Cu mesh increased the methane production by 61% and 59%, respectively. Community and metabolic analysis found that the introduction of metal meshes significantly enhanced the metabolic process of VFAs by promoting the growth of the syntrophic microorganisms. The microorganisms involved in the DIET and synergistic methanogenesis process were enriched, including *Bacteroidetes_vadinHA17*, *Anaerolineaceae*, *Synergistaceae*, *Rhodocyclaceae*, and *Methanosaetaceae*. Comparing the effects of different metal meshes on AD, it was found that biocompatibility was more important than electrical conductivity. The engineering implication of metal meshes in anaerobic fermentation is that they can be recycled and reused. This was a preliminary discussion, and more specific optimization experiments should be carried out continuously to obtain more valuable results.

Author Contributions: Writing and original draft preparation, L.W.; software, C.L.; methodology, X.F.; review and editing, C.Y.; validation, X.Z.; conceptualization, Z.G. All authors have read and agreed to the published version of the manuscript.

Funding: This research was funded by Open Project of State Key Laboratory of Urban Water Resource and Environment, Harbin Institute of Technology (No. HC202148), the National Natural Science Foundation of China, China (No. 52100148 and No. 52000090), the Natural Science Foundation of Shandong Province, China (No. ZR2021QE161), the Natural Science Foundation of Jiangsu Province (No. BK20190980), and the China Postdoctoral Science Foundation (No. 2021M701511).

Conflicts of Interest: The authors declare no conflict of interest.

Abbreviations

WAS	Waste activated sludge	DIET	Direct electron transfer process
AD	Anaerobic digestion	WWTPs	Wastewater treatment plants
CMs	Conductive materials	EAB	Electrochemically active bacteria
VFAs	Volatile fatty acids	VSS	Volatile suspended solids
TSS	Total suspended solids	COD	Chemical oxygen demand
SS	Stainless steel	PCoA	Principal component analysis
PBS	Phosphate buffered saline	LCFAs	Long-chain fatty acids
SAO	Synergistic acetic acid oxidative		

References

1. He, Z.W.; Zou, Z.S.; Sun, Q.; Jin, H.Y.; Yao, X.Y.; Yang, W.J.; Tang, C.C.; Zhou, A.J.; Liu, W.Z.; Ren, Y.X.; et al. Freezing-low temperature treatment facilitates short-chain fatty acids production from waste activated sludge with short-term fermentation. *Bioresour. Technol.* **2022**, *347*, 126337. [CrossRef] [PubMed]
2. Pang, H.L.; Jiao, Q.Q.; An, L.; Yang, T.; He, J.G.; Xie, B.H.; Yan, Z.S.; Lu, J.S. New insight into selective Na^+ stress on acidogenic fermentation of waste activated sludge from microbial perspective: Hydrolase secretion, fermentative bacteria screening, and metabolism modification. *Chem. Eng. J.* **2022**, *442*, 136098. [CrossRef]
3. Xie, S.H.; Li, X.; Wang, C.D.; Kulandaivelu, J.; Jiang, G.M. Enhanced anaerobic digestion of primary sludge with additives: Performance and mechanisms. *Bioresour. Technol.* **2020**, *316*, 123970. [CrossRef] [PubMed]
4. Abbas, Y.; Yun, S.N.; Wang, Z.Q.; Zhang, Y.W.; Zhang, X.M.; Wang, K.J. Recent advances in bio-based carbon materials for anaerobic digestion: A review. *Renew. Sustain. Energy Rev.* **2021**, *135*, 110378. [CrossRef]
5. Park, J.H.; Kang, H.J.; Park, K.H.; Park, H.D. Direct interspecies electron transfer via conductive materials: A perspective for anaerobic digestion applications. *Bioresour. Technol.* **2018**, *254*, 300–311. [CrossRef]
6. Yin, Q.D.; Gu, M.Q.; Wu, G.X. Inhibition mitigation of methanogenesis processes by conductive materials: A critical review. *Bioresour. Technol.* **2020**, *317*, 123977. [CrossRef]
7. Kang, H.J.; Lee, S.H.; Lim, T.G.; Park, J.H.; Kim, B.; Buffiere, P.; Park, H.D. Recent advances in methanogenesis through direct interspecies electron transfer via conductive materials: A molecular microbiological perspective. *Bioresour. Technol.* **2021**, *322*, 124587. [CrossRef]
8. Wu, Y.; Wang, S.; Liang, D.H.; Li, N. Conductive materials in anaerobic digestion: From mechanism to application. *Bioresour. Technol.* **2020**, *298*, 122403. [CrossRef]
9. Abdelsalam, E.; Samer, M.; Attia, Y.A.; Abdel-Hadi, M.A.; Hassan, H.E.; Badr, Y. Influence of zero valent iron nanoparticles and magnetic iron oxide nanoparticles on biogas and methane production from anaerobic digestion of manure. *Energy* **2017**, *120*, 842–853. [CrossRef]
10. Abdelsalam, E.; Samer, M.; Attia, Y.A.; Abdel-Hadi, M.A.; Hassan, H.E.; Badr, Y. Effects of Co and Ni nanoparticles on biogas and methane production from anaerobic digestion of slurry. *Energy Convers. Manag.* **2017**, *141*, 108–119. [CrossRef]
11. Ajay, C.M.; Mohan, S.; Dinesha, P.; Rosen, M.A. Review of impact of nanoparticle additives on anaerobic digestion and methane generation. *Fuel* **2020**, *277*, 118234. [CrossRef]
12. Amen, T.W.M.; Eljamal, O.; Khalil, A.M.E.; Matsunaga, N. Biochemical methane potential enhancement of domestic sludge digestion by adding pristine iron nanoparticles and iron nanoparticles coated zeolite compositions. *J. Environ. Chem. Eng.* **2017**, *5*, 5002–5013. [CrossRef]
13. Amen, T.W.M.; Eljamal, O.; Khalil, A.M.E.; Sugihara, Y.; Matsunaga, N. Methane yield enhancement by the addition of new novel of iron and copper-iron bimetallic nanoparticles. *Chem. Eng. Process. Process Intensif.* **2018**, *130*, 253–261. [CrossRef]
14. Liu, H.Y.; Xu, Y.; Li, L.; Dai, X.H.; Dai, L.L. A review on application of single and composite conductive additives for anaerobic digestion: Advances, challenges and prospects. *Resour. Conserv. Recycl.* **2021**, *174*, 105844. [CrossRef]
15. Pang, H.L.; Jiao, Q.Q.; He, J.G.; Zhang, Z.Q.; Wang, L.; Yan, Z.S.; Lu, J.S. Enhanced short-chain fatty acids production through a short-term anaerobic fermentation of waste activated sludge: Synergistic pretreatment of alkali and alkaline hydrolase blend. *J. Clean. Prod.* **2022**, *342*, 130954. [CrossRef]
16. Pang, H.L.; He, J.G.; Ma, Y.Q.; Pan, X.L.; Zheng, Y.S.; Yu, H.R.; Yan, Z.S.; Nan, J. Enhancing volatile fatty acids production from waste activated sludge by a novel cation-exchange resin assistant strategy. *J. Clean. Prod.* **2021**, *278*, 123236. [CrossRef]
17. Wang, L.; He, Z.W.; Guo, Z.C.; Sangeetha, T.; Yang, C.X.; Gao, L.; Wang, A.J.; Liu, W.Z. Microbial community development on different cathode metals in a bioelectrolysis enhanced methane production system. *J. Power Sources* **2019**, *444*, 227306. [CrossRef]
18. Wang, A.J.; Shi, K.; Ning, D.L.; Cheng, H.Y.; Wang, H.C.; Liu, W.Z.; Gao, S.H.; Li, Z.L.; Han, J.L.; Liang, B.; et al. Electrical selection for planktonic sludge microbial community function and assembly. *Water Res.* **2021**, *206*, 117744. [CrossRef]
19. Wang, H.; Liu, W.Z.; Haider, M.R.; Ju, F.; Yu, Z.; Shi, Y.J.; Cai, W.W.; Wang, A.J. Waste activated sludge lysate treatment: Resource recovery and refractory organics degradation. *J. Hazard. Mater.* **2021**, *416*, 126206. [CrossRef]
20. Wang, B.; Liu, W.Z.; Zhang, Y.F.; Wang, A.J. Natural solar intermittent-powered electromethanogenesis towards green carbon reduction. *Chem. Eng. J.* **2022**, *432*, 134369. [CrossRef]

21. Yang, C.X.; Zhao, S.; Guo, Z.C.; Liu, W.Z.; Wang, L.; Yu, S.P.; Liu, B.L.; Cong, X. Alkaline aided thermophiles pretreatment of waste activated sludge to increase short chain fatty acids production: Microbial community evolution by alkaline on hydrolysis and fermentation. *Environ. Res.* **2020**, *186*, 109503. [CrossRef] [PubMed]
22. Lin, X.Q.; Li, Z.L.; Liang, B.; Zhai, H.L.; Cai, W.W.; Nan, J.; Wang, A.J. Accelerated microbial reductive dechlorination of 2,4,6-trichlorophenol by weak electrical stimulation. *Water Res.* **2019**, *162*, 236–245. [CrossRef] [PubMed]
23. Su, L.; Shi, X.; Guo, G.; Zhao, A.; Zhao, Y. Stabilization of sewage sludge in the presence of nanoscale zero-valent iron (nZVI): Abatement of odor and improvement of biogas production. *J. Mater. Cycles Waste Manag.* **2013**, *15*, 461–468. [CrossRef]
24. Tsapekos, P.; Alvarado-Morales, M.; Tong, J.; Angelidaki, I. Nickel spiking to improve the methane yield of sewage sludge. *Bioresour. Technol.* **2018**, *270*, 732–737. [CrossRef] [PubMed]
25. Hassanein, A.; Naresh Kumar, A.; Lansing, S. Impact of electro-conductive nanoparticles additives on anaerobic digestion performance—A review. *Bioresour. Technol.* **2021**, *342*, 126023. [CrossRef] [PubMed]
26. Yu, Z.; Liu, W.Z.; Shi, Y.J.; Wang, B.; Huang, C.; Liu, C.S.; Wang, A.J. Microbial electrolysis enhanced bioconversion of waste sludge lysate for hydrogen production compared with anaerobic digestion. *Sci. Total Environ.* **2021**, *767*, 144344. [CrossRef] [PubMed]
27. Baek, G.; Rossi, R.; Saikaly, P.E.; Logan, B.E. The impact of different types of high surface area brush fibers with different electrical conductivity and biocompatibility on the rates of methane generation in anaerobic digestion. *Sci. Total Environ.* **2021**, *787*, 147683. [CrossRef] [PubMed]
28. Nakasaki, K.; Koyama, M.; Maekawa, T.; Fujita, J. Changes in the microbial community during the acclimation process of anaerobic digestion for treatment of synthetic lipid-rich wastewater. *J. Biotechnol.* **2019**, *306*, 32–37. [CrossRef]
29. Lv, Z.P.; Chen, Z.B.; Chen, X.; Liang, J.Z.; Jiang, J.H.; Loake, G.J. Effects of various feedstocks on isotope fractionation of biogas and microbial community structure during anaerobic digestion. *Waste Manag.* **2019**, *84*, 211–219. [CrossRef]
30. Zamorano-Lopez, N.; Borras, L.; Seco, A.; Aguado, D. Unveiling microbial structures during raw microalgae digestion and co-digestion with primary sludge to produce biogas using semi-continuous AnMBR systems. *Sci. Total Environ.* **2020**, *699*, 134365. [CrossRef]
31. Yamada, T.; Imachi, H.; Ohashi, A.; Harada, H.; Hanada, S.; Kamagata, Y.; Sekiguchi, Y. *Bellilinea caldifistulae* gen. nov., sp nov and *Longilinea arvoryzae* gen. nov., sp nov., strictly anaerobic, filamentous bacteria of the phylum *Chloroflexi* isolated from methanogenic propionate-degrading consortia. *Int. J. Syst. Evol. Microbiol.* **2007**, *57*, 2299–2306. [CrossRef] [PubMed]
32. Zhang, D.J.; Shen, J.Y.; Shi, H.F.; Su, G.Y.; Jiang, X.B.; Li, J.S.; Liu, X.D.; Mu, Y.; Wang, L.J. Substantially enhanced anaerobic reduction of nitrobenzene by biochar stabilized sulfide-modified nanoscale zero-valent iron: Process and mechanisms. *Environ. Int.* **2019**, *131*, 105020. [CrossRef] [PubMed]
33. Pascual, J.; Wust, P.K.; Geppert, A.; Foesel, B.U.; Huber, K.J.; Overmann, J. Novel isolates double the number of chemotrophic species and allow the first description of higher taxa in *Acidobacteria* subdivision 4. *Syst. Appl. Microbiol.* **2015**, *38*, 534–544. [CrossRef] [PubMed]
34. Lv, L.Y.; Li, W.G.; Zheng, Z.J.; Li, D.H.; Zhang, N. Exogenous acyl-homoserine lactones adjust community structures of bacteria and methanogens to ameliorate the performance of anaerobic granular sludge. *J. Hazard. Mater.* **2018**, *354*, 72–80. [CrossRef] [PubMed]
35. Guo, B.; Zhang, Y.D.; Zhang, L.; Zhou, Y.; Liu, Y. RNA-based spatial community analysis revealed intra-reactor variation and expanded collection of direct interspecies electron transfer microorganisms in anaerobic digestion. *Bioresour. Technol.* **2020**, *298*, 122534. [CrossRef] [PubMed]
36. Yang, Z.M.; Xu, X.H.; Guo, R.B.; Fan, X.L.; Zhao, X.X. Accelerated methanogenesis from effluents of hydrogen-producing stage in anaerobic digestion by mixed cultures enriched with acetate and nano-sized magnetite particles. *Bioresour. Technol.* **2015**, *190*, 132–139. [CrossRef]
37. Xing, L.Z.; Wang, Z.F.; Gu, M.Q.; Yin, Q.D.; Wu, G.X. Coupled effects of ferroferric oxide supplement and ethanol co-metabolism on the methanogenic oxidation of propionate. *Sci. Total Environ.* **2020**, *723*, 137992. [CrossRef]
38. Zhao, Z.S.; Li, Y.; Yu, Q.L.; Zhang, Y.B. Ferroferric oxide triggered possible direct interspecies electron transfer between *Syntrophomonas* and *Methanosaeta* to enhance waste activated sludge anaerobic digestion. *Bioresour. Technol.* **2018**, *250*, 79–85. [CrossRef]
39. Poirier, S.; Dejean, S.; Chapleur, O. Support media can steer methanogenesis in the presence of phenol through biotic and abiotic effects. *Water Res.* **2018**, *140*, 24–33. [CrossRef]
40. Wang, C.Q.; Liu, Y.; Jin, S.; Chen, H.; Xu, X.Y.; Wang, Z.R.; Xing, B.; Zhu, L. Responsiveness extracellular electron transfer (EET) enhancement of anaerobic digestion system during start-up and starvation recovery stages via magnetite addition. *Bioresour. Technol.* **2019**, *272*, 162–170. [CrossRef]

water

MDPI

Article

Nitrate Removal from Groundwater by Heterotrophic and Electro-Autotrophic Denitrification

Shuangshuang Yao, Lei Liu, Shiyang Zhang and Xinhua Tang *

School of Civil Engineering and Architecture, Wuhan University of Technology, No.122 Luoshi Road, Hongshan District, Wuhan 430070, China; yaoshuangshuang@whut.edu.cn (S.Y.); 13125177038@163.com (L.L.); zhangshiyang7@126.com (S.Z.)
* Correspondence: tangxinhua@whut.edu.cn

Abstract: A heterotrophic and autotrophic denitrification (HAD) system shows satisfactory performance for groundwater with nitrate contamination. In this study, an HAD system combining solid-phase heterotrophic denitrification and electrochemical hydrogen autotrophic denitrification (SHD-EHD) was developed for the treatment of nitrate-contaminated groundwater, in which polycaprolactone (PCL) was used as the carbon source to enhance the nitrate removal performance and prevent secondary pollution of the electrochemical hydrogen autotrophic denitrification (EHD) system. The denitrification performance, microbial community structure and nitrogen metabolism were investigated. The results showed that a high nitrate removal rate of 99.04% was achieved with an influent nitrate concentration of 40 mg/L, a current of 40 mA and a hydraulic retention time (HRT) of 4 h. By comparing the performance with the EHD system, it was found that the HAD system with PCL promoted the complete denitrification and reduced the accumulation of NO_2^--N. Analysis of the microbial community structure identified the key denitrifying bacteria: *Dechloromonas*, *Thauera* and *Hydrogenophaga*. A comparison of microbial communities from SHD-EHD and solid-phase heterotrophic denitrification (SHD) demonstrated that electrical stimulation promoted the abundance of the dominant denitrifying bacteria and the electroactive bacteria. Analysis of the nitrogen metabolic pathway revealed that the conversion of NO to N_2O was the rate-limiting step in the overall denitrification pathway. The SHD-EHD developed in this study showed great potential for groundwater nitrate removal.

Keywords: nitrate; biofilm electrode reactor; hydrogen autotrophic denitrification; solid-phase heterotrophic denitrification; microbial community

1. Introduction

Groundwater represents 99% of the world's liquid freshwater and is the source of one quarter of all the water used by humans [1]. Two billion people still lack access to safe drinking water, and groundwater plays a significant role in sustainable development according to the Sustainable Development Goals (SDGs) from the United Nations [2]. Human production activities are the main reason for the contamination of nitrate in groundwater. Excessive application of chemical fertilizers, uncontrolled discharge of industrial/domestic wastewater and infiltration of waste leachate all contribute to nitrate pollution in groundwater [3]. These contaminations not only affect water systems worldwide [4], but also threaten human health. Nitrate can be transformed to nitrite in the human body, which may lead to methemoglobinemia; nitrite can be further converted to nitrosamine, increasing the risk of cancer [5]. There have been many reports of nitrate contamination around the world, such as in eastern Nebraska in USA and in the Yellow River Delta in China [6]. The maximum concentration of nitrate and nitrite regulated by China are 10 mg/L NO_3^--N and 1 mg/L NO_2^--N, respectively (GB5749-2022).

Nitrogen in water undergoes migration and transformation under biological and chemical processes. Denitrification is a significant process of the nitrogen cycle, as well as in the biological treatment of nitrate in groundwater [7]. Among various methods for nitrate removal, biofilm electrode reactor (BER) has attracted increasing attention due to its satisfactory performance [8,9]. However, heterotrophic denitrification has the problem of secondary pollution caused by added organic carbon source [10]. In situ electrochemical hydrogen production realized by BER overcomes the disadvantages of low H_2 mass transfer rate of conventional autotrophic denitrification and eliminates the risks of H_2 during transportation and operation. However, the processing capacity of BER is affected by the reactor configuration [11] and the supply of inorganic carbon sources [12]. The BER also consumes a lot of electrical energy to produce CO_2 and H_2. In addition, the acclimatization of autotrophic microorganisms is time-consuming and the biofilm is easy to shed. The difference in current intensity (I) and hydraulic retention time (HRT) affects the number, species, activity and metabolism of microorganisms on the biofilm, thus influencing nitrate removal rate and accumulation of nitrite [13].

To further improve the processing capacity of BER, researchers increased the cathode surface area and adapted reactor operating conditions, such as dicyclic-type electrode based biofilm reactor [14], upflow BER [15], and added microbial carriers such as fiber threads [12,16] and wheat-rice stone powder [17]. In addition, researchers developed an enhanced biofilm electrode reactor that combined autotrophic and heterotrophic denitrification, which provided adequate inorganic carbon for autotrophic denitrification [18,19]. Moreover, it has been demonstrated that heterotrophic and autotrophic denitrification (HAD) can be realized synergistically [16]. In an HAD system, heterotrophic denitrification combined with autotrophic denitrification reduces the consumption of organic carbon to some extent [18], overcoming the weaknesses of secondary pollution and blockage in the reactor. External carbon sources currently used for HAD are mainly liquid-phase carbon sources such as glucose, ethanol, methanol and acetate [20]. Although liquid-phase carbon sources are effective for heterotrophic denitrification, the dosing of carbon source is easily influenced by fluctuations of influent quality, and the system requires complicated online monitoring and control instruments [21]. To solve the above problems, the recently developed solid-phase carbon source denitrification technology, which uses biodegradable polymers as carbon sources and biofilm carriers as well, has shown a promising potential in the removal of nitrate from groundwater. These biodegradable polymers are polymeric compounds formed by covalent bonding of monomer substances that can easily be hydrolyzed by enzymes of bacteria or fungi [22,23]. Many biodegradable polymers, such as polyhydroxyalkanoates (PHAs) [24,25], polycaprolactone (PCL) [26,27] and polybutylene succinate (PBS) [28,29], have been used for solid-phase denitrification and have exhibited satisfactory performance. However, few studies on the combination of heterotrophic denitrification using biodegradable polymers and electrochemical hydrogen autotrophic denitrification have been reported.

Considering the above-mentioned aspects, an HAD system with a solid-phase heterotrophic denitrification and electrochemical hydrogen autotrophic denitrification (SHD-EHD) was constructed for the treatment of nitrate-contaminated groundwater. PCL was used as the solid carbon source for denitrification. PCL is insoluble in water and has good hydrophobicity, but it can be hydrolyzed by microbial enzymes. The amount of hydrolyzed PCL is determined by the number of microorganisms in the reactor, so there is no need to artificially control the dosing of PCL. In addition, the hydrolyzed products of PCL are CO_2 and H_2O, which can continuously provide a carbon source for autotrophic denitrification.

A comparison between the start-up of the SHD-EHD and the BER was studied. The effects of current intensity (I) and HRT on nitrate removal from groundwater by SHD-EHD were investigated to optimize the system. Under the optimal conditions, the water quality indices and microbial communities were studied to evaluate the transformation and synergistic degradation mechanism of nitrate in SHD-EHD. This study aimed to provide an effective method to remove nitrate from groundwater, which was in line with the target of "improving water quality by reducing pollution" in the SDGs.

2. Materials and Methods

2.1. Materials

The physical characteristics of PCL granules (Capa 6800, Solvay Corporation, Pasadena, CA, USA) were as follows: shape (ellipsoidal), diameter (2–3 mm), average weight (0.0123 g per piece), mass density (1.15 g/cm^3), molecular weight (80,000 Da).

The simulated nitrate-contaminated groundwater (1 L deionized water) contained 0.243 g $NaNO_3$ (40 mg/L NO_3^--N). Appropriate amounts of mineral elements (23 mg/L $MgSO_4{\cdot}7H_2O$, 131 mg/L $NaHCO_3$, 13 mg/L Na_2HPO_4, 158 mg/L $MgCl_2{\cdot}6H_2O$) were added to the simulated groundwater for microbial growth.

2.2. Experimental Apparatus

The experimental apparatus consisted of a laboratory-scale SHD-EHD, a DC regulated power supply, a peristaltic pump and two storage tanks (Figure 1). The SHD-EHD was made by a PVC cylinder (9 cm in diameter and 29 cm in height) with an effective volume of 2 L, in which two baffles were set at the bottom and the top of the reactor to fix the electrodes and achieve uniform water distribution. A graphite rod (1 cm in diameter and 30 cm in length) was placed in the center of the cylindrical reactor to serve as the anode, while a 3 mm thick carbon felt was in a circle against the inner wall of the reactor as the cathode. Carbon felt has been widely used as electrode material due to its good biocompatibility, satisfactory conductivity and low cost [30]. The utilization of the carbon rod anode was ascribed to its good conductivity and biochemical stability. The flow rate of influent and effluent were controlled by a multi-channel peristaltic pump (BT600S, Leadfluid, Baoding, China). A DC power supply (IT6302, Itech, Nanjing, China) was applied to provide constant DC power for different operating conditions. The lower section of the reactor was filled with plastic multifaceted hollow spheres to support PCL. As the control, SHD was the same as SHD-EHD without current application.

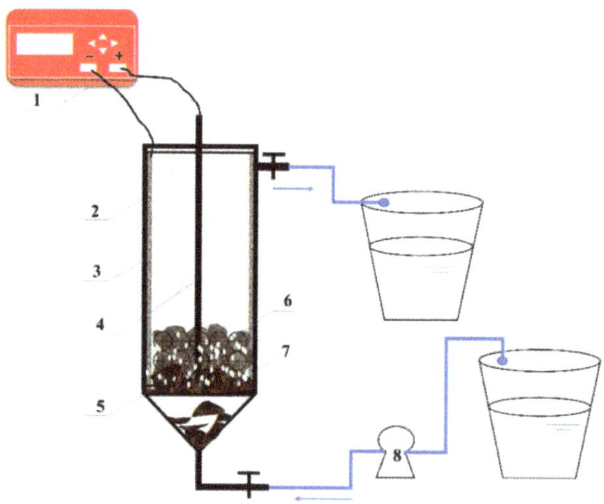

Figure 1. Experimental apparatus. 1 DC Power supply; 2 Reactor; 3 Cathode: Carbon felt; 4 Anode: Graphite rod; 5 Supporting board; 6 PCL; 7 Biofilm; 8 Peristaltic pump.

2.3. Experimental Start-Up and Operation

The experiments were divided into the following parts: EHD start-up and operation (stage I); SHD-EHD start-up and operation (stage II). Detailed operations are presented below. All experiments were conducted under room temperature.

EHD start-up: First, 0.5 L of activated sludge from Beihu wastewater treatment plant (Wuhan, China) was inoculated into the EHD. In the early stage of start-up, the EHD was operated under the sequence batch mode without applying current. The synthetic groundwater containing 30 mg/L NO_3^--N was replaced every 12 h and the pH was adjusted to neutral. After 15 days of anaerobic cultivation, the current was applied to acclimatize the microorganisms in the reactor. The current intensity gradually increased from 1 mA to 100 mA, and the feed mode was changed to continuous flow with an HRT of 12 h. The start-up was basically considered complete when nitrate removal rate reached 75%.

Stage I: The denitrification performances under different current intensities, HRT, influent pH and influent NO_3^--N concentrations were tested to obtain optimal operating conditions of the EHD, which also provided guidance for the SHD-EHD operating. The experimental conditions and performance of the EHD were summarized in Figure S1.

SHD-EHD start-up: The collected activated sludge was first inoculated into the SHD-EHD. Different from the EHD, the NO_3^--N concentration of the synthetic groundwater was 40 mg/L and replaced every 10 h during the start-up. Other procedures were the same as EHD start-up except that the applied current intensity increased from 10 mA to 40 mA.

Stage II: The influences of current intensity and HRT on the denitrification performance of the SHD-EHD were studied. The influent pH was controlled at about 7.5. Detailed influent conditions were shown in Table 1.

Table 1. Operation parameters of SHD-EHD.

Days	I (mA)	HRT (h)	inf-NO_3^--N (mg/L)	Days	I (mA)	HRT (h)	inf-NO_3^--N (mg/L)
1~8	5	10	40	1~7	40	2.5	40
9~16	10	10	40	8~14	40	3.0	40
17~25	20	10	40	15~22	40	3.5	40
26~34	30	10	40	23~30	40	4.0	40
35~43	40	10	40	31~38	40	4.5	40

The nitrate removal rate (η, %) and the denitrification rate (R, mg NO_3^--N/(cm^2·d)) were calculated as follows:

$$\eta = \frac{C_0 - C_t}{C_0} \times 100\% \quad (1)$$

$$R = \frac{Q \times (C_0 - C_t) \times 24}{A} \quad (2)$$

where C_0 is the initial NO_3^--N concentration (mg/L); C_t is the NO_3^--N concentration at time t (mg/L); Q is the influent flow rate (L/h); A is the cathode surface area (cm^2).

2.4. Analytical Methods

Influent and effluent samples were collected and analyzed according to standard methods. All samples were filtered using 0.45 μm filter before analyses. Concentrations of NO_3^--N, NO_2^--N and TN were analyzed using a UV spectrophotometer (HACH, DR 5000, Loveland, CO, USA). pH was measured by a pH meter (SIN-CT6321, Sinomeasure, Hangzhou, China). High-throughput Sequencing was conducted at Majorbio Technology CO., LTD. (Shanghai, China) and the detailed procedures were presented in Supplementary Materials.

2.5. Chemical Equations

Chemical equations related to material conversion and mechanism were as follows:

$$5CH_2O + 4NO_3^- \rightarrow 2N_2 + 4OH^- + 5CO_2 + 3H_2O \quad (3)$$

$$O_2 + 2H_2O + 4e^- \rightarrow 4OH^- \quad (4)$$

$$2H_2O + 2e^- \rightarrow H_2 + 2OH^- \quad (5)$$

$$2NO_3^- + 5H_2 \rightarrow N_2 + 4H_2O + 2OH^- \quad (6)$$

$$C + 2H_2O \rightarrow CO_2 + 4H^+ + 4e^- \quad (7)$$

3. Results and Discussion

3.1. Start-Up of the EHD and SHD-EHD

As showed in Figure 2, the nitrate removal rate was positively correlated with the current intensity during the start-up period of the EHD. When the current intensity was adjusted to 70 mA and maintained for 7 days at an HRT of 12 h, the nitrate removal rate could be higher than 75%, and tawny biofilm was observed on the cathode, which was regarded as the completion of the start-up. The whole process took about 60 days.

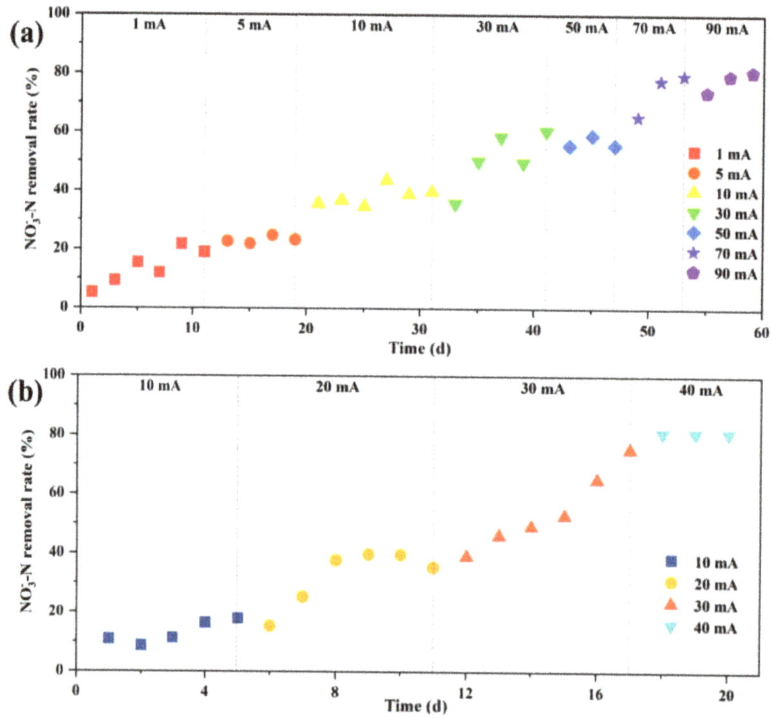

Figure 2. Nitrate removal rate during the start-up of (**a**) EHD; (**b**) SHD–EHD.

For the start-up of the SHD-EHD, when the current intensity was adjusted to 30 mA and maintained for 7 days at an HRT of 10 h, the nitrate removal rate could be more than 75% and tawny biofilm was observed on both PCL and the cathode (Figure S2), which was regarded as the completion of the start-up. The whole process took about 22 days. The time required for the start-up of the SHD-EHD was considerably shorter compared with the EHD, while the required current was smaller. These results indicate that the SHD-EHD not only achieved fast start-up to overcome the disadvantage of the long start-up of the EHD, but also realized high removal rate in treating simulated groundwater.

3.2. Optimization of the EHD Operating Parameters

In stage I, the operating parameters such as the current intensity, HRT, influent pH and influent NO_3^--N concentration of EHD were optimized. The influent NO_3^--N concentration was set to 30 mg N/L. The operating parameters and overall performance of the EHD were presented in Figure S1.

The nitrate removal rate increased with the increase of current intensity in the range of 10~86 mA. When the current intensity was set to 86 mA, the nitrate removal rate reached a maximum of 90.80%. A decreasing trend was observed after the current intensity was greater than 86 mA. This might be due to the occurrence of "hydrogen inhibition" caused by high concentration of H_2 generated under high current intensity, which reduced the microbial activity [31]. The NO_2^--N accumulation kept growing as the current intensity increased from 10 mA to 90 mA (Figure S3).

When HRT < 9 h, the nitrate removal rate was positively correlated with HRT, rising from 43.33% to 90.12%. When HRT > 9 h, the influence of HRT on the nitrate removal rate was gradually weakened. The nitrate removal rate and the effluent NO_2^--N concentration remained stable (Figure S4).

The pH range suitable for the EHD was weakly alkaline (pH 7.45–8.1) (Figure S5). The EHD had a buffering capacity for pH variations due to CO_2 production by electrolysis of the carbon rod anode (Figure S6).

When the influent NO_3^--N concentration was lower than 40 mg/L, the removal amount of NO_3^--N increased markedly with the increase of influent NO_3^--N concentration. When the NO_3^--N concentration was 40~60 mg/L, the removal amount of NO_3^--N still increased but the increase rate was much smaller. The removal amount of NO_3^--N decreased when the concentration of influent NO_3^--N was higher than 60 mg/L (Figure S7).

To sum up, the optimal operating parameters for the EHD in this study were as follows: current intensity of 86 mA~96 mA; HRT of 9 h; influent pH of 7.45~8.1; influent NO_3^--N concentration of 40 mg N/L. Under these conditions, the nitrate removal rate ranged from 80% to 90%, while both the effluent NO_3^--N concentration and the effluent NO_2^--N concentration was 4~6 mg/L. This provided a reference for the following study of the range of the current and HRT for SHD-EHD.

3.3. Effect of Current Intensity on Performance of the SHD-EHD

As shown in Figure 3, the NO_3^--N and TN concentrations in the effluent gradually declined with the growth of current intensity under an HRT of 9 h. When the current intensity increased from 5 mA to 40 mA, the nitrate removal rate increased from 50.30% to 86.53% and the TN removal rate increased from 15.30% to 78.72%. The main reason for the growth of the nitrogen removal rate was that the boost of current intensity promoted the generation of H_2 on the cathode and CO_2 on the anode, which served as the electron donor and inorganic carbon source for the autotrophic microorganisms, respectively. In addition, the increase of current intensity could facilitate the activity of denitrification microorganisms [16] and stimulate the activity of nitrate reductase in autotrophic microorganisms [32], thus enhancing the utilization of H_2 by autotrophic microorganisms and improving the denitrification rate of the SHD-EHD. When the current intensity increased from 30 mA to 40 mA, the increase in the nitrate removal rate was not obvious. Further increase of current intensity might cause a waste of electrical energy. Moreover, the applied current intensity should not be too high to inhibit the bacterial activity; it was confirmed that high current density (>320 mA/m^2) would inhibit the activity of microorganisms in a SHD-EHD system [16]. With the rising current, the concentration of NO_2^--N in the effluent gradually increased from 0.2 mg/L to 0.62 mg/L. The possible reasons were as follows: short HRT and insufficient electron donor lead to incomplete denitrification; nitrite reductase was less capable of competing for electrons than nitrate reductase. Therefore, in an unstable environment, denitrifying bacteria tended to use NO_3^--N as the terminal electron acceptor instead of NO_2^--N, leading to the accumulation of NO_2^--N.

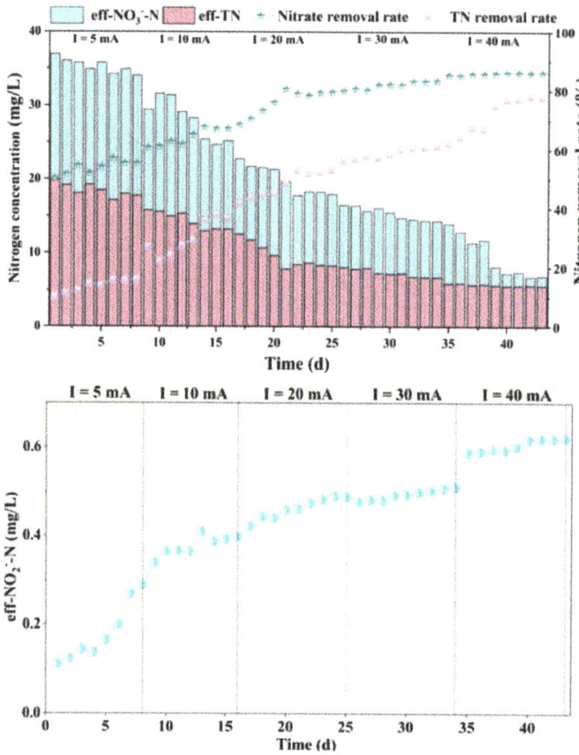

Figure 3. Concentrations of NO_3^--N, TN and NO_2^--N, nitrate removal rate in the SHD−EHD at different current intensities.

3.4. *Effect of HRT on Performance of the SHD-EHD*

As shown in Figure 4, the effluent concentrations of NO_3^--N, NO_2^--N and TN all declined with the increase of HRT. When the HRT increased to 4 h, the effluent NO_2^--N concentration was low and the nitrate removal rate reached 90%. With the extension of HRT, the nitrogen removal rate of the SHD-EHD improved obviously.

HRT is one of the key parameters of an SHD-EHD system, which determines the treatment capacity of the reactor at a certain influent flow rate. The longer the HRT, the more the bacteria can remove NO_3^--N. In the study of the EHD, 92.36% of NO_3^--N could be removed at an HRT of 9 h when the influent NO_3^--N concentration was 30 mg/L. The nitrate removal rate of the SHD-EHD achieved 94.02% at an HRT of 4 h when the influent NO_3^--N concentration was 40 mg/L. Obviously, the SHD-EHD could enhance the nitrate removal rate.

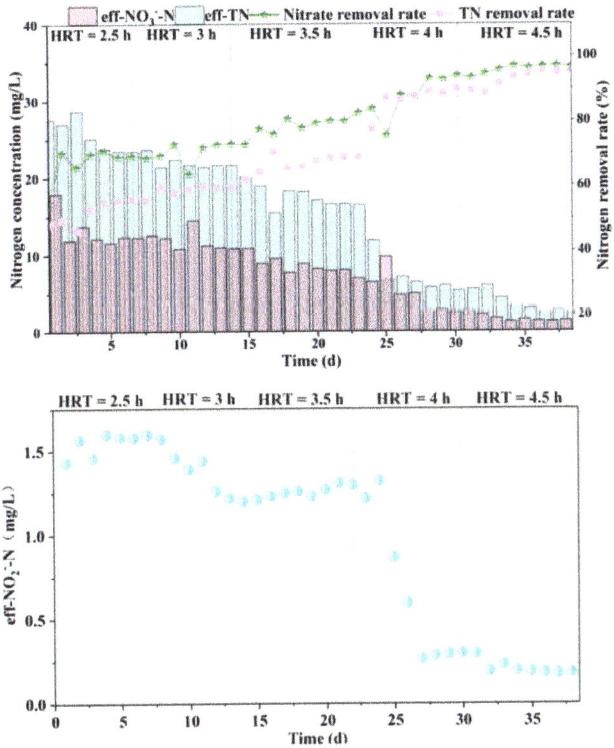

Figure 4. Concentrations of NO_3^--N, TN and NO_2^--N, nitrate removal rate in the SHD−EHD at different HRT.

3.5. pH Variation

Figure 5 shows the pH variation of the SHD-EHD under different current intensity and HRT. It could be seen that the effluent pH remained between 7 and 8.25. The effluent pH reduced from 8.22 to 7.05 when the current intensity increased from 5 mA to 40 mA. The effluent pH increased from 7.44 to 8.25 when the HRT boosted from 2.5 h to 4.5 h.

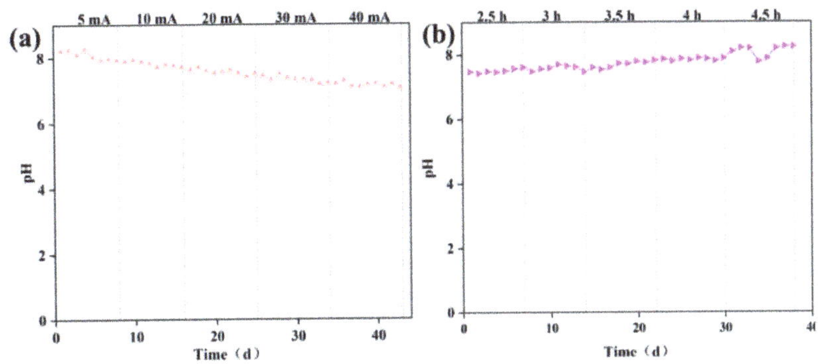

Figure 5. Variation of pH under different (**a**) Current intensities; (**b**) HRT.

According to the calculation, the removal of 1 mol NO_3^--N produced 3.57 mg of alkalinity ($CaCO_3$). Therefore, the pH in the reactor would increase as the treatment proceeded. However, the effluent pH showed small fluctuations and finally stabilized between 7 and 8.25, indicating that the SHD-EHD had good pH buffering capacity. Possible reasons were as follows:

- H^+ and CO_2 were generated by the electrochemical reaction between the carbon rod anode and H_2O. The generated H^+ neutralized the alkalinity produced by denitrification, while the CO_2 combined with OH^- in water and existed in the form of HCO_3^-, which endowed the reactor with good pH buffering capacity;
- Solid-phase heterotrophic denitrification could also produce CO_2, which reacted with OH^- to reduce the alkalinity. This also provided a stable pH favorable for denitrification.

3.6. Denitrification Performance of the SHD-EHD under Optimal Conditions

From the above nitrate removal and NO_2^--N accumulation study, it could be concluded that the optimal condition of the SHD-EHD was as follows: current intensity of 40 mA, HRT of 4 h, influent pH of 7.4~7.5 and influent NO_3^--N concentration of 40 mg/L. The SHD-EHD was operated under this condition for 43 days, and the results are shown in Figure 6. Under the optimal condition, the nitrate removal rate of the SHD-EHD was above 90% and the highest nitrate removal rate reached 99.04%. The effluent NO_3^--N concentration ranged from 1.41 mg/L to 3. 76 mg/L and the effluent NO_2^--N concentration ranged from 0.27 mg/L to 0.62 mg/L. Compared with the EHD, the SHD-EHD was able to obtain a higher nitrate removal rate in a shorter HRT. The SHD-EHD reactor also increased the denitrification rate and reduced the NO_2^--N concentration in the effluent (Figure 7). It indicated that the addition of PCL facilitated the removal of NO_3^--N and enhanced the complete denitrification in groundwater. These results demonstrate that the SHD-EHD could effectively remove NO_3^--N.

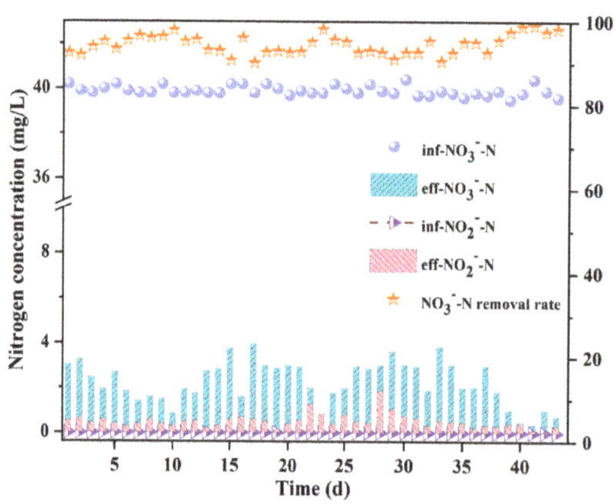

Figure 6. Denitrification performance of SHD−EHD under optimal conditions (I = 40 mA, HRT = 4 h, pH = 7.4~7.5, inf-NO_3^--N = 40 mg/L).

A comparison of nitrate removal rate in different HAD systems was presented in Table 2. The SHD-EHD in this study achieved a satisfactory nitrate removal rate within a much shorter HRT when the influent NO_3^--N concentrations were similar. Tong et al. [33] filled the reactor with pine sawdust and haycite as solid carbon sources and biocarriers; the HRT required to achieve satisfactory nitrate removal rate was longer than the SHD-EHD in this study. This might be due to the fact that some natural biopolymer materials were difficult to be utilized by microorganisms. A previous study suggested that denitrification rates using lignocellulosic materials as electron donors were usually lower than those of conventional liquid-phase organics (such as glucose and ethanol) [34] while PCL could reach a similar performance to liquid-phase organics. In most reported studies, biodegradable polymers as carbon sources achieved higher denitrification performance than that of natural biopolymer materials [26]. Compared with the results of Tong et al. [16], PCL as the carbon source of HAD required much shorter HRT and lower current intensity, which might be due to a shortage of methanol, or because the liquid-phase carbon source dosage was difficult to control. However, PCL could provide a sufficient carbon source for denitrification via continuous hydrolysis by microorganisms. Furthermore, PCL also acted as a biological carrier to increase the biomass in the reactor, which was conducive to the denitrification.

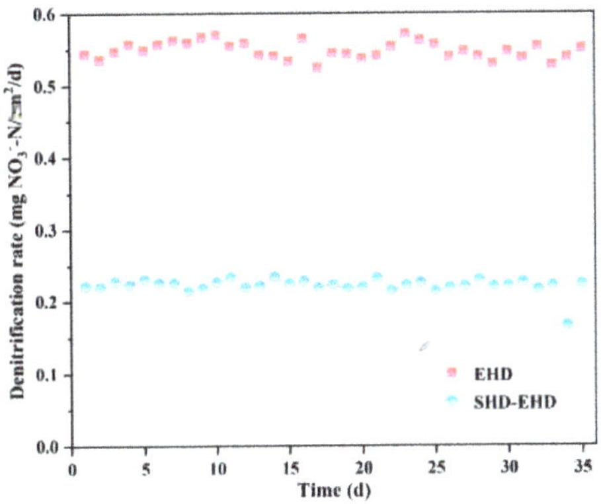

Figure 7. Comparison of denitrification rates of SHD–EHD and EHD under optimal conditions (I = 40 mA, HRT = 4 h, pH = 7.4~7.5, inf-NO_3^--N = 40 mg/L).

Table 2. Comparison with other denitrification systems.

Reactors	inf-NO_3^--N (mg/L)	Current (mA)	HRT (h)	Removal Rate (%)	External Carbon Source	Cathode and Anode Material	Reference
HEAD-PBR	25	100	18	99.70	pine sawdust	cathode: stainless steel mesh anode: Ti/RuO_2 biological carrier: haycite	[33]
HAD-BER	50	60	8	99.90	methanol	cathode: stainless steel mes hanode: stainless steel bar biological carrier: cotton fibers	[16]
SHD-EHD	40	40	4	99.04	PCL	cathode: carbon felt anode: graphite rods biological carrier: PCL	This study
3DBER-SAD	35	60	12	92.00	-	cathode/anode: carbon rods biological carrier: activated carbon electron donor: sulfur	[35]

Compared with the SHD-EHD constructed in this study (Table 3), the conventional hydrogen autotrophic denitrification reactors required more energy consumption; the energy consumption of the SHD-EHD was reduced by 71.15–95.41% compared to them [36–38], which was 44.09% lower than the EHD in this study. These results indicated that the SHD-EHD could significantly reduce the energy consumption. In addition, although the energy consumption of the SHD-EHD was slightly higher than that of the HAD-BER with liquid carbon source [16], the SHD-EHD with PCL as a solid carbon source could be operated continuously without supplementing the carbon source, reducing the maintenance and operation costs.

Table 3. Comparison of energy consumption with previous denitrification systems.

Reactor Type	Denitrification Type	Energy Consumption (kWh/kg NO_3^--N)	Reference
multi-electrode BER	Hydrogen autotrophic	70	[36]
3D-BER	Hydrogen autotrophic	440	[37]
two-chambe BES	Hydrogen autotrophic	200	[38]
EHD	Hydrogen autotrophic	36.11	This study
HAD-BER	Heterotrophic/ hydrogen autotrophic	20	[16]
SHD-EHD	Heterotrophic/ hydrogen autotrophic	20.19	This study

3.7. Analysis of Microbial Community Structure

The 16S rRNA high-throughput sequencing was performed on the samples collected from the SHD-EHD (b1) and the solid-phase heterotrophic denitrification reactor (SHD) (b2), respectively, to investigate the abundance and diversity of the microbial communities. The bacterial community structures were also identified at the phylum and genus levels. The Venn diagram (Figure 8a) shows that the operational taxonomic units of the two samples b1 and b2 were 175 and 533, respectively, indicating that a large number of species in the SHD-EHD system were eliminated. The abundance indices (Ace and Chao) of the SHD were much higher than that of the SHD-EHD (Figure 8b), which suggests the similar conclusion that the microbial diversity homogeneity of the SHD reactor was significantly higher than that of the SHD-EHD. Nevertheless, the denitrification performance of the SHD-EHD was better than the SHD, which might be due to the electrical stimulation that promoted the proliferation of dominant denitrifying bacteria and electroactive bacteria.

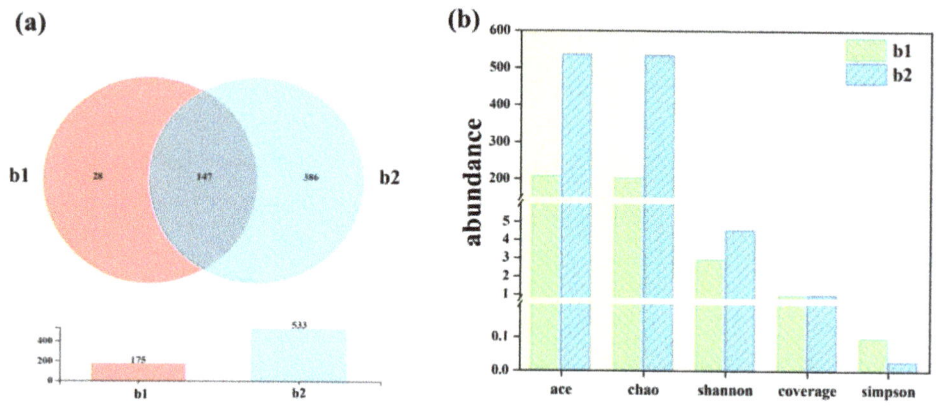

Figure 8. (**a**)Venn diagram; (**b**) Diversity indexes of samples from SHD-EHD and SHD.

At the phylum level, microbial community composition and abundance were exhibited in Figure 9a. *Proteobacteria* was the dominant phylum (86.81%) in the SHD-EHD reactor. As previously reported, many types of denitrifying bacteria, such as *Thauera*, *Hydrogenophaga*, *Alcaligenes* and *Dechloromonas*, were affiliated with *Proteobacteria* [33]. In addition, as *Proteobacteria* in the SHD-EHD consisted of as high as 86.81% of the abundance; other phyla had a low abundance. In contrast, the phyla distribution in the SHD was more uniform, in which the abundance of the dominant phyla *Proteobacteria*, *Bacteroidota* and *Chloroflexi* reached 44.48%, 22.22% and 17.83%, respectively. It could be concluded that denitrifying bacteria could effectively adapt to a electrical environment and kept proliferating while bacteria without such ability was finally eliminated by electrical selection and lost in the competition with denitrifying bacteria. Therefore, high nitrate removal rate in the SHD-

EHD was achieved by high abundance of electrically enriched denitrifying bacteria. The results demonstrated that the bacterial community composition in the SHD-EHD could bring better nitrate removal rate, which reflects the significance of enriching denitrifying bacteria through electrical stimulation.

At the class level, most of the sequences in the SHD-EHD and the SHD belonged to 16 classes, among which *Gammaproteobacteria*, *Bacteroidia*, *Alphaproteobacteria*, *Anaerolineae* and *Chloroflexia* were the main ones (Figure 9b). *Gammaproteobacteria* had the largest proportion of all classes in both samples. The relative abundance of *Gammaproteobacteria* in SHD-EHD and SHD were 72.93% and 39.04%, respectively. *Gammaproteobacteria* included many denitrifying bacteria [39]; therefore, the abundance of *Gammaproteobacteria* was positively correlated with the denitrification performance. This suggested that the addition of PCL not only provided carbon source for denitrification, but also promoted the enrichment of denitrifying bacteria.

At the genus level, the main denitrifying bacteria in the SHD-EHD mainly included *Dechloromonas*, *Thauera*, *Hydrogenophag* and *Acidovorax* (Figure 9c). *Dechloromonas* was the dominant genus with the highest abundance in the SHD-EHD, which could use organic carbon sources as electron donors to reduce O_2, nitrate and nitrite under anoxic condition [40,41]. The relative abundance of *Dechloromonas* in the SHD-EHD increased compared to the SHD, probably due to the electrical stimulation that promoted its activity and improved its ability to utilize organic carbon sources. *Thauera* could not only utilize organic matters as carbon sources for heterotrophic denitrification, but also could use H_2 as an electron donor for hydrogen autotrophic denitrification [42]. The relative abundance of *Thauera* in the SHD-EHD and the SHD were 14.56% and 3.82%, respectively, indicating that electrical stimulation facilitated the enrichment of *Thauera*. Therefore, the SHD-EHD constructed in this study achieved autotrophic–heterotrophic synergistic denitrification. *Hydrogenophaga* in the SHD-EHD was from the *Comamonadacea* family, which was able to use H_2 for autotrophic growth [43]. H_2 generated at the cathode could be utilized by *Hydrogenophaga*. The inorganic carbon generated during the heterotrophic denitrification provided carbon sources for the growth of *Hydrogenophaga*. In consequence, the heterotrophic denitrifying bacteria could facilitate the growth of *Hydrogenophaga* in the SHD-EHD. *Acidovorax* with a relative abundance of 4.29% in the SHD-EHD had the ability to oxidize H_2 for denitrification [44]. In addition, small amounts of *Rhodopseudomonas* (3.21%) and *Alishewanella* (1.77%) were also found in the biofilm sample from the SHD-EHD. *Rhodopseudomonas* was able to fix nitrogen, produce H_2 and fix CO_2 [45]; it could remove NO_3^--N probably by biological nitrogen fixation in denitrification system [46]. It was reported that *Alishewanella* had the ability to reduce NO_3^--N [47]. In this study, the relative abundance of *Stenotrophomonas* was 0.04% in the SHD and 2.47% in the SHD-EHD, which demonstrated that H_2 generated by electrolysis promoted the enrichment of *Stenotrophomonas*. Moreover, it was reported that *Stenotrophomonas* could shuttle electrons generated during the oxidation of substrates via electron mediators or direct contact with the electrode. *Denitratisoma* was also found in the two reactors. A previous study showed that the accumulation of NO_2^--N was low in an HAD system, owing to the high abundance of *Denitratisoma* in which the nitrite reductase could facilitated the reduction of NO_2^--N [43].

In addition, there were some genera that had the abilities of degrading biodegradable polymer and denitrification. The denitrifying genus *Acidovorax* in the SHD-EHD was capable of both autotrophic denitrification and metabolism of PCL because of its potential to hydrolyze ester bonds [48,49]. This enabled *Acidovorax* to be enriched by an order of magnitude in the SHD-EHD, where PCL served as the solid carbon source. Moreover, the enrichment of *Bacteroidetes* might facilitate the hydrolysis of PCL, thus releasing small molecule carbon sources more favorable to denitrifying bacteria [50].

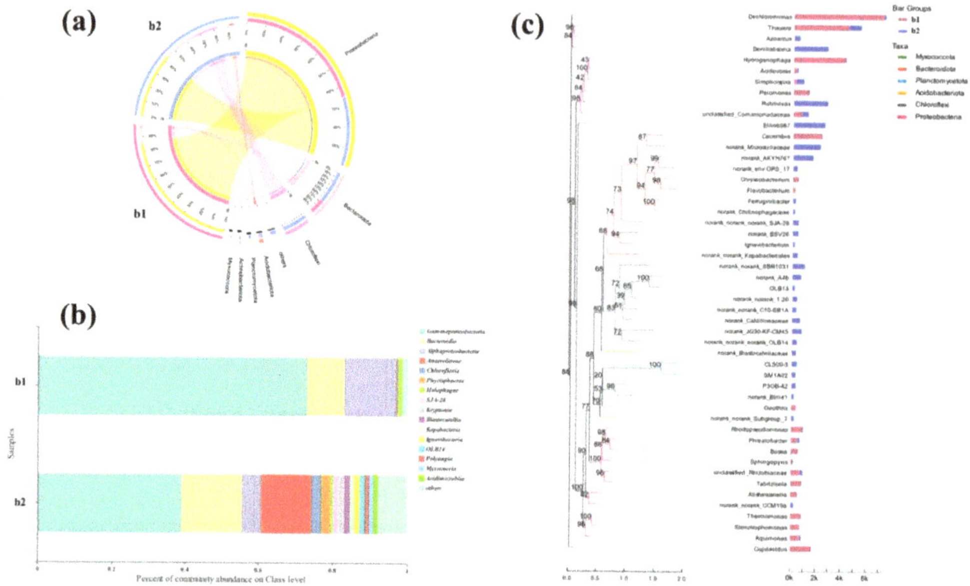

Figure 9. Relative abundance of samples at (**a**) Phylum level; (**b**) Class level; (**c**) Genus level.

3.8. Analysis of Metabolic Pathway

The nitrogen metabolic pathway was predicted by combining the gene data from the metagenome sequencing results with the nitrogen metabolism information from the KEGG database (Figure 10). Four related denitrifying enzymes appeared in both samples, which were iso-nitrate reductase (narG, narH and napA), nitrite reductase (nirK and nirS), nitric oxide reductase (norB and norC) and nitrous oxide reductase (nosZ). The abundance of enzymes in the SHD-EHD was higher than in the SHD, representing more active denitrification in the SHD-EHD. According to previous results, the SHD-EHD displayed a better nitrate removal rate. Therefore, electrical stimulation increased the activity of nitrate reductase and nitrite reductase, which was consistent with former study [13]. Nitrate reductases were represented by membrane-bound nitrate reductase (narG and narH) and periplasmic nitrate reductase (napA). Although napA was a peritroplasmic nitrate reductase, its distance from the outer membrane was shorter than that of narG and narH, and it could be connected to external electron flow [51]. Therefore, the expression of napA was expected to increase at a high current intensity. However, the expression of napA decreased while narG and narH expression increased at high current intensity. This might be due to the gene regulation, which had not evolved for adapting to the change of carbon source [52], or narG and narH increased the electron utilization for the reduction of NO_3^--N, rather than for respiration.

NosZ was proposed to be the rate-limiting step in the whole denitrification pathway. The low expression of nosZ might lead to the accumulation of N_2O. The accumulation of denitrifying intermediates except N_2O was not observed. The gene expression data of nitrite reductase was in correspondence with the effluent NO_2^--N concentration. The accumulation of NO_2^--N was not observed after NO_3^--N was reduced to NO_2^--N at all current intensities, representing that the expression of nirK and nirS gradually increased to saturation with the increase of current intensity. The nitrite reductase byproduct NO might not accumulate because it was unstable and was easily converted to N_2O or other forms through abiotic processes. In addition, the expression data of norB and norC suggested that bioconversion of NO to N_2O might have been greatly activated. Finally, from the accumulation of N_2O

and the low expression of nosZ, it was concluded that the transformation of NO to N_2O was the rate-limiting step in the whole denitrification pathway.

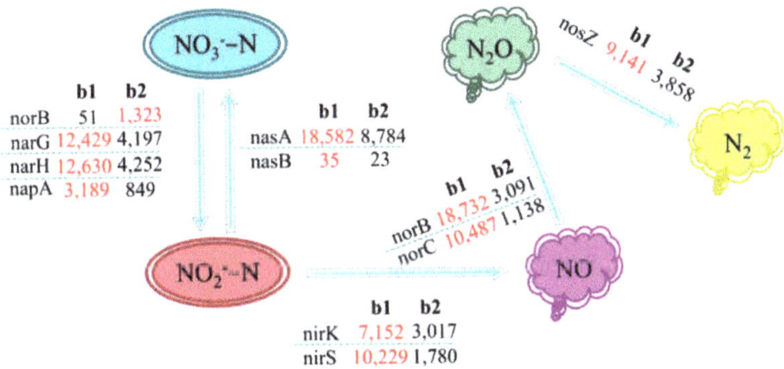

Figure 10. Nitrogen metabolism pathway analysis.

3.9. Material Conversion and Mechanism

As a biodegradable polymer, PCL could be decomposed by microbial enzymes. In the SHD-EHD, the PCL acted as an electron donor for heterotrophic denitrification Equation (3) and the carrier for the attachment of microorganisms as well. The decomposition product CO_2 could be used as the carbon source for autotrophic denitrification.

In the lower part of the SHD-EHD, dissolved oxygen was consumed by electrochemical reaction Equation (4), creating an anoxic environment for denitrifying bacteria. At the same time, H_2 generated through electrolysis served as the electron donor for autotrophic denitrification Equations (5) and (6). The anodic carbon rod was oxidized to generate CO_2 Equation (7), offering the carbon source required for autotrophic denitrification for denitrifying bacteria. Autotrophic denitrification was the main process of inorganic carbon consumption.

As described above, the SHD-EHD was a collaborative system combining multiple processes. The possible pathway of material transformation in the SHD-EHD was listed as follows (the transformation approach in the SHD-EHD was shown in Figure 11):

1. *In situ* H_2 production at the cathode was consumed for autotrophic denitrification.
2. NO_3^- and NO_2^- produced by electrochemical denitrification and incomplete denitrification were transformed into N_2 mainly by heterotrophic and autotrophic denitrification.
3. Part of NO_3^--N was assimilated by microorganisms and transformed to organic nitrogen.
4. PCL could be hydrolyzed by microorganisms into an organic carbon source that could be easily utilized by heterotrophic microorganisms.
5. Organic carbon was transformed into inorganic carbon by microbial hydrolysis, metabolism and heterotrophic denitrification, which provided a carbon source for autotrophic denitrification, thus achieving synergistic heterotrophic and autotrophic denitrification in the SHD-EHD.

Efficient material transformation was achieved by the combination of solid-phase heterotrophic denitrification and electrochemical hydrogen autotrophic denitrification. Satisfactory nitrate removal performance and low concentrations of by-products were realized, providing a novel and efficient method for nitrate removal.

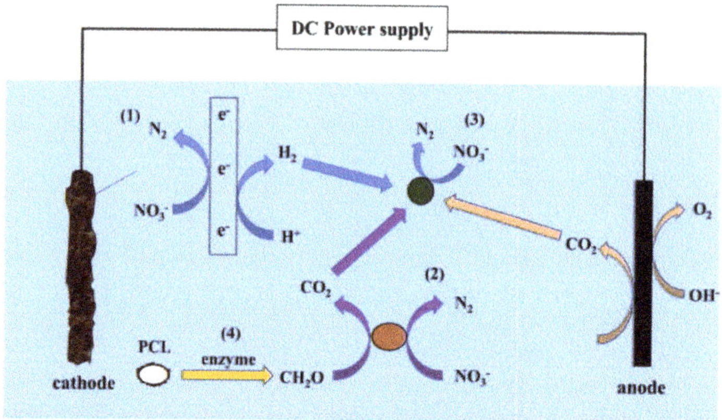

Figure 11. Denitrification mechanism in the SHD−EHD: (1) Electrochemical denitrification reduction; (2) Heterotrophic denitrification; (3) Hydrogen autotrophic denitrification; (4) Hydrolysis of PCL.

4. Conclusions

In summary, a heterotrophic denitrification coupled with electro-autotrophic denitrification system was developed for treating nitrate-contaminated groundwater, in which PCL served as the carbon source. Under the optimal condition (influent nitrate concentration of 40 mg/L, current intensity of 40 mA and HRT of 4 h), a high nitrate removal rate of 99.04% could be realized. Compared with the EHD, the SHD-EHD with PCL facilitated the complete denitrification and reduced the accumulation of NO_2^--N. *Dechloromonas*, *Thauera* and *Hydrogenophaga* were identified as the key denitrifying bacteria in the SHD-EHD. Comparison of microbial communities from the SHD-EHD and the SHD revealed that electrical stimulation boosted the abundance of the dominant denitrifying bacteria and the electroactive bacteria. The conversion of NO to N_2O was the rate-limiting step in the overall denitrification pathway, according to the analysis of the nitrogen metabolic pathway. The SHD-EHD developed in this study shows promising potential for the removal of nitrate from groundwater. However, further study on the generation of intermediate gases and synergistic removal with other pollutants in the SHD-EHD is required to explain the detailed nitrate removal mechanisms of the SHD-EHD.

Supplementary Materials: The following supporting information can be downloaded at: https://www.mdpi.com/article/10.3390/w14111759/s1, Figure S1: Denitrification performance of EHD under optimal conditions; DNA extraction and Illumina MiSeq sequencing; Figure S2: SEM of carbon felt in SHD-EHD: (**a**) bare carbon felt; (**b**) carbon felt after reaction; Figure S3: Effect of different current on NO_3^--N and NO_2^--N concentrations; Figure S4: Effect of different HRT on NO_3^--N and NO_2^--N concentrations; Figure S5: Effect of different pH on NO_3^--N and NO_2^--N; Figure S6: Effect of influent pH on anode, cathode and effluent pH of EHD; Figure S7: Effect of different inf-NO_3^--N on NO_3^--N and NO_2^--N concentrations.

Author Contributions: Conceptualization, X.T.; methodology, S.Y.; investigation, S.Y.; resources, X.T. and S.Z.; writing—original draft preparation, S.Y. and L.L.; writing—review and editing, X.T. and L.L.; supervision, X.T.; funding acquisition, X.T. All authors have read and agreed to the published version of the manuscript.

Funding: This research was funded by the National Natural Science Foundation of China (No. 21806126) and the Fundamental Research Funds for the Central Universities (WUT.2019IVB031).

Informed Consent Statement: Not applicable.

Data Availability Statement: Not applicable.

Acknowledgments: The advice of microbial analysis from Hao-Jin Peng in School of Civil Engineering and Architecture, Wuhan University of Technology, China is gratefully acknowledged.

Conflicts of Interest: The authors declare no conflict of interest.

References

1. United Nations. The United Nations World Water Development Report 2022: Groundwater: Making the Invisible Visible. Available online: https://digitallibrary.un.org/record/3967679?ln=zh_CN (accessed on 23 March 2022).
2. Arjen, H.; Ashok, C.; Pieter, V.O. Advancing Water Footprint Assessment Research: Challenges in Monitoring Progress towards Sustainable Development Goal 6. *Water* **2017**, *9*, 438.
3. Ghafari, S.; Hasan, M.; Aroua, M.K. Effect of carbon dioxide and bicarbonate as inorganic carbon sources on growth and adaptation of autohydrogenotrophic denitrifying bacteria. *J. Hazard. Mater.* **2009**, *162*, 1507–1513. [CrossRef] [PubMed]
4. Ballantine, K.A.; Groffman, P.M.; Lehmann, J.; Schneider, R.L. Stimulating nitrate removal processes of restored wetlands. *Environ. Sci. Technol.* **2014**, *48*, 7365–7373. [CrossRef] [PubMed]
5. Garcia-Segura, S.; Lanzarini-Lopes, M.; Hristovski, K.; Westerhoff, P. Electrocatalytic reduction of nitrate: Fundamentals to full scale water treatment applications. *Appl. Catal. B: Environ.* **2018**, *236*, 546–568. [CrossRef]
6. Wang, H.; Chen, N.; Feng, C.; Deng, Y. Insights into heterotrophic denitrification diversity in wastewater treatment systems: Progress and future prospects based on different carbon sources. *Sci. Total Environ.* **2021**, *780*, 146521. [CrossRef]
7. Liang, J.; Chen, N.; Tong, S.; Liu, Y.; Feng, C. Sulfur autotrophic denitrification (SAD) driven by homogeneous composite particles containing CaCO3-type kitchen waste for groundwater remediation. *Chemosphere* **2018**, *212*, 954–963. [CrossRef]
8. Feng, H.; Huang, B.; Zou, Y.; Li, N.; Wang, M.; Yin, J.; Cong, Y.; Shen, D. The effect of carbon sources on nitrogen removal performance in bioelectrochemical systems. *Bioresour. Technol.* **2013**, *128*, 565–570. [CrossRef]
9. Molognoni, D.; Devecseri, M.; Cecconet, D.; Capodaglio, A.G. Cathodic groundwater denitrification with a bioelectrochemical system. *J. Water Process Eng.* **2017**, *19*, 67–73. [CrossRef]
10. Tang, Y.; Ziv-El, M.; Meyer, K.; Zhou, C.; Shin, J.H.; Ahn, C.H.; McQuarrie, J.; Candelaria, D.; Swaim, P.; Scott, R.; et al. Comparing heterotrophic and hydrogen-based autotrophic denitrification reactors for effluent water quality and post-treatment. *Water Supply* **2012**, *12*, 227–233. [CrossRef]
11. Kiss, I.; Szekeres, S.; Bejerano, T.T.; Soares, M.I.M. Hydrogen-dependent denitrification: Preliminary assessment of two bio-electrochemical systems. *Water Sci. Technol.* **2000**, *42*, 373–379. [CrossRef]
12. Zhao, Y.X.; Feng, C.P.; Wang, Q.H.; Yang, Y.N.; Zhang, Z.Y.; Sugiura, N. Nitrate removal from groundwater by cooperating heterotrophic with autotrophic denitrification in a biofilm-electrode reactor. *J. Hazard. Mater.* **2011**, *192*, 1033–1039. [CrossRef] [PubMed]
13. Liu, H.Y.; Tong, S.; Chen, N.; Liu, Y.; Feng, C.P.; Hua, Q.L. Effect of electro-stimulation on activity of heterotrophic denitrifying bacteria and denitrification performance. *Bioresour. Technol.* **2015**, *196*, 123–128. [CrossRef] [PubMed]
14. Zhai, S.Y.; Ji, M.; Zhao, Y.X.; Pavlostathis, S.G.; Zhao, Q. Effects of salinity and COD/N on denitrification and bacterial community in dicyclic-type electrode based biofilm reactor. *Chemosphere* **2018**, *192*, 328–336. [CrossRef] [PubMed]
15. Ghafari, S.; Hasan, M.; Aroua, M.K. Nitrate remediation in a novel upflow bio-electrochemical reactor (UBER) using palm shell activated carbon as cathode material. *Electrochim. Acta* **2009**, *54*, 4164–4171. [CrossRef]
16. Tong, S.; Zhang, B.; Feng, C.; Zhao, Y.; Chen, N.; Hao, C.; Pu, J.; Zhao, L. Characteristics of heterotrophic/biofilm-electrode autotrophic denitrification for nitrate removal from groundwater. *Bioresour. Technol.* **2013**, *148*, 121–127. [CrossRef]
17. Liu, H.Y.; Hu, Q.L. Analysis of denitrification performance and microbial community structure in a bio-electrochemical reactor under different current densities with wheat-rice stone powder. *Water Reuse* **2022**, *12*, 66–77. [CrossRef]
18. Zhao, Y.X.; Zhang, B.G.; Feng, C.P.; Huang, F.Y.; Zhang, P.; Zhang, Z.Y.; Yang, Y.N.; Sugiura, N. Behavior of autotrophic denitrification and heterotrophic denitrification in an intensified biofilm-electrode reactor for nitrate-contaminated drinking water treatment. *Bioresour. Technol.* **2012**, *107*, 159–165. [CrossRef]
19. Zhang, J.M.; Feng, C.P.; Hong, S.Q.; Hao, H.L.; Yang, Y.N. Behavior of solid carbon sources for biological denitrification in groundwater remediation. *Water Sci. Technol.* **2012**, *65*, 1696–1704. [CrossRef]
20. Ribera-Guardia, A.; Kassotaki, E.; Gutierrez, O.; Pijuan, M. Effect of carbon source and competition for electrons on nitrous oxide reduction in a mixed denitrifying microbial community. *Process Biochem.* **2014**, *49*, 2228–2234. [CrossRef]
21. Wang, J.L.; Chu, L.B. Biological nitrate removal from water and wastewater by solid-phase denitrification process. *Biotechnol. Adv.* **2016**, *34*, 1103–1112. [CrossRef]
22. Zhong, Y.; Godwin, P.; Jin, Y.; Xiao, H. Biodegradable polymers and green-based antimicrobial packaging materials: A mini-review. *Adv. Ind. Eng. Polym. Res.* **2020**, *3*, 27–35. [CrossRef]
23. RameshKumar, S.; Shaiju, P.; O'Connor, K.E. Bio-based and biodegradable polymers—State-of-the-art, challenges and emerging trends. *Curr. Opin. Green Sustain. Chem.* **2020**, *21*, 75–81. [CrossRef]
24. Mergaert, J.; Boley, A.; Cnockaert, M.C.; Muller, W.R.; Swings, J. Identity and potential functions of heterotrophic bacterial isolates from a continuous-upflow fixed-bed reactor for denitrification of drinking water with bacterial polyester as source of carbon and electron donor. *Syst. Appl. Microbiol.* **2001**, *24*, 303–310. [CrossRef] [PubMed]
25. Chu, L.B.; Wang, J.L. Denitrification of groundwater using PHBV blends in packed bed reactors and the microbial diversity. *Chemosphere* **2016**, *155*, 463–470. [CrossRef] [PubMed]
26. Li, P.; Zuo, J.E.; Wang, Y.J.; Zhao, J.; Tang, L.; Li, Z.X. Tertiary nitrogen removal for municipal wastewater using a solid-phase denitrifying biofilter with polycaprolactone as the carbon source and filtration medium. *Water Res.* **2016**, *93*, 74–83. [CrossRef]

27. Zhang, Q.; Ji, F.; Xu, X. Effects of physicochemical properties of poly-ε-caprolactone on nitrate removal efficiency during solid-phase denitrification. *Chem. Eng. J.* **2016**, *283*, 604–613. [CrossRef]
28. Shen, Z.Q.; Yin, Y.N.; Wang, J.L. Biological denitrification using poly(butanediol succinate) as electron donor. *Applied Microbiol. Biotechnol.* **2016**, *100*, 6047–6053. [CrossRef]
29. Wu, W.Z.; Yang, L.H.; Wang, J.L. Denitrification using PBS as carbon source and biofilm support in a packed-bed bioreactor. *Environ. Sci. Pollut. Res.* **2013**, *20*, 333–339. [CrossRef]
30. Cui, Y.; Lai, B.; Tang, X.H. Microbial Fuel Cell-Based Biosensors. *Biosensors* **2019**, *9*, 18. [CrossRef]
31. Singh, B.K.; Quince, C.; Macdonald, C.A.; Khachane, A.; Thomas, N.; Abu Al-Soud, W.; Sorensen, S.J.; He, Z.L.; White, D.; Sinclair, A.; et al. Loss of microbial diversity in soils is coincident with reductions in some specialized functions. *Environ. Microbiol.* **2014**, *16*, 2408–2420. [CrossRef]
32. Jian-Cha, W. Enhanced Removal of Nitrogen and Phosphorus in 3DBER Filled with Sponge Iron and Activated Carbon. *China Water Wastewater* **2014**, *209*, 57–64. [CrossRef]
33. Peng, T.; Feng, C.P.; Hu, W.W.; Chen, N.; He, Q.C.; Dong, S.S.; Xu, Y.X.; Gao, Y.; Li, M. Treatment of nitrate-contaminated groundwater by heterotrophic denitrification coupled with electro-autotrophic denitrifying packed bed reactor. *Biochem. Eng. J.* **2018**, *134*, 12–21. [CrossRef]
34. Shen, Z.; Zhou, Y.; Wang, J. Comparison of denitrification performance and microbial diversity using starch/polylactic acid blends and ethanol as electron donor for nitrate removal. *Bioresour. Technol.* **2013**, *131*, 33–39. [CrossRef] [PubMed]
35. Hao, R.X.; Meng, C.C.; Li, J.B. An integrated process of three-dimensional biofilm-electrode with sulfur autotrophic denitrification (3DBER-SAD) for wastewater reclamation. *Appl. Microbiol. Biotechnol.* **2016**, *100*, 7339–7348. [CrossRef] [PubMed]
36. Sakakibara, Y.; Nakayama, T. A novel multi-electrode system for electrolytic and biological water treatments: Electric charge transfer and application to denitrification. *Water Res.* **2001**, *35*, 768–778. [CrossRef]
37. Zhou, M.; Fu, W.; Gu, H.; Lei, L. Nitrate removal from groundwater by a novel three-dimensional electrode biofilm reactor. *Electrochim. Acta* **2007**, *52*, 6052–6059. [CrossRef]
38. Pous, N.; Koch, C.; Vila-Rovira, A.; Balaguer, M.D.; Colprim, J.; Muehlenberg, J.; Mueller, S.; Harnisch, F.; Puig, S. Monitoring and engineering reactor microbiomes of denitrifying bioelectrochemical systems. *Rsc Adv.* **2015**, *5*, 68326–68333. [CrossRef]
39. Feng, T.; Wu, J.; Chai, K.; Liu, F. Effect of Pseudomonas sp. on the degradation of aluminum/epoxy coating in seawater. *J. Mol. Liq.* **2018**, *263*, 248–254. [CrossRef]
40. Horn, M.A.; Ihssen, J.; Matthies, C.; Schramm, A.; Acker, G.; Drake, H.L. Dechloromonas denitrificans sp nov., Flavobacterium denitrificans sp nov., Paenibacillus anaericanus sp. nov and Paenibacillus terrae strain MH72, N2O-producing bacteria isolated from the gut of the earthworm Aporrectodea caliginosa. *Int. J. Syst. Evol. Microbiol.* **2005**, *55*, 1255–1265. [CrossRef]
41. Coates, J.D.; Chakraborty, R.; Lack, J.G.; O'Connor, S.M.; Cole, K.A.; Bender, K.S.; Achenbach, L.A. Anaerobic benzene oxidation coupled to nitrate reduction in pure culture by two strains of Dechloromonas. *Nature* **2001**, *411*, 1039–1043. [CrossRef]
42. Zhang, Y.W.; Wei, D.Y.; Morrison, L.; Ge, Z.B.; Zhan, X.M.; Li, R.H. Nutrient removal through pyrrhotite autotrophic denitrification: Implications for eutrophication control. *Sci. Total Environ.* **2019**, *662*, 287–296. [CrossRef] [PubMed]
43. Xu, Z.S.; Dai, X.H.; Chai, X.L. Biological denitrification using PHBV polymer as solid carbon source and biofilm carrier. *Biochem. Eng. J.* **2019**, *146*, 186–193. [CrossRef]
44. Vasiliadou, I.A.; Pavlou, S.; Vayenas, D.V. Dynamics of a chemostat with three competitive hydrogen oxidizing denitrifying microbial populations and their efficiency for denitrification. *Ecol. Model.* **2009**, *220*, 1169–1180. [CrossRef]
45. Karpinets, T.V.; Pelletier, D.A.; Pan, C.L.; Uberbacher, E.C.; Melnichenko, G.V.; Hettich, R.L.; Samatova, N.F. Phenotype Fingerprinting Suggests the Involvement of Single-Genotype Consortia in Degradation of Aromatic Compounds by Rhodopseudomonas palustris. *PLoS ONE* **2009**, *4*, e4615. [CrossRef] [PubMed]
46. Du, L.; Trinh, X.; Chen, Q.R.; Wang, C.; Wang, H.H.; Xia, X.; Zhou, Q.H.; Xu, D.; Wu, Z.B. Enhancement of microbial nitrogen removal pathway by vegetation in Integrated Vertical-Flow Constructed Wetlands (IVCWs) for treating reclaimed water. *Bioresour. Technol.* **2018**, *249*, 644–651. [CrossRef]
47. Morris, J.M.; Jin, S.; Crimi, B.; Pruden, A. Microbial fuel cell in enhancing anaerobic biodegradation of diesel. *Chemical Eng. J.* **2009**, *146*, 161–167. [CrossRef]
48. Nalcaci, O.O.; Boke, N.; Ovez, B. Potential of the bacterial strain Acidovorax avenae subsp avenae LMG 17238 and macro algae Gracilaria verrucosa for denitrification. *Desalination* **2011**, *274*, 44–53. [CrossRef]
49. McIlroy, S.J.; Kirkegaard, R.H.; McIlroy, B.; Nierychlo, M.; Kristensen, J.M.; Karst, S.M.; Albertsen, M.; Nielsen, P.H. MiDAS 2.0: An ecosystem-specific taxonomy and online database for the organisms of wastewater treatment systems expanded for anaerobic digester groups. *Database J. Biol. Databases Curation* **2017**. [CrossRef]
50. Zhang, D.Y.; Vahala, R.K.; Wang, Y.; Smets, B.F. Microbes in biological processes for municipal landfill leachate treatment: Community, function and interaction. *Int. Biodeterior. Biodegrad.* **2016**, *113*, 88–96. [CrossRef]
51. Simpson, P.J.L.; Richardson, D.J.; Codd, R. The periplasmic nitrate reductase in Shewanella: The resolution, distribution and functional implications of two NAP Isoforms, NapEDABC and NapDAGHB. *Microbiology. SGM* **2010**, *156*, 302–312. [CrossRef]
52. Doan, T.V.; Lee, T.K.; Shukla, S.K.; Tiedje, J.M.; Park, J. Increased nitrous oxide accumulation by bioelectrochemical denitrification under autotrophic conditions: Kinetics and expression of denitrification pathway genes. *Water Res.* **2013**, *47*, 7087–7097. [CrossRef] [PubMed]

water

MDPI

Article

Electrode Microbial Communities Associated with Electron Donor Source Types in a Bioelectrochemical System Treating Azo-Dye Wastewater

Zechong Guo [1,2,3], Lu Zhang [1], Min-Hua Cui [2,4,*] and Aijie Wang [2,5,*]

1 School of Environmental and Chemical Engineering, Jiangsu University of Science and Technology, Zhenjiang 212100, China; guozechong@just.edu.cn (Z.G.); avavip@126.com (L.Z.)
2 State Key Laboratory of Urban Water Resource and Environment, Harbin Institute of Technology, Harbin 150090, China
3 Jiangxi Jindalai Environmental Protection Co., Ltd., Nanchang 330100, China
4 School of Environmental and Civil Engineering, Jiangnan University, Wuxi 214122, China
5 Key Laboratory of Environmental Biotechnology, Research Center for Eco-Environmental Sciences, Chinese Academy of Sciences, Beijing 100085, China
* Correspondence: cuiminhua@jiangnan.edu.cn (M.-H.C.); waj0578@hit.edu.cn (A.W.)

Abstract: Bioelectrochemical systems (BESs) have been acknowledged to be an efficient technology for refractory pollution treatment. An electron donor is as an indispensable element of BES, and domestic wastewater (DW) has been proved as a cost-efficient and accessible alternative option to expensive carbon sources (such as acetate and glucose), yet its effect on microbial community evolution has not been thoroughly revealed. In this study, the electrode microbial communities from BESs treating azo dye wastewater fed by DW (R_{DW}), acetate (R_{Ac}), and glucose (R_{Glu}) were systematically revealed based on 16S rRNA Illumina MiSeq sequencing platform. It was found that there were significant differences between three groups in microbial community structures. *Desulfovibrio*, *Acinetobacter*, and *Klebsiella* were identified as the predominant bacterial genera in R_{DW}, R_{Ac}, and R_{Glu}, respectively. *Methanosaeta*, the most enriched methanogen in all reactors, had a relative lower abundance in R_{DW}. Microbial communities in R_{Ac} and R_{Glu} were sensitive to electrode polarity while R_{DW} was sensitive to electrode position. Compared with pure substrates, DW increased the diversity of microbial community and, thus, may enhance the stability of electrode biofilm. This study provides an insight into the microbial response mechanism to the electron donors and provides engineering implications for the development of BES.

Keywords: bioelectrochemical system (BES); electron donor source; canonical correspondence analysis; microbial community structure

1. Introduction

Electrochemical technologies are proved as an efficient method for contamination remediation by electrochemical oxidation and electrochemical reduction processes [1,2]. However, the high cost and extreme operating conditions impede wide applications in the practical scene. Considering the cost-efficient and feasibility of large-scale capability, biological treatment is still the most widely used wastewater treatment technology. However, it was difficult to satisfy the stricter discharging standard due to the low efficiency and complicated manipulation, especially in treating refractory industrial wastewaters, such as azo dye wastewater, chemical wastewater, etc. Introducing electrochemical technology into conventional biological facilities seems to be a potential strategy to deal with wastewater issues [3]. The bioelectrochemical system (BES) is a rapidly developing technology that inherits both the advantages of electrochemical and biological processes. The feasibility and superiority of implementing BES to remediate pollution have been verified in covering azo dyes, nitro compounds and metals reduction, and dehalogenation [4]. Among them, azo dye decolorization is one of the most successful applications of BES in industrial wastewater treatments. Due to the presence of external circuit, the electron transfer process was obviously enhanced and resulted in fast decolorization compared to traditional anaerobic process. For now, several pioneer works have confirmed the feasibility of the practical application of BES technology [4]. However, more efforts are required to promote the industrialized application of BES, such as evaluating the performance under practical scenario, exploring the mechanism of microbial functional stabilization, reducing the operational cost of BES technology, etc.

In a BES, electrochemical active microorganisms (EAMs) play an important role by involving extracellular electron transfer. Electrons are produced by anode respiration bacteria oxidizing organic matters and transferred to the cathode to drive reduction actions. These EAMs can serve as a biocatalyst to decrease applied voltage and increase reaction efficiency. Thus, to maintain a stable microbial community structure capable of achieving efficient electron transfer, pollutant degradation is crucial to the successful operation of a BES. In addition to some common physical and chemical influencing factors [5–10] in BES, the microbial community structure might also be influenced by some special operation parameters, such as electrode position and polarity [4,11], salinity and COD/N [12], applied voltage [13], organic loading rates, and electron donors. In previous studies, simple and pure carbon sources such as acetate, glucose, and yeast extract are commonly employed to support BES's operation, but they comprise diseconomy and are impracticable to access in the large-scale application scene. To decrease operating costs, various organic wastes/wastewater were proposed to serve as alternative electron donors to drive bioelectrochemical reactions. Among them, domestic wastewater (DW) has been proved to be a cost-effective yet highly efficient electron donor source in a BES treating azo dye wastewater in our previous study [14]; nevertheless, the characteristics of the microbial community structure of electrode biofilm with DW as a carbon source has not been thoroughly revealed. Therefore, this study systematically analyzed the effects of various electron donors (including acetate, glucose, and DW) on microbial community structure by conducting 16S rRNA Illumina Miseq sequencing and evaluated the feasibility and superiority of DW as electron donor at the level of microbial community evolution.

2. Materials and Methods

2.1. Reactor Configuration and Operation

Three identical cylindrical reactors were manufactured with plexiglass [14]. Each one had a working volume of approximately 1.25 L (ID × L = 8 cm × 25 cm) (see Figure 1). Two pairs of electrodes were installed into the reactor, from the bottom to the top, which was arranged as down-cathode, down-anode, up-cathode, and up-anode with the distance of 2.5 cm between each electrode. Both anodes and cathodes were constructed by granular graphite (diameter from 3 to 6 mm) with 8 cm in diameter and 4 cm in height, producing the total electrode volume of 200 cm^3. Before use, granular graphite was washed by soaking in 32% HCl four times to remove foreign materials [15]. The graphite rod (Φ = 4 mm) penetrated the electrode zone and worked as an electron collector. A saturated calomel electrode (SCE, +247 mV vs. standard hydrogen electrode) was used as the reference electrode. During all experiments, a voltage of 0.5 V was supplied between the anode and cathode by a DC power supply. A 10 Ω resistor was connected in series into the electric circuit. The voltages across this resistor were recorded every 10 min by a data acquisition system (Keithley 2700), which were automatically converted to the current according to Ohm law.

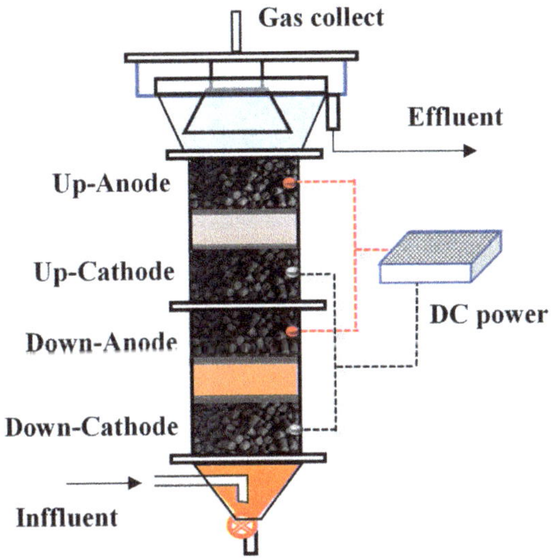

Figure 1. Schematic representation of the reactor [14].

All three reactors were startup in batch mode. Effluent collected from a long-term operated single chamber BES was amended with sodium acetate (NaAc, 1000 mg/L) and acid orange 7 (AO7, 200 mg/L) as the inoculum solution. In the first batch operation cycle, 50 mL anaerobic activated sludge (SS of 35.57 mg/L and VSS of mg/L) was also added into the reactors to strengthen the biomass. After that, the inoculum solution was replaced every two days. When a stable current output was observed, the reactors were considered to be successfully started up. Subsequently, the operation of reactors was changed to continuous flow mode. The hydraulic retention times for all reactors were set at 6 h throughout the experiment. Domestic wastewater (DW) collected from a local sewage well was filtered by a 400-mesh sieve to remove particles that might block the interspace of electrodes. The COD of the filtrated DW was detected as 309 ± 18 mg/L in all experiments, total nitrogen of 59.38 ± 1.54 mg/L, ammonium of 49.68 ± 5.29 mg/L, total phosphorus of 13.14 ± 1.43 mg/L, sulfate of 88.90 ± 11.33 mg/L, chloride of 67.41 ± 18.04 mg/L, alkalinity of 217 ± 75 mg-$CaCO_3$/L, and pH of 7.39 ± 0.07. DW was then amended with AO7 (200 mg/L) as the influent of one reactor (R_{DW}). As the controls, the other two reactors were fed with synthetic wastewater with acetate (R_{Ac}) and glucose (R_{Glu}) as the electron donor sources, respectively, which contributed the same COD as that in DW. In addition to the electron donor sources, synthetic wastewater comrpised AO7 (200 mg/L) and other minerals, as reported previously [16].

2.2. Chemicals and Analytical Method

The liquid samples were collected every day. AO 7 was used as the model azo dye (purity > 95%, Shanghai Sangon Biotech Co., Ltd., Shanghai, China). Liquid samples taken from reactors were immediately filtered through the 0.45 μm filters (Tianjin Jinteng Experiment Equipment Co., Ltd., Tianjin, China). AO7 concentration was quantified by a UV-Vis spectrophotometer (UV-1800, Shanghai Meipuda instrument Co., Ltd., Shanghai, China) at a wavelength of 484 nm. COD was quantified by the HACH method.

2.3. Biofilm Sampling and High-Throughput 16S rRNA Gene Illumina MiSeq Sequencing

After operating those reactors for more than 200 days, biofilms along with graphite particles were collected from anodes and cathodes, respectively. Graphite particles of each electrode were sampled from at least 6 positions in different heights and combined for DNA extraction.

Amplicon libraries were constructed by Illumina Miseq PE 250 using universal primers 515F (5′-GTGCCAGCMGCCGCGGTAA-3′) and 806R (5′-GGACTACHVGGGTWTCTAAT-3′). Both forward and reverse primers were added with barcodes. PCR amplification, products purification and quantification, and sequencing were carried out by using the Illumina MiSeq platform in The Beijing Genomics Institute (BGI).

3. Results and Discussion

3.1. Performance of BESs Served with Different Electron Donor Source

The BESs fed with domestic wastewater (R_{DW}), glucose (R_{Glu}), and acetate (R_{Ac}) presented similar and commendable decolorization efficiency (>98%) with azo dye AO 7 loading rates increased from 200 to 800 $g/(m^3 \cdot d)$. The COD consumptions among the three reactors were significantly different despite influent COD concentrations being controlled at a parallel level. The highest COD removal efficiency of R_{Ac} was 39.89 ± 1.95% followed by R_{Glu} of 34.01 ± 2.20%, and lowest COD removal efficiency was recorded in R_{DW} (19.56 ± 3.21%). Given the different yet similar electrons from COD oxidation used for azo dye decolorization, R_{DW} showed the highest electrons utilization efficiency. In R_{Glu} and R_{Ac}, more electrons were lost in the unwanted parallel routes, such as biomass production and methanogenesis. By decreasing electron donor concentrations from 300 to 80 mg-COD/L, the decolorization efficiencies of R_{Glu} and R_{Ac} obviously deteriorated to 85.85 ± 2.33% and 72.41 ± 1.37%, respectively, which was still kept at 94.91 ± 1.55% in R_{DW}. The preferable performance of R_{DW} under lower electron donation conditions emphasized that DW could serve as a cost-efficient electron donor source to drive BES for implementing the decolorization of azo dye [14].

3.2. Overall Microbial Community Structures

Illumina sequencing platform was employed to analyze microbial communities from 12 electrode biofilms of three reactors, the nomenclature and alpha diversity were shown in Table 1. Approximate 30,000 high-quality sequences were obtained from each sample (Figure 2). Significantly more operational taxonomic units (OTUs) were observed in samples fed with DW, followed by glucose and acetate, respectively. It indicated that microbial diversity highly depended on the complexity of the electron donor source. The estimated species richness abundance-based coverage estimator (ACE) and Chao1 were also consistent with this trend, which was significantly higher in R_{DW} samples than that in R_{Ac} and R_{Glu} ($p < 0.05$). It implied that more species were found in R_{DW}. Shannon index was employed to evaluate the evenness of the samples and reflected that the microbial communities from R_{DW} showed better evenness. It was also confirmed by the lowest Simpson index obtained in R_{DW} samples.

Principal component analysis (PCA) demonstrated that samples from the same reactor clustered together and presented high similarity of the microbial communities (Figure 3). It is reasonable considering the process of varied electron donor source utilization. Acetate is the simplest one and can be directly utilized as an electron donor. Glucose would be fermented into volatile fatty acids and then be utilized by microbes. DW metabolic processes involve hydrolysis, acidogenesis, acetogenesis, and methanogenesis, which are operated by diverse consortia.

Table 1. Nomenclature and alpha diversity of 12 microbial communities.

Mark	Sample Name	Number of Sequences	OTU	Chao 1	ACE	Shannon	Simpson
C1	R_{DW} down-cathode	29,514	649	699	715	4.29	0.04
C2	R_{DW} down-anode	28,786	625	662	682	4.23	0.04
C3	R_{DW} up-cathode	31,281	539	592	609	3.67	0.07
C4	R_{DW} up-anode	30,224	577	629	644	3.79	0.06
C5	R_{Ac} down-cathode	33,033	335	356	373	2.80	0.17
C6	R_{Ac} down-anode	34,760	346	400	408	3.10	0.11
C7	R_{Ac} up-cathode	35,182	398	468	455	3.12	0.15
C8	R_{Ac} up-anode	33,810	351	388	415	3.30	0.08
C9	R_{Glu} down-cathode	32,182	376	421	438	3.23	0.13
C10	R_{Glu} down-anode	30,776	387	464	469	3.63	0.06
C11	R_{Glu} up-cathode	31,687	354	383	393	3.30	0.12
C12	R_{Glu} up-anode	31,257	402	463	483	3.29	0.10

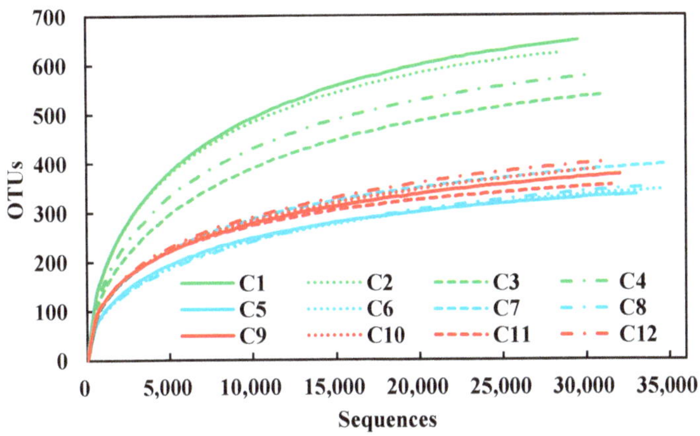

Figure 2. Rarefaction curves of the microbial communities.

Hierarchical cluster analysis was used to identify the differences of 12 microbial communities, as shown in Figure 4. Overall, microbial communities from R_{Ac} and R_{Glu} were relatively similar and separated from the R_{DW} group, suggesting clear distinctions of community structure between the complex feeding reactor and simple one. This was consistent with PCA results. In R_{Ac} and R_{Glu}, the microbial communities from the same polarity of electrode biofilm clustered together even though they were located in a different position. It indicated that the electrode polarity was the major influence factor on the microbial community structures from electrode biofilms in R_{Ac} and R_{Glu}. It was different in R_{DW} where position rather than polarity seems to be the dominant influence factor on the microbial communities. It can be deduced that the electron donor source was a dominant factor that influenced the microbial community structure of electrode biofilms. Simple carbon source (NaAc and glucose) was more likely to be utilized by electricigens, which led to a decrease in diversity and an increase in similarity of biofilm microbial community structure. While serving a complex carbon source (DW), electron production requires a long and complicated metabolic pathway constructed by a more diversified microbial community structure.

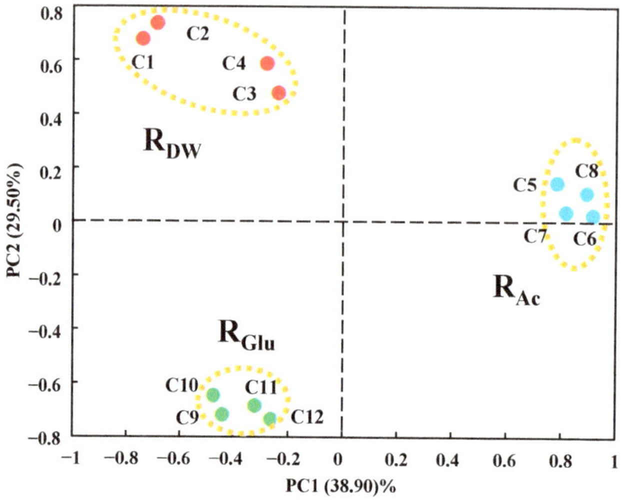

Figure 3. Principal component analysis (PCA) of microbial communities from different electron donor sources feeding BESs.

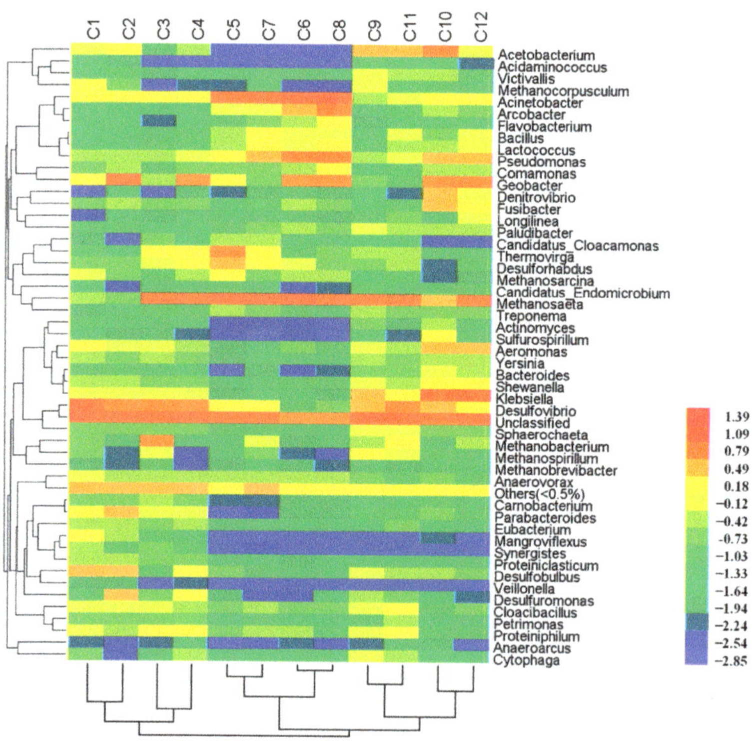

Figure 4. Hierarchical cluster analysis of microbial communities 12 biofilm samples.

3.3. Microbial Community Structures at Phylum, Class, and Family Levels

Microbial community structures from 12 samples were compared at phylum, class, and family levels (Figure 5). As it can be seen, the carbon source significantly altered the microbial community structure. A total of 11 phyla with relative abundance >1% were identified as shown in Figure 5A. Bacteroidetes, Firmicutes, and Proteobacteria were the dominant member. Bacteroidetes was the most enriched phylum in R_{DW} with an average abundance of 31.92 ± 3.31%, which was followed by 7.15 ± 1.99% and 11.52 ± 3.22% in R_{Ac} and R_{Glu}, respectively. Bacteroidetes were widely reported in BESs [17]. Firmicutes were enriched in R_{Glu} with a relative abundance of 32.65 ± 7.18%. Firmicutes were frequently identified from BES that fed glucose as a carbon source [18,19]. Proteobacteria were the absolute dominant phylum in R_{Ac} (47.22 ± 18.13%), which were 29.69 ± 11.45% and 29.67 ± 17.59% in R_{DW} and R_{Glu}, respectively. In addition, Synergistota was the other relative enriched phylum with the abundances of 3.96 ± 0.98, 2.64 ± 2.84, and 1.31 ± 1.12% in R_{DW}, $R_{Ac,}$ and R_{Glu}, respectively. It is worth noting that Synergistota seems to be enriched in the anode biofilm. Especially in R_{Ac} and R_{Glu} that feed with a simple carbon source, the relative abundance of Synergistota in the anode biofilm was one order of magnitude higher than the corresponding cathode. It seems to imply that the phylum of Synergistetes was relevant to the anodic respiration and extracellular electron transfer; however, there was no solid evidence to confirm this, and a deep investigation is required for illumination. A total of 18 classes were identified from 12 biofilm samples (relative abundance >1%, as shown in Figure 5B). Δ-proteobacteria with the relative abundance of 23.23 ± 11.12% was the dominant member and followed by γ-proteobacteria (5.48 ± 1.06%) in biofilms from R_{DW}, and γ-proteobacteria was the dominant class in R_{Ac} (33.46 ± 12.95%). While Bacilli, γ- and δ-proteobacteria were found to be the dominant member in biofilms in R_{Glu} with a relative abundance of 22.40 ± 9.67%, 18.23 ± 13.33%, and 9.89 ± 2.93%, respectively.

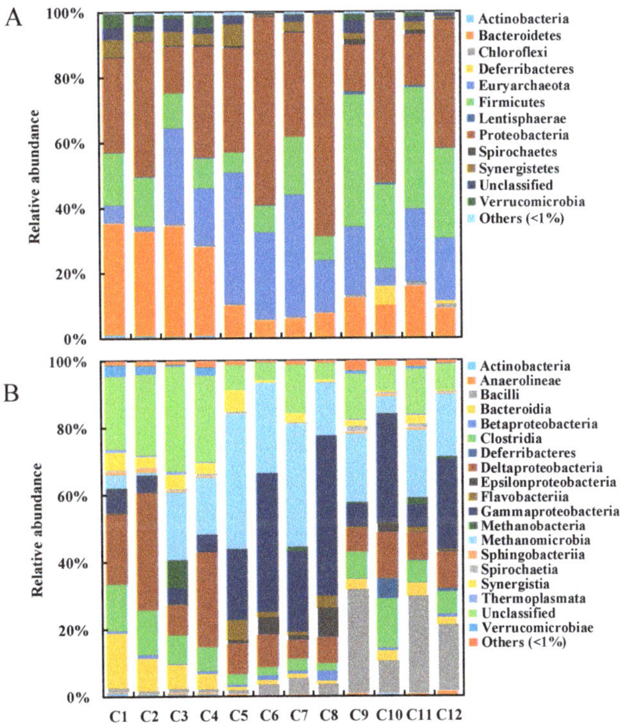

Figure 5. Microbial community comparison at phylum (**A**) and class (**B**) level.

3.4. Potential Function of Dominant Genera

An in-depth characterization of microbial community was performed at a genus level. Thirty-one genera with relative abundance >1% in the twelve microbial communities are shown in Table 2. *Geobacter*, a typical exoelectrogen, was found to be the dominant member in all anodic biofilms with the relative abundance in a range of 6.75% and 14.15%, which was higher than their corresponding cathodic biofilms. *Geobacter sulfurreducens* was the first reported strain in which microbial electricity production occurred solely by cells attached to an electrode [20]. In addition, the azo dye reduction capability of *Geobacter* has also been discovered [21].

Table 2. Microbial community comparison at genus level.

Genus	C1	C2	C3	C4	C5	C6	C7	C8	C9	C10	C11	C12
Methanosaeta	0.70	0.27	**19.00**	**16.46**	**37.25**	**26.26**	**35.83**	**15.43**	**17.29**	4.84	**18.94**	**18.51**
Acinetobacter	1.53	1.67	1.68	2.03	**18.76**	**32.89**	**18.26**	**38.51**	0.74	2.57	1.38	2.50
Pseudomonas	0.55	0.48	0.51	0.86	2.05	**7.80**	**5.21**	**8.35**	0.59	4.99	1.46	3.54
Geobacter	1.65	**14.15**	0.46	**11.84**	1.71	**8.35**	0.35	**6.75**	0.26	**8.02**	0.12	**7.68**
Arcobacter	0.04	0.08	0.04	0.08	1.09	**5.14**	1.23	**8.76**	0.07	0.75	0.11	0.32
Thermovirga	0.16	0.03	2.16	1.86	**6.34**	0.16	1.98	0.29	0.47	0.09	0.52	0.07
Desulforhabdus	0.19	0.32	1.49	1.73	4.90	0.76	2.70	0.28	0.31	0.01	0.39	0.14
Desulfovibrio	**13.96**	**11.99**	**6.30**	**8.31**	1.88	0.16	2.00	0.26	4.71	4.84	**6.50**	2.23
Desulfobulbus	4.34	3.82	0.15	2.16	0.29	0.07	0.18	0.07	1.00	0.55	0.51	0.36
Methanobacterium	0.26	0.04	**8.07**	0.03	0.05	0.02	1.04	0.04	0.43	0.07	1.81	0.45
Parabacteroides	2.19	3.51	0.88	1.32	-	0.02	0.01	0.01	0.15	0.36	0.10	0.13
Desulfuromonas	0.12	3.93	0.36	2.90	0.02	0.00	0.01	0.04	0.48	0.05	0.24	0.01
Klebsiella	1.24	0.82	1.07	0.78	0.02	0.04	0.04	0.01	3.24	**17.43**	2.44	**15.33**
Acetobacterium	1.30	1.65	0.17	0.51	-	-	0.00	-	4.87	**7.27**	3.28	1.22
Aeromonas	2.52	1.04	0.82	0.69	0.02	0.21	0.05	0.13	0.89	4.03	0.62	3.16
Denitrovibrio	-	0.02	-	0.02	0.01	0.03	0.01	0.02	0.04	5.94	0.01	1.01
Fusibacter	0.31	0.54	0.29	0.20	0.04	0.70	0.17	0.21	0.08	3.41	0.12	2.94
Proteiniphilum	1.00	0.52	2.35	1.86	0.30	0.43	0.48	0.13	1.01	0.18	2.22	0.17
Cloacibacillus	1.40	1.18	1.04	0.57	0.19	0.32	0.71	0.20	1.17	0.16	1.91	0.23
Mangroviflexus	2.58	0.39	0.10	0.06	-	-	-	-	-	0.01	-	-
Anaerovorax	1.22	0.56	0.73	0.51	0.52	0.38	0.60	0.48	0.23	0.19	0.14	0.12
Methanocorpusculum	2.02	0.29	-	0.01	0.01	0.00	0.04	0.00	1.12	0.10	0.18	0.09
Methanosarcina	1.05	0.20	0.17	0.63	2.76	0.49	0.29	0.56	0.06	0.01	0.10	0.02
Shewanella	0.53	0.50	0.32	0.22	0.03	0.14	0.09	0.13	0.80	2.38	0.62	1.50
Comamonas	0.31	0.43	0.13	0.31	0.32	0.56	0.26	2.08	0.08	0.49	0.11	0.65
Methanospirillum	0.24	0.01	0.89	0.00	0.04	0.01	0.25	0.00	1.82	0.19	0.97	0.06
Lactococcus	0.09	0.07	0.10	0.15	0.47	1.44	2.11	1.62	0.14	0.41	0.60	0.93
Bacteroides	0.10	0.15	0.03	0.03	-	0.00	0.02	0.01	0.29	1.16	0.21	0.65
Bacillus	0.04	0.05	0.04	0.13	0.42	1.31	2.00	1.44	0.06	0.34	0.79	1.03
Sulfurospirillum	0.07	0.02	0.02	0.01	0.00	0.00	-	-	0.03	1.59	0.01	0.09
Flavobacterium	0.04	0.03	0.01	0.03	0.12	0.69	0.33	2.89	0.02	0.08	0.03	0.06
Unclassified	49.70	44.25	43.67	36.64	15.93	8.01	18.17	7.63	49.76	21.83	47.64	28.93
Others # (<1%)	8.53	6.99	6.95	7.09	4.48	3.59	5.58	3.66	7.80	5.69	5.89	5.85

Dominant genera (>5%) in electrode biofilms were bolded; #: genus relative abundance less than 1% were classified as "Others"; "0.00": genus that identified yet low relative abundance; "-": genus that were undetected in electrode biofilms.

In the R_{DW} group, *Desulfovibrio* was enriched (13.96, 11.99, 6.30, and 8.31%) compared to R_{Ac} and R_{Glu}. *Desulfovibrio* has been proved as a functional genus for extracellular electron transfer (EET) and azo dye reduction [22,23]. It was interesting that the *Desulfovibrio* relative abundance of the microbial community was more sensitive to the electrode position rather than polarity, which was selectively enriched in the biofilms collected from down position electrodes. It is reasonable considering that this genus can grow with a broad range and as complex organic matters with electron donors [24], and it is adaptable to the complex composition of DW.

Acinetobacter is non-fermentive and capable of growth in mineral media with acetate as the sole carbon source [25]. It seems to be the reason that *Acinetobacter* was the dominant member in all samples from the R_{Ac} group. *Acinetobacter* was selectively enriched in anodic biofilm with relative abundances of 32.89 and 38.51%, which decreased to 18.76 and 18.26% in cathodic ones. It strongly implied that *Acinetobacter* was involved in electrode respiration; however, we failed to find out an explicit report to confirm the direct link between

Acinetobacter and EET. *Acinetobacter* presented function in azo dye decolorization [26]. *Pseudomonas*, a reported exoelectrogen as well as an azo dye decolorizing genus, was selected enriched in anodic biofilms (7.80% and 8.35%) versus cathodes (2.05% and 5.21%), which can produce phenazine-based metabolites to stimulate EET [27,28]. In addition, *Arcobacter* was another exoelectrogen that had high relative abundances in R_{Ac} electrodes (1.09%, 5.14%, 1.23%, and 8.76%). It has been verified that they were neither fermented nor oxidized carbohydrates, yet organic and amino acids could be utilized as carbon sources [29,30].

In the R_{Glu} group, *Klebsiella* was the dominant member with relative abundances of 17.43% and 15.33% in anodes and one order of magnitude lower (3.24% and 2.44%) in cathodes. It is a functional genus of EET and azo dye reduction [31,32]. *Acetobacterium* was an enriched genus in R_{Glu} group, with relative abundances of 4.87%, 7.27%, 3.28%, and 1.22%. *Acetobacterium* was not proved for its function in EET or azo dye reduction; however, it was an acetogenic microbe and can produce acetate by oxidizing glucose [33].

In terms of archaea, *Methanosaeta* was the predominant methane producer that specializes in only using acetate for methane production. No growth on or methane production from H_2/CO_2, formate, methanol, ethanol, trimethylamine, isobutanol, or isopropanol [34] was observed. This explains the enrichment of *Methanosaeta* on the R_{Ac} biofilms (37.25%, 26.26%, 35.83%, and 15.43%). The relative abundances of *Methanosaeta* were lower in the R_{DW} and R_{Glu} groups and relatively enriched in the down electrode biofilms. Both DW and glucose should be fermented in acetate to create a favorable condition for *Methanosaeta* metabolism so that they are relatively enriched on the up electrodes. The recent literature indicated *Methanosaeta* is able to acquire electrons from exoelectrogen (e.g., *Geobacter*) by direct interspecies electron transfer process [35,36]. In addition, *Methanosaeta* was found to be a dominant member in a microbial community from an anaerobic baffled reactor treating dyes wastewater, which implied it could tolerate the toxicity of the dyes and possibly be involved in dye removal. Other methane producers, *Methanobacterium*, *Methanocorpusculum*, *Methanosarcina*, and *Methanospirillum*, were found but with quite lower relative abundances in 12 biofilms.

Canonical correspondence analysis (CCA) was employed to evaluate the effect of electron donor source types on the microbial community structures, as shown in Figure 6. The arrow direction of acetate was positively correlated with the horizontal canonical axis, and DW and glucose were located in the middle of the second and third quadrant of the cartesian coordinate. The separation of three electron donor sources indicated the relatively independent selective pressure. *Mangroviflexus* stood in the DW direction and far away from the origin, and its enrichment was highly sensitive to DW. It was a dominant genus in the rural household biogas digesters and played a crucial role in reducing and oxidizing reactions to bio-degrade organic matters [37]. *Parabacteroides* was an obligately anaerobic species from wastewater of a paper mill [38]. *Desulfuromonas* was able to oxidize complex organics such as long-chain fatty acids and use tetrachloroethylene, trichloroethylene, and Fe(III) as electron acceptors [39,40]. *Klebsiella* had a function of EET from glucose oxidation [41]; thus, it was enriched on the anodes from R_{Glu}. *Acinetobacter*, *Arcobacter*, and *Flavobacterium* were highly sensitive to the acetate.

The comparisons of the dominant genera that were identified from BESs fed with various electron donor sources are summarized in Table S1 [42–45]. It is quite clear that the genera with a function of extracellular electron transfer, such as *Geobacter*, *Pseudomonas*, *Desulfovibrio*, and *Klebsiella*, were enriched in the electrode microbial community. Those genera seemed to endow an electrochemical activity of the electrode biofilms. Specific genus was enriched based on the characteristics of the substrate. *Comamonadaceae* was enriched in a BES with phenol as the substrate [45], and *Klebsiella* was enriched in this work with domestic wastewater as the electron donor.

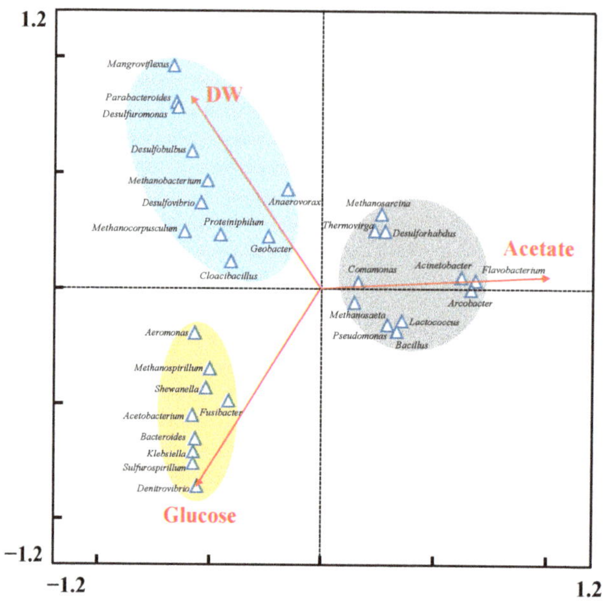

Figure 6. Canonical correspondence analysis for electron donor sources and functional genera from electrode microbial communities.

4. Conclusions

Employing domestic wastewater as electron donor source to drive bioelectrochemical reaction was feasible and cost-efficient. In this study, the electrode microbial communities from bioelectrochemical systems fed with domestic wastewater (R_{DW}), acetate (R_{Ac}), and glucose (R_{Glu}) were systematically revealed. Similar decolorization performances were observed among three BESs, while the microbial community structures were quite different. Microbial diversity highly depended on the complexity of the electron donor source, and significantly more OTUs were observed in R_{DW} compared to R_{Glu} and R_{Ac}. Bacteroidetes, Firmicutes, and Proteobacteria were identified as the dominated phyla in R_{DW}, R_{Glu}, and R_{Ac}, respectively. *Desulfovibrio*, *Acinetobacter*, and *Klebsiella* were identified as the predominant bacterial genera in R_{DW}, R_{Ac}, and R_{Glu}, respectively. Typical exoelectrogen *Geobacter* was found to be enriched in the anodes' biofilms among three BES reactors. *Methanosaeta* was the most enriched methanogen in all reactors. This study provides an insight in the microbial response mechanism to the electron donor sources and provides great inspiration to bring bioelectrochemical technology closer to applications.

Supplementary Materials: The following supporting information can be downloaded at: https://www.mdpi.com/article/10.3390/w14091505/s1, Table S1: Comparison of dominant genera fed with various electron donor sources.

Author Contributions: Z.G.: Investigation, funding acquisition, and writing—review and editing. L.Z.: Investigation and writing—original draft. M.-H.C.: Conceptualization, funding acquisition, and writing—review and editing. A.W.: Project administration and supervision. All authors have read and agreed to the published version of the manuscript.

Funding: This research was funded by the Natural Science Foundation of Jiangsu Province, No. BK20190980; the National Natural Science Foundation of China, No. 52000090 and 52000088; the China Postdoctoral Science Foundation, No. 2021M701511; the Open Project of Key Laboratory of Environmental Biotechnology, CAS, Grant No. kf2020010.

Institutional Review Board Statement: Not applicable.

Informed Consent Statement: Not applicable.

Data Availability Statement: Not applicable.

Conflicts of Interest: The authors declare no conflict of interest.

References

1. Suprun, E.V.; Radko, S.P.; Khmeleva, S.A.; Mitkevich, V.A.; Archakov, A.I.; Makarov, A.A.; Shumyantseva, V.V. Electrochemical oxidation of amyloid-beta peptide isoforms on carbon screen printed electrodes. *Electrochem. Commun.* **2017**, *75*, 33–37. [CrossRef]
2. Kong, D.; Liang, B.; Yun, H.; Cheng, H.; Ma, J.; Cui, M.; Wang, A.; Ren, N. Cathodic degradation of antibiotics: Characterization and pathway analysis. *Water Res.* **2015**, *72*, 281–292. [CrossRef] [PubMed]
3. Cui, M.-H.; Liu, W.-Z.; Tang, Z.-E.; Cui, D. Recent advancements in azo dye decolorization in bio-electrochemical systems (BESs): Insights into decolorization mechanism and practical application. *Water Res.* **2021**, *203*, 117512. [CrossRef]
4. Cui, M.-H.; Cui, D.; Gao, L.; Cheng, H.-Y.; Wang, A.-J. Analysis of electrode microbial communities in an up-flow bioelectrochemical system treating azo dye wastewater. *Electrochim. Acta* **2016**, *220*, 252–257. [CrossRef]
5. Tang, C.-C.; Zhang, X.-Y.; Wang, R.; Wang, T.-Y.; He, Z.-W.; Wang, X.C. Calcium ions-effect on performance, growth and extracellular nature of microalgal-bacterial symbiosis system treating wastewater. *Environ. Res.* **2022**, *207*, 112228. [CrossRef] [PubMed]
6. Zieliński, M.; Dębowski, M.; Kazimierowicz, J. Microwave Radiation Influence on Dairy Waste Anaerobic Digestion in a Multi-Section Hybrid Anaerobic Reactor (M-SHAR). *Processes* **2021**, *9*, 1772. [CrossRef]
7. Westerholm, M.; Crauwels, S.; Van Geel, M.; Dewil, R.; Lievens, B.; Appels, L. Microwave and ultrasound pre-treatments influence microbial community structure and digester performance in anaerobic digestion of waste activated sludge. *Appl. Microbiol. Biotechnol.* **2016**, *100*, 5339–5352. [CrossRef]
8. Dębowski, M.; Zieliński, M.; Kisielewska, M.; Kazimierowicz, J. Evaluation of Anaerobic Digestion of Dairy Wastewater in an Innovative Multi-Section Horizontal Flow Reactor. *Energies* **2020**, *13*, 2392. [CrossRef]
9. Lyu, W.; Song, Q.; Shi, J.; Wang, H.; Wang, B.; Hu, X. Weak magnetic field affected microbial communities and function in the A/O/A sequencing batch reactors for enhanced aerobic granulation. *Sep. Purif. Technol.* **2021**, *266*, 118537. [CrossRef]
10. Zhao, B.; Sha, H.; Li, J.; Cao, S.; Wang, G.; Yang, Y. Static magnetic field enhanced methane production via stimulating the growth and composition of microbial community. *J. Clean. Prod.* **2020**, *271*, 122664. [CrossRef]
11. Daghio, M.; Gandolfi, I.; Bestetti, G.; Franzetti, A.; Guerrini, E.; Cristiani, P. Anodic and cathodic microbial communities in single chamber microbial fuel cells. *New Biotechnol.* **2015**, *32*, 79–84. [CrossRef] [PubMed]
12. Zhai, S.; Ji, M.; Zhao, Y.; Pavlostathis, S.G.; Zhao, Q. Effects of salinity and COD/N on denitrification and bacterial community in dicyclic-type electrode based biofilm reactor. *Chemosphere* **2017**, *192*, 328–336. [CrossRef] [PubMed]
13. Zhu, X.; Yates, M.D.; Hatzell, M.C.; Rao, H.A.; Saikaly, P.E.; Logan, B.E. Microbial Community Composition Is Unaffected by Anode Potential. *Environ. Sci. Technol.* **2013**, *48*, 1352–1358. [CrossRef] [PubMed]
14. Cui, M.-H.; Cui, D.; Gao, L.; Wang, A.-J.; Cheng, H.-Y. Azo dye decolorization in an up-flow bioelectrochemical reactor with domestic wastewater as a cost-effective yet highly efficient electron donor source. *Water Res.* **2016**, *105*, 520–526. [CrossRef]
15. Mu, Y.; Rabaey, K.; Rozendal, R.A.; Yuan, Z.; Keller, J. Decolorization of Azo Dyes in Bioelectrochemical Systems. *Environ. Sci. Technol.* **2009**, *43*, 5137–5143. [CrossRef]
16. Cheng, H.-Y.; Liang, B.; Mu, Y.; Cui, M.-H.; Li, K.; Wu, W.; Wang, A.-J. Stimulation of oxygen to bioanode for energy recovery from recalcitrant organic matter aniline in microbial fuel cells (MFCs). *Water Res.* **2015**, *81*, 72–83. [CrossRef]
17. Quan, X.-C.; Quan, Y.-P.; Tao, K. Effect of anode aeration on the performance and microbial community of an air–cathode microbial fuel cell. *Chem. Eng. J.* **2012**, *210*, 150–156. [CrossRef]
18. Liang, B.; Cheng, H.; Van Nostrand, J.; Ma, J.; Yu, H.; Kong, D.; Liu, W.; Ren, N.; Wu, L.; Wang, A.; et al. Microbial community structure and function of Nitrobenzene reduction biocathode in response to carbon source switchover. *Water Res.* **2014**, *54*, 137–148. [CrossRef]
19. Cui, M.-H.; Cui, D.; Lee, H.-S.; Liang, B.; Wang, A.-J.; Cheng, H.-Y. Effect of electrode position on azo dye removal in an up-flow hybrid anaerobic digestion reactor with built-in bioelectrochemical system. *Sci. Rep.* **2016**, *6*, 25223. [CrossRef]
20. Bond, D.R.; Lovley, D.R. Electricity Production by *Geobacter sulfurreducens* Attached to Electrodes. *Appl. Environ. Microbiol.* **2003**, *69*, 1548–1555. [CrossRef]
21. Liu, G.; Zhou, J.; Chen, C.; Wang, J.; Jin, R.; Lv, H. Decolorization of azo dyes by *Geobacter metallireducens*. *Appl. Microbiol. Biotechnol.* **2012**, *97*, 7935–7942. [CrossRef] [PubMed]
22. Kang, C.S.; Eaktasang, N.; Kwon, D.-Y.; Kim, H.S. Enhanced current production by *Desulfovibrio desulfuricans* biofilm in a mediator-less microbial fuel cell. *Bioresour. Technol.* **2014**, *165*, 27–30. [CrossRef]
23. Kim, S.-Y.; An, J.-Y.; Kim, B.-W. Improvement of the decolorization of azo dye by anaerobic sludge bioaugmented with *Desulfovibrio desulfuricans*. *Biotechnol. Bioprocess Eng.* **2007**, *12*, 222–227. [CrossRef]
24. Sass, H.; Ramamoorthy, S.; Yarwood, C.; Langner, H.; Schumann, P.; Kroppenstedt, R.M.; Spring, S.; Rosenzweig, R.F. *Desulfovibrio idahonensis* sp. nov., sulfate-reducing bacteria isolated from a metal(loid)-contaminated freshwater sediment. *Int. J. Syst. Evol. Microbiol.* **2009**, *59*, 2208–2214. [CrossRef] [PubMed]

25. Li, W.; Zhang, D.; Huang, X.; Qin, W. *Acinetobacter harbinensis* sp. nov., isolated from river water. *Int. J. Syst. Evol. Microbiol.* **2014**, *64 Pt 5*, 1507–1513. [CrossRef] [PubMed]
26. Cai, Z.; Zhang, W.; Ma, J.; Cai, J.; Li, S.; Zhu, X.; Yang, G.; Zhao, X. Biodegradation of Azo Dye Disperse Orange S-RL by a Newly Isolated Strain *Acinetobacter* sp. SRL8. *Water Environ. Res.* **2015**, *87*, 516–523. [CrossRef]
27. Pham, T.H.; Boon, N.; De Maeyer, K.; Höfte, M.; Rabaey, K.; Verstraete, W. Use of Pseudomonas species producing phenazine-based metabolites in the anodes of microbial fuel cells to improve electricity generation. *Appl. Microbiol. Biotechnol.* **2008**, *80*, 985–993. [CrossRef]
28. Qiao, Y.; Qiao, Y.-J.; Zou, L.; Ma, C.-X.; Liu, J. Real-time monitoring of phenazines excretion in Pseudomonas aeruginosa microbial fuel cell anode using cavity microelectrodes. *Bioresour. Technol.* **2015**, *198*, 1–6. [CrossRef]
29. Vandamme, P.; Falsen, E.; Rossau, R.; Hoste, B.; Segers, P.; Tytgat, R.; De Ley, J. Revision of *Campylobacter*, *Helicobacter*, and *Wolinella* Taxonomy: Emendation of Generic Descriptions and Proposal of *Arcobacter* gen. nov. *Int. J. Syst. Bacteriol.* **1991**, *41*, 88–103. [CrossRef]
30. Toh, H.; Sharma, V.K.; Oshima, K.; Kondo, S.; Hattori, M.; Ward, F.B.; Free, A.; Taylor, T.D. Complete Genome Sequences of *Arcobacter butzleri* ED-1 and *Arcobacter* sp. Strain L, Both Isolated from a Microbial Fuel Cell. *J. Bacteriol.* **2011**, *193*, 6411–6412. [CrossRef]
31. Lee, Y.-Y.; Kim, T.G.; Cho, K.-S. Enhancement of electricity production in a mediatorless air–cathode microbial fuel cell using *Klebsiella* sp. IR21. *Bioprocess Biosyst. Eng.* **2016**, *39*, 1005–1014. [CrossRef] [PubMed]
32. Elizalde-González, M.; Fuentes, L.; Guevara-Villa, M. Degradation of immobilized azo dyes by *Klebsiella* sp. UAP-b5 isolated from maize bioadsorbent. *J. Hazard. Mater.* **2009**, *161*, 769–774. [CrossRef] [PubMed]
33. Balch, W.E.; Schoberth, S.; Tanner, R.S.; Wolfe, R.S. *Acetobacterium*, a New Genus of Hydrogen-Oxidizing, Carbon Dioxide-Reducing, Anaerobic Bacteria. *Int. J. Syst. Bacteriol.* **1977**, *27*, 355–361. [CrossRef]
34. Ma, K. *Methanosaeta harundinacea* sp. nov., a novel acetate-scavenging methanogen isolated from a UASB reactor. *Int. J. Syst. Evol. Microbiol.* **2006**, *56 Pt 1*, 127–131. [CrossRef]
35. Liu, F.; Rotaru, A.-E.; Shrestha, P.M.; Malvankar, N.S.; Nevin, K.P.; Lovley, D.R. Promoting direct interspecies electron transfer with activated carbon. *Energy Environ. Sci.* **2012**, *5*, 8982–8989. [CrossRef]
36. Rotaru, A.-E.; Shrestha, P.M.; Liu, F.; Shrestha, M.; Shrestha, D.; Embree, M.; Zengler, K.; Wardman, C.; Nevin, K.P.; Lovley, D.R. A new model for electron flow during anaerobic digestion: Direct interspecies electron transfer to *Methanosaeta* for the reduction of carbon dioxide to methane. *Energy Environ. Sci.* **2013**, *7*, 408–415. [CrossRef]
37. Han, R.; Yuan, Y.; Cao, Q.; Li, Q.; Chen, L.; Zhu, D.; Liu, D. PCR–DGGE Analysis on Microbial Community Structure of Rural Household Biogas Digesters in Qinghai Plateau. *Curr. Microbiol.* **2017**, *75*, 541–549. [CrossRef]
38. Tan, H.-Q.; Li, T.-T.; Zhu, C.; Zhang, X.-Q.; Wu, M.; Zhu, X.-F. *Parabacteroides chartae* sp. nov., an obligately anaerobic species from wastewater of a paper mill. *Int. J. Syst. Evol. Microbiol.* **2012**, *62 Pt 11*, 2613–2617. [CrossRef]
39. Coates, J.D.; Lonergan, D.J.; Philips, E.J.P.; Jenter, H.; Lovley, D.R. *Desulfuromonas palmitatis* sp. nov., a marine dissimilatory Fe(III) reducer that can oxidize long-chain fatty acids. *Arch. Microbiol.* **1995**, *164*, 406–413. [CrossRef]
40. Krumholz, L.R. *Desulfuromonas chloroethenica* sp. nov. Uses Tetrachloroethylene and Trichloroethylene as Electron Acceptors. *Int. J. Syst. Bacteriol.* **1997**, *47*, 1262–1263. [CrossRef]
41. Xia, X.; Cao, X.-X.; Liang, P.; Huang, X.; Yang, S.-P.; Zhao, G.-G. Electricity generation from glucose by a *Klebsiella* sp. in microbial fuel cells. *Appl. Microbiol. Biotechnol.* **2010**, *87*, 383–390. [CrossRef] [PubMed]
42. Yu, Y.-Y.; Zhen, S.-H.; Chao, S.-L.; Wu, J.; Cheng, L.; Li, S.-W.; Xiao, X.; Zhou, X. Electrochemistry of newly isolated Gram-positive bacteria Paenibacillus lautus with starch as sole carbon source. *Electrochim. Acta* **2022**, *411*, 140068. [CrossRef]
43. Mai, Q.; Yang, G.; Cao, J.; Zhang, X.; Zhuang, L. Stratified microbial structure and activity within anode biofilm during electrochemically assisted brewery wastewater treatment. *Biotechnol. Bioeng.* **2020**, *117*, 2023–2031. [CrossRef] [PubMed]
44. Wang, Y.; Pan, Y.; Zhu, T.; Wang, A.; Lu, Y.; Lv, L.; Zhang, K.; Li, Z. Enhanced performance and microbial community analysis of bioelectrochemical system integrated with bio-contact oxidation reactor for treatment of wastewater containing azo dye. *Sci. Total Environ.* **2018**, *634*, 616–627. [CrossRef] [PubMed]
45. Yang, L.-H.; Cheng, H.-Y.; Ding, Y.-C.; Su, S.-G.; Wang, B.; Zeng, R.; Sharif, H.M.A.; Wang, A.-J. Enhanced treatment of coal gasification wastewater in a membraneless sleeve-type bioelectrochemical system. *Bioelectrochemistry* **2019**, *129*, 154–161. [CrossRef]

water

MDPI

Article

Mechanism and Kinetic Analysis of the Degradation of Atrazine by O_3/H_2O_2

Yixin Lu [1,2,3,4], Chenghan Tang [3], Yujie Liu [3] and Jiao Chen [1,2,3,*]

[1] School of Materials and Environmental Engineering, Chengdu Technological University, Chengdu 611730, China; yxlu61@163.com
[2] State Environmental Protection Key Laboratory of Synergetic Control and Joint Remediation for Soil & Water Pollution, Chengdu 610059, China
[3] Faculty of Geosciences and Environmental Engineering, Southwest Jiao Tong University, Chengdu 611756, China; 15201822120@163.com (C.T.); lyjenter.163.com (Y.L.)
[4] Sichuan Haitianshenghuan Technology Co., Ltd., Chengdu 610200, China
* Correspondence: chjicn@foxmail.com

Abstract: In phosphate buffer, the degradation of ATZ by ozone/(O_3/H_2O_2) under various circumstance was explored and the degradation mechanism and dynamics were probed. The findings revealed that when maintaining the reaction temperature at 25 °C, the H_2O_2 concentration and the O_3 concentration were 20 mol/L and 20 mol/L, respectively. Moreover, the degradation rate of 5 mol/L ATZ under the influence of O_3/H_2O_2 was 92.59% in phosphate buffer at pH7. The mechanism analysis showed that HO• and O_3 underwent co-oxidized degradation and that the HO• and O_3 oxidation degradation ratios were close to 1:1 under acidic conditions. Furthermore, HO• oxidative degradation dominated the ATZ degradation process. The kinetics analysis showed that the ATZ kinetics of O_3/H_2O_2 degradation were more compatible with quasi-second-order reaction kinetics under different temperatures, pH values, and H_2O_2 concentrations.

Keywords: ozone; hydrogen peroxide; atrazine; degradation mechanism; kinetics

1. Introduction

Atrazine (ATZ, common molecular formula: $C_8H_{14}ClN_5$) is now the world's second most extensive herbicide, with cheap cost, low toxicity, and good weeding properties [1,2]. Since it is considered to be one of the most polluting pesticides, seven European countries, including Germany, France and Sweden, banned the use of atrazine in 2007 [3]. Atrazine consumption is reported to be between 80,000 tons on average per year worldwide [4]. Nevertheless, China's yearly atrazine consumption has been estimated to be greater than 108 tons [5]. Although not very toxic, ATZ is structurally stable and is often found in water resources due to its long half-life (30–100 days) [6]. A survey has shown that the residual concentration of atrazine in most bodies of water in China is 1–20 μg/L, and the mass concentration of atrazine detected in some lakes and streams in many EU countries is between 0.1–30.3 μg/L [7]. Buster et al. investigated 18 lakes in Sweden, and the test results show that atrazine exists in the waters, and the highest mass concentration reaches 4 μg/L [8]. ATZ can migrate and be transformed within different environmental media, such as water, atmosphere, and soil. Overland runoff, lixiviating, and dry and wet deposition are the most typical methods by which ATZ reaches the aquatic environment [9,10].

The harmful effects of ATZ on various organisms in nature are mainly manifested in the interference of ATZ in the normal operation of the endocrine system and its causing organisms to produce toxic reactions. From the viewpoint of Sun et al. [11], ATZ with a concentration of 10 mg/L or above inhibited the germination rate of rice seeds. Fu et al. [12] found that an ATZ concentration of 0.125 mg/cm^2 inhibited the embryonic development of the red-eared turtle. The research of Bellona et al. [13] suggests that ATZ has considerable effects on the growth of the mature F1 mouse brain during a vital and sensitive stage, resulting in a change in behavioral development trajectories. Stradtman et al. [14] reported that atrazine use in the United States and the rest of the world constitutes a substantial threat to life, health, and the environment due to its neurotoxicity properties and to its being an endocrine-disrupting compound and a reproductive toxin. Sagarkar et al. [15] studied how a battery of nuclear and mitochondrial-encoded genes involved in oxidative phosphorylation (OXPHOS) were affected by short-term atrazine treatment in human liver (HepG2) and rat muscle (L6) cell lines. The results showed that the EC50 amount of atrazine for mitotoxicity in HepG2 and L6 cells was around 0.162 and 0.089 mM, respectively. Mitochondrial dysfunction could be introduced by atrazine induction. According to the study by Marcus et al. [16], as a result of exposure to a dose of atrazine, the proportion of pups that emerged, the proportion of adults that emerged, and adult survival all decreased. Atrazine sped up the development of the flies, causing them to pupate and emerge earlier than the controls. Therefore, the residual ATZ in the environment has attracted the attention of many scholars [17,18].

In recent years, studies have found that SO_4^{-} (2.5~3.1 V) has a higher redox potential than HO• (2.3~2.7 V), since it can efficiently remove organic pollutants and it is less affected by the water quality environment. Therefore, persulfate (PS) has attracted the attention of many researchers. The most important of its forms are peroxymonosulfate (PMS) and peroxodisulfate (PDS). Phosphate is an excellent buffer in the degradation of ATZ by O_3/PMS. In addition, pH has a significant effect on the degradation of ATZ by O_3/PMS. The degradation efficiency under alkaline conditions is much better than that under acidic conditions. Phosphate can stimulate PMS to produce SO_4^{-}, which shows that the excitation effect under acidic conditions is better than that under alkaline conditions. Furthermore, the phosphate system alone has no degradation effect on ATZ. Studies have shown that O_3 has a better degradation effect on ATZ in the phosphate system, and PMS alone has a degradation effect on ATZ in the phosphate buffer system. Under the same conditions, using NaOH to adjust the pH, PMS has no degradation effect on ATZ [19].

At the moment, various ozone (O_3)-based AOPs have been shown to be able to effectively degrade ATZ in water, including ozonation [20], ferrosilite/O_3 [21], UV/O_3/ultrasound [22], peroxide [23], photo-fenton and photo-fenton-like processes [24], and photocatalysis [25]. Among these technologies, the UV/O_3-based ATZ water treatment method is more eco-friendly and more suitable for many kinds of wastewater [20]. Wen et al. indicated that the UV/O_3/US process can effectively degrade atrazine in manufacturing wastewater [26]. However, there are few reports on the degradation of ATZ by O_3/H_2O_2, even though some studies have shown that the O_3/H_2O_2 system has been used in the treatment of water pollution and that it has shown excellent degradation ability for pollutants in actual treatment [27,28]. Therefore, this paper investigates the effect of O_3/H_2O_2 on ATZ in a phosphate buffer system under different conditions. Through the obtained experimental results, the mechanism of ATZ degradation in the system and the kinetic model of ATZ degradation are discussed. This study thereby contributes to the further enrichment of the chemical processing technology of ATZ.

2. Materials and Methods

2.1. Reagents and Instruments

Reagents: All reagents used in the experiments were of analytical grade unless otherwise specified: methyl alcohol for chromatography, caustic soda, sodium phosphate monobasic, sodium nitrite, TBA (tertiary butanol), hydrogen peroxide ($H_2O_2 \geq 30\%$), ATZ purchased from Aladdin Co., Ltd. (Shanghai, China), and oxygen with a purity greater than 99%.

Instruments: HPLC (Waters 2695–2996); electronic balance; laboratory pH-meter of Shanghai Lei Magnetic Co., Ltd. (Shanghai, China) (PHSJ-3F); KH5 200 DB CNC ultrasonic cleaner; Ultrapure Water Polishing System from ULUPURE Co., Ltd. (Chengdu, China); intelligent constant temperature bath (DC-1030); constant temperature magnetic stirrer

(78HW-1); and a laboratory O_3 generator from Harbin Jiujiu Electrochemical Engineering Technology Co., Ltd. (Harbin, China) (DHX-SS-1G).

2.2. Experimental Scheme

2.2.1. Solution Preparation

In this experiment, ultrapure water (resistivity 18.24 MΩ·cm) was used to prepare ATZ mother liquor with a concentration of 100 μmol/L, 0.2 mol/L of NaH_2PO4 solution, 0.2 mol/L of NaOH solution, 0.01 mol/L of $NaNO_2$ solution, 0.01 mol/L of H_2O_2 solution, and 8 g/L of tert-butanol solution. Table 1 shows the preparation methods for the pH6, pH7, and pH8 phosphate buffers (both to 1 L). The phosphate buffer (PB) solution was used to adjust the pH of the reaction, so that the pH value was constant at 7 (except for the single factor of pH). We compared the pH of the reaction system to 6.26 and 6.58 after the reaction was completed when the initial pH of the reaction system was adjusted by NaOH to 7 and 8, respectively. The main reason for the pH change was that the initiation and participation of O_3 by OH- decomposes the chain reaction, and the chain reaction consumes the amount of OH- in the reaction system, causing the pH of the reaction system to drop and become acidic. The chain reaction is then terminated, and PB has the function of maintaining the pH of the solution.

Table 1. The preparation method for the NaH_2PO_4-NaOH buffer.

pH	0.2 mol/L NaH_2PO_4-NaOH (mL)	0.2 mol/L NaOH (mL)
6	250	28.50
7	250	148.15
8	250	244.00

2.2.2. Experimental Scheme of ATZ Degradation by O_3/H_2O_2

The degradation efficiency of O_3/H_2O_2 for ATZ in a 12.5 mmol/L phosphate buffer solution was investigated under the following conditions: different H_2O_2 concentrations of 5, 10, 15, and 20 μmol/L; pH values of 6, 7, and 8; and different temperatures of 10, 15, 20, and 25 °C. In a water bath at 20 °C, the concentration of O_3 was 20 μmol/L, the concentration of phosphate was 12.5 mmol/L, the concentration of H_2O_2 was 20 μmol/L, and the concentration of ATZ was 5 μmol/L. The mechanism of the degradation of ATZ by O_3/H_2O_2 was investigated by adding different concentrations of tert-butanol and adjusting the pH. The reaction was stopped using a 0.01 mol/L solution of $NaNO_2$.

In the actual experiment, three parallel samples were set for each group of experiments, and the experimental results were taken as the average of three parallel experiments.

2.3. Analytical Method

2.3.1. Liquid Chromatography Analysis

To detect ATZ, the Symmetry® C18 StableBond was used, and the particular test procedure was as follows: a 60:40 methyl alcohol to ultrapure water mobile phase ratio, a flow velocity of 0.8 mL/min, operating temperatures of 40 °C, and a determinate wavelength of 225 nm. O_3 concentration was determined through the UV_{258} method [29].

2.3.2. Kinetic Analysis of ATZ Degradation by O_3/H_2O_2

The pseudo-first-order kinetic equation was first introduced by Lagergren, and it is usually used in the form proposed by Ho and McKay [30].

$$Ln(q_e - q_t) = Lnq_e - K_1 t \tag{1}$$

Combine formula:

$$q_e = \frac{(C_{0-}C_e)V}{m} \tag{2}$$

Deduced:

$$\frac{C_t - C_e}{C_O - C_e} = e^{-K_1 t} \tag{3}$$

Assuming that ATZ can be completely degraded in the O_3/H_2O_2 system, that is, that the concentration of ATZ in the solution is 0 (Ce = 0) at equilibrium, we can obtain:

$$\frac{C_t}{C_O} = e^{-K_1 t} \tag{4}$$

Therefore, the kinetic model of the O_3 oxidative degradation of ATZ is established according to the following kinetic equation, and the first-order reaction kinetic equation is:

$$Ln\frac{C_t}{C_O} = -K_1 t \tag{5}$$

where C/C_0 represents the surplus rate of ATZ in the reactor, and the elimination rate of ATZ in the reactor $\beta = 1 - C/C_0$; thus, $C/C_0 = 1 - \beta$.

A linear second-order rate equation was then introduced [31]:

$$\frac{1}{C_t} = K_2 t + \frac{1}{C_0} \tag{6}$$

Substitute $C/C_0 = 1 - \beta$ into Equation (2) to obtain the following result:

$$\frac{1}{(1-\beta)} = K_2 C_0 t \tag{7}$$

q_e: Adsorption capacity value at equilibrium, mg/g;
q_t: Adsorption capacity value at time t, mg/g;
K_1: Pseudo-first-order rate constant, 1/min;
K_2: Pseudo-second-order rate constant, L/(μmol·min);
β: Represents the elimination rate of ATZ;
m: Mass of the adsorbent, g;
C_0: Initial concentration of ATZ;
C_e: Reaction ATZ concentration at equilibrium;
C_t: ATZ concentration at any time, mg/L.

3. Results and Discussion

3.1. Effect of Temperature on ATZ Degradation by O_3/H_2O_2

When the concentrations of ATZ, O_3, and H_2O_2 were 5 μmol/L, 20μmol/L, and 20 μmol/L, respectively, the impact of the temperature changes on ATZ decomposition by O_3/H_2O_2 in phosphate buffer at pH7 is illustrated in Figure 1. According to Figure 1, the effect of ATZ degradation by O_3/H_2O_2 was enhanced when the temperature of the reaction increased. This was mainly due to the increase in the temperature: the percentage of activated O_3 increased and the decomposition of O_3 was accelerated to generate HO•. At the same time, increasing the temperature speeds up the movement of molecules, thereby increasing the collision frequency between ATZ and O_3 and HO•, thereby accelerating the degradation rate. The ATZ removal rate increased from 83.96 to 95.52% when the temperature of the reaction system was gradually adjusted from 10 °C to 25 °C, and although there was no obvious difference in the removal effect, the reaction rate was considerably different. The degradation rate of ATZ at 25 °C changed slowly after five minutes, and basically tended to be stable. At five minutes, the degradation rate of ATZ by O_3/H_2O_2 quickly reached 92.59%. The elimination rate of ATZ decreased slightly at 20 °C compared to the other temperatures and increased at 25 °C. The reason for this is that in the ozonation process, on the one hand, increasing the temperature can increase the rate constants of activated molecules and chemical reaction rates, thereby improving the degradation efficiency. On the other hand, the increase in temperature reduces the solubility of ozone in aqueous solution, resulting in a lower degradation efficiency. When discussing the utilization efficiency of ozone at different temperatures, Lei Zhao et al. [32] mentioned that in a separate ozone system, with the increase in temperature, the utilization of ozone first increased, then gradually became stable, and finally dropped. At the same time, the presence of heterogeneous catalysts can reduce the negative effects of decreasing ozone solubility in water with increasing temperature, resulting in an increase in ozone utilization, and contributing to the production of •OH during ozonolysis [33]. Therefore, in the combined result of the two opposite effects, the removal rate of ATZ by O_3/H_2O_2 decreased at 20 °C. At 25 °C, due to the increase in the temperature of the reaction solution, the chemical reaction and mass transfer rate were accelerated, and more •OH was generated, thereby increasing the removal rate of ATZ. It is easy to observe that when the temperature increased from 15 °C to 20 °C, the removal rate of ATZ was significantly improved, indicating that the temperature change in the normal temperature range has a greater impact on the degradation of ATZ by O_3.

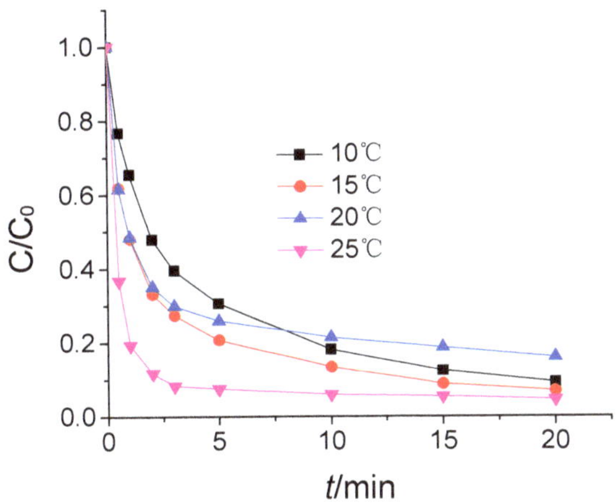

Figure 1. The ATZ removal rate under different temperatures.

In addition, with the progress of the reaction, different temperatures caused different dissolved concentrations of O_3 in the system, resulting in different contents of HO•; the removal of ATZ in aqueous solution was also gradually different. Within the same time period, the increase in the reaction temperature can speed up the reaction rate of ATZ with O_3, HO•, and other oxidants, but at the same time it will accelerate the ineffective decomposition of O_3 and reduce the solubility of O_3 in water; it is thus not conducive to the degradation of ATZ in solution. These two contradictory reactions lead to the process of degrading ATZ by O_3/H_2O_2, and the removal rate of ATZ showed a trend of increasing first and then leveling off with the increase in reaction temperature. Within a certain range of conditions, the increase in temperature promotes the improvement of the reaction rate, which is beneficial to the degradation of ATZ in the O_3/H_2O_2 system [34]. Therefore, under the experimental conditions, the optimal temperature for degrading ATZ by O_3/H_2O_2 is 25 °C.

3.2. The Effect of H_2O_2 Concentration on ATZ Degradation by O_3/H_2O_2

In Figure 2, the effect of the amount of H_2O_2 in the reaction system concentration on the degradation of ATZ by O_3/H_2O_2 is shown. As shown in the figure, there was no obvious change in ATZ degradation by O_3/H_2O_2 when the quantity of H_2O_2 in the reaction process rose. This is mainly because a small amount of deprotonated H_2O_2 could give a strong boost to O_3 to generate more HO• [35], whose reaction rate with ATZ (2.54×10^9/Ms [36]) is far greater than that of O_3 with ATZ. Therefore, even a small amount of H_2O_2 input could result in distinct ATZ degradation when the O_3 concentration maintains a certain value in the O_3/H_2O_2 system, indicating that the O_3/H_2O_2 system is suitable for the treatment of high concentrations of ATZ solution. The ATZ removal rate reached its highest level at 92.92% when the H_2O_2 concentration was 10μmol/L in the system; that is, ATZ degradation by O_3/H_2O_2 achieved its highest efficiency when the ratio of nH_2O_2 to nO_3 was 1:2.

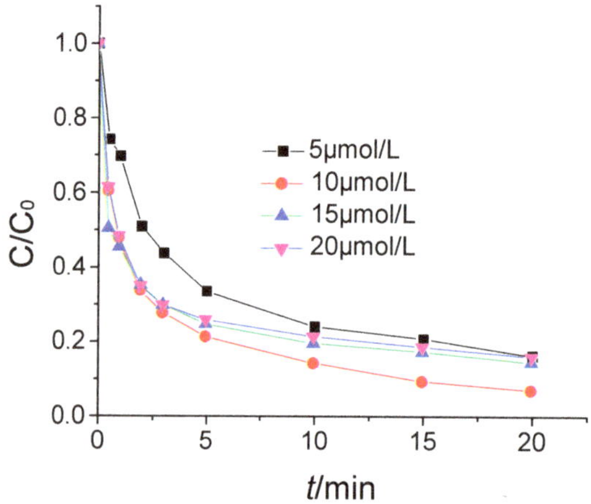

Figure 2. The ATZ removal rate under different H_2O_2 densities.

When only O_3 is used for oxidation, the degradation process for organic pollutants usually includes two aspects: one is the direct oxidation of pollutants by molecular O_3, and the other is the indirect oxidation of organic pollutants by HO• produced by O_3 conversion [37]. When the O_3/H_2O_2 system is used to degrade ATZ, there is not only the direct oxidation of ATZ by O_3 and H_2O_2, but more importantly, due to the existence of H_2O_2, H_2O_2 is rapidly dissociated to form HO_2^-, which can react with O_3 to promote its rapid and efficient production. A large amount of HO• is generated, which in turn strengthens the indirect oxidation of ATZ [38–40]. Therefore, compared with O_3 oxidation alone, the use of an O_3/H_2O_2 advanced oxidation process can greatly improve the ability to generate HO•. The reaction is shown in Formulas (4)–(8) [41,42].

$$H_2O_2 \rightarrow HO_2^- + H^+ \quad (8)$$

$$O_3 + OH^- \rightarrow HO_2^- + O_2 \quad (9)$$

$$O_3 + HO_2^- \rightarrow HO\cdot + O_2^- + O_2 \quad (10)$$

$$O_3 + O_2^- \rightarrow O_3^- + O_2 \quad (11)$$

$$O_3^- + H_2O \rightarrow HO\cdot + OH^- + O_2 \quad (12)$$

When the concentration of H_2O_2 is low, the excitation effect of HO• is poor, which is not conducive to the improvement of ATZ removal efficiency. However, excess H_2O_2 molecules can react quickly with HO•. The reaction formula is: H_2O_2+HO•→HO_2• + H_2O. The presence of excess H_2O_2 will thus also reduce the degradation efficiency of HO• for ATZ. In addition, the high concentration of H_2O_2 will also cause unnecessary waste and the risk of secondary pollution. Therefore, choosing the appropriate concentration of H_2O_2 is of great practical significance for improving the efficiency of ATZ degradation in the O_3/H_2O_2 system. When using the O_3/H_2O_2 oxidation process to treat wastewater, the optimal concentration ratio of O_3 and H_2O_2 should be determined through practical experiments [43,44].

3.3. The Effect of pH Value on ATZ Degradation by O_3/H_2O_2

In phosphate buffer solution at 20 °C, the degradation results of ATZ at different pH values were provided in Figure 3. The quantity of ATZ, O_3, and H_2O_2 were 5 μmol/L, 20 μmol/L, and 20 μmol/L, respectively. From the figure, the results of the decomposition of ATZ by O_3/H_2O_2 showed a rapid enhancement as the pH value increased in the reaction process. When the pH was raised from 6 to 8, the clearance rate of ATZ improved from 52.95 to 89.03%. After gradually adjusting the pH value from 6 to 7, the ratio of the decomposition amount of ATZ to the initial amount increased significantly from 52.95 to 83.96%. Furthermore, the ratio of the decomposition amount of ATZ increased from 83.96 to 89.03% when the pH value was adjusted from 7 to 8; the removal rate did not rise significantly, but the reaction rate increased notably. The removal rate reached 89% at 2 min under pH8, then tended to be stable. When the pH of the reaction system increased from 6 to 8, the removal rate of ATZ was significantly enhanced. This was because when the pH was acidic, O_3 mainly relied on selective direct oxidation to degrade the target, and the oxidation ability was low; on the other hand, when the pH is alkaline, O_3 mainly relies on the chain reaction to generate non-selective strong oxidizing HO• to oxidatively degrade the target, and HO• has a significantly stronger oxidative ability for ATZ than O_3. It can be seen that the pH of the reaction system has a very significant effect on the degradation of ATZ by O_3. Yang et al. [45] investigated the ozonation-induced degradation of atrazine in the presence of hydroxylamine, and discovered that pH had a substantial impact on atrazine degradation.

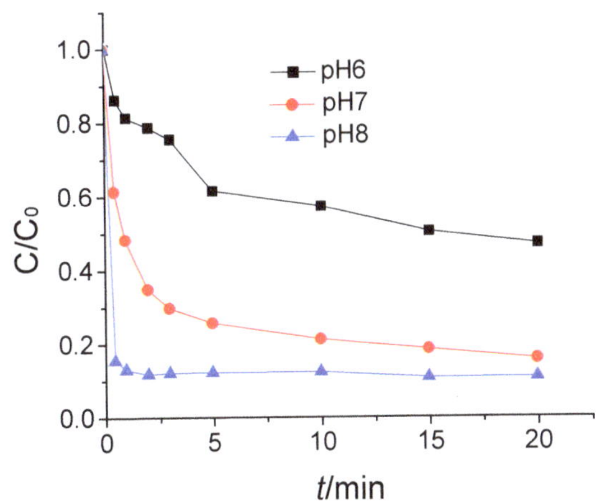

Figure 3. The ATZ removal rate under different pH values.

The dissociation of H_2O_2 is a key step in the chain reaction of the O_3/H_2O_2 reaction system. The conjugate base HO_2^- is the active substance in the reaction, and its concentration is closely related to the pH value. Under acidic conditions, the reaction rate of H_2O_2 and O_3 slowed down, and when the pH value was greater than 5, the decomposition rate of O_3 and H_2O_2 increased rapidly. With the increase in pH value, the amount of HO• produced by the combined action of H_2O_2 and O_3 gradually increased, which triggered a free-radical chain reaction, more fully oxidized ATZ, and an improved degradation effect. Under higher pH conditions, the initiator HO_2^-, which can effectively promote the decomposition of O_3, can be dissociated even in a lower concentration of H_2O_2 solution [46,47].

Feng et al. [48] used the O_3/H_2O_2 process to treat the mother liquor of gas field wastewater, and found that pH had the most critical effect on the treatment effect. The best operating conditions can be obtained when the pH value is 10.9, the flow rate of O_3 is 0.8 L/min, and the amount of H_2O_2 is 6.2 mL. Ochir et al. [49] used the O_3/H_2O_2 system to treat tebuconazole, carbendazim, and pyrimethanil in turn with an initial concentration of 2 μmol/L. When the reaction temperature was 25 °C, the reaction time was 0.5 h, and the pH value increased from 5 to 9; the removal rates of tebuconazole, carbendazim, and pyrimethanil also increased from 17.0 ± 1.0%, 66.4 ± 2.2%, and 90.9 ± 1.6% to 30.3 ± 2.0%, 80.5 ± 1.4%, and 99.5 ± 0.4%, respectively. A higher pH value is not always better. Farzaneh et al. [50] studied the effect of pH value on the removal of ibuprofen by the O_3/H_2O_2 advanced oxidation process. After 5 min of ozone oxidation at pH7.6, the residual ozone concentration was analyzed and found to have increased from 1.5 mg/L drops to about 0.15 mg/L, which means that at pH 7.6 and above, ozone is almost completely decomposed within 5 min, and other compounds will quickly consume the free radicals. Therefore, even at higher pH, HO• is not sufficient for achieving greater ibuprofen removal. From the existing research, for different water bodies or different pollutants, the pH value required by O_3/H_2O_2 is different, but most of them are in the range of 6~12.

3.4. Mechanism Analysis of ATZ Degradation by O_3/H_2O_2

3.4.1. Mechanism Analysis of ATZ Degradation by O_3/H_2O_2 in Phosphate Buffer at pH6

In phosphate buffer at pH6, different amounts of tertiary butyl alcohol were added to the O_3 alone and O_3/H_2O_2 systems to capture HO• in the reaction system, so as to investigate the O_3/H_2O_2 degradation mechanism of ATZ [51,52]. The results are shown in Figure 4. It was found that H_2O_2 alone had virtually no degradation effect on ATZ. The degradation amount of ATZ by O_3 alone was basically the same when the tertiary butyl concentration was 8 and 16 mg/L, respectively, proving that all HO• in the O_3 alone system can be captured when the concentration of tertiary butyl is 8 mg/L.

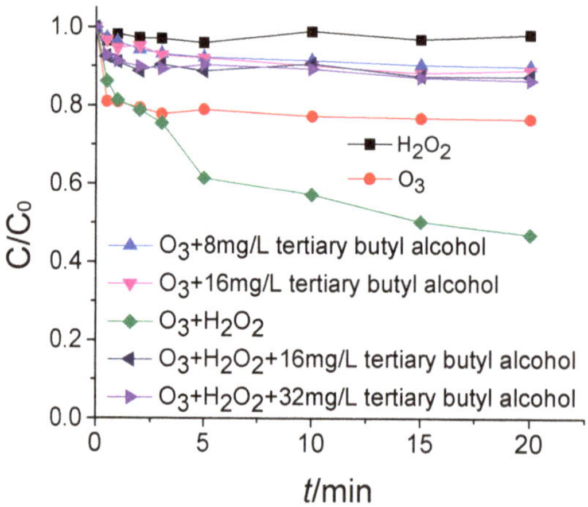

Figure 4. The effect of free radical scavenger on the degradation of ATZ by O_3/H_2O_2 in pH6 phosphate buffer.

The percentage of ATZ elimination dropped from 23.38% to 11% when excessive tertiary butanol was added to the O_3 reaction system, indicating that only 10% of ATZ in the system was directly oxidatively removed by O_3. The amount of ATZ directly oxidized by O_3 accounted for 47.05% of the total removal rate, while ATZ degraded by HO• made up the other 52.95%, showing that O_3 and HO• each accounted for half of the ATZ degradation in phosphate buffer at pH6.

The degradation effect of O_3/H_2O_2 on ATZ was basically the same when the tertiary butanol concentration was 16 and 32 mg/L in the O_3/H_2O_2 reaction system, revealing that all the HO• in the system was captured when the concentration of tertiary butanol was 16 mg/L. The percentage of ATZ elimination dropped from 52.95% to around 13% when excessive amounts of tertiary butanol were added to the O_3/H_2O_2 system, indicating that only 13% of ATZ was directly oxidized by O_3. The O_3 oxidation accounted for 24.55% of the overall ATZ removal rate, whereas HO• degradation accounted for the remaining 75.45%, indicating that HO• was superior for degrading ATZ in phosphate buffer at pH6.

The ATZ degradation efficiency by O_3 alone and H_2O_2 alone was 23.38% and 2%, respectively, in phosphate buffer at pH6, while the degradation efficiency of O_3/H_2O_2 was 27.57% higher than the sum of the two, proving that O_3 and H_2O_2 can excite each other to produce more HO• than each would produce separately.

3.4.2. Mechanism Analysis of ATZ Degradation by O_3/H_2O_2 in Phosphate Buffer at pH7

In the phosphate buffer at pH7, different amounts of tertiary butyl alcohol were added to the O_3 alone and O_3/H_2O_2 systems to capture HO• in the reaction system, so as to investigate the O_3/H_2O_2 degradation mechanism of ATZ. The results are shown in Figure 5. According to early research, when the concentration of tertiary butyl in the O_3 alone environment was 16 mg/L, all of the HO• in the system could be collected.

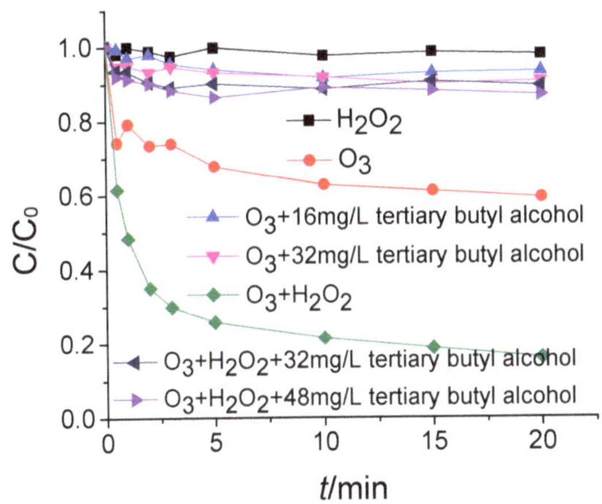

Figure 5. The effect of free radical scavenger on the degradation of ATZ by O_3/H_2O_2 in pH7 phosphate buffer.

From Figure 5, the degradation rate of ATZ by O_3 alone decreased from 40.76% to 9.4% when the concentration of tertiary butanol was 16 mg/L in the system, suggesting that only 9.4% of ATZ was directly oxidized by O_3. The O_3 oxidation accounted for 23.06% of the total ATZ removal rate, while the amount of ATZ degraded by HO• accounted for the other 76.94%, reflecting the dominant role of HO• in the ATZ degradation of O_3 alone in phosphate buffer at pH 7.

The degradation effect of O_3/H_2O_2 on ATZ was basically the same when the tertiary butanol concentration was 32 and 48 mg/L in the O_3/H_2O_2 reaction system, revealing that all the HO• in the system was captured when the concentration of tertiary butanol was 32 mg/L. The rate of ATZ degradation fell from 83.96% to roughly 12% when excessive amounts of tertiary butanol were added to the O_3/H_2O_2 system, indicating that only 12% of ATZ was directly oxidized by O_3. The O_3 oxidation contributed a 14.29% degradation rate of ATZ, while HO• removal accounted for the remaining 85.71%. In phosphate buffer at pH7, O_3 and HO• both played a role in ATZ degradation and HO• played the leading role.

3.4.3. Mechanism Analysis of ATZ Degradation by O_3/H_2O_2 in Phosphate Buffer at pH8

In phosphate buffer at pH8, different amounts of tertiary butyl alcohol were added to the O_3 alone and O_3/H_2O_2 systems to capture HO• in the reaction system, so as to investigate the O_3/H_2O_2 degradation mechanism of ATZ. Figure 6 depicts the experimental consequences. According to prior tests, all the HO• in the system can be captured when the tertiary butyl concentration inside this O_3 alone system is 32 mg/L.

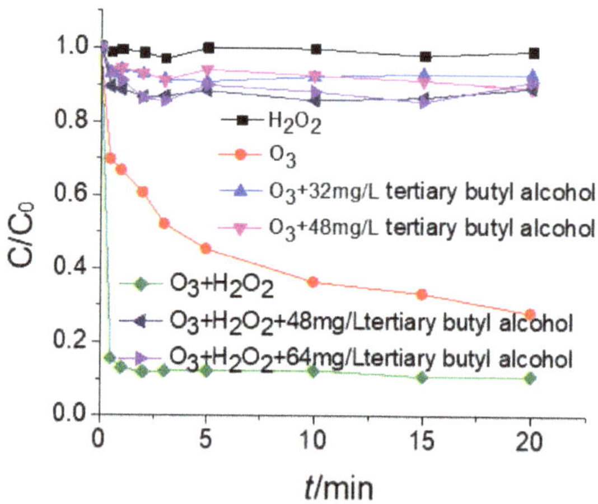

Figure 6. The influence of free radical scavenger on the degradation of ATZ by O_3/H_2O_2 in pH8 phosphate buffer.

From Figure 6, the degradation rate of ATZ by O_3 alone decreased from 71.83% to 8% when the tertiary butyl concentration inside this O_3 alone system was 32 mg/L, suggesting that only 8% of ATZ was directly oxidized by O_3. The O_3 oxidation accounted for 11.14% of the total ATZ removal rate, while the amount of ATZ degraded by HO• accounted for the other 88.86%, indicating the dominant role of HO• in the ATZ degradation of O_3 alone in phosphate buffer at pH8.

The degradation effect of O_3/H_2O_2 on ATZ was basically the same when the tertiary butanol concentration was 48 and 64 mg/L in the O_3/H_2O_2 reaction system, revealing that the HO• in the system was basically captured when the concentration of tertiary butanol was 48 mg/L. The removal efficiency of ATZ dropped from 89.03% to roughly 10% when excessive amounts of tertiary butanol were added to the O_3/H_2O_2 system, indicating that only 10% of ATZ was directly oxidized by O_3. The O_3 oxidation was responsible for a 11.23% ATZ elimination rate, while HO• degradation was responsible for the remaining 88.77%. In phosphate buffer at pH8, O_3 and HO• both played a role in ATZ degradation and HO• played the leading role [53,54].

3.4.4. Intermediate Products and Degradation Mechanism of ATZ Degradation

The results of the product analysis with HPLC-ESI-MS show that, as shown in Figure 7, the degradation pathways of ATZ mainly include the following four pathways: (I) Hydroxylation: During the degradation process of ATZ, the methyl group was replaced by a hydroxyl group, and the mass-to-charge ratio became $m/z = 218$; (II) De-isopropylation: Under the action of oxidant, the ATZ de-isopropyl group is generated to form de-isopropyl ATZ, and the mass-to-charge ratio becomes $m/z = 174$ and $m/z = 146$; (III) De-ethylation: During this process, ATZ removes the ethyl group to form de-ethylated ATZ, and the mass-to-charge ratio becomes $m/z = 188$ and $m/z = 146$; (IV) De-chlorination: ATZ removes Cl- and combines with -OH to form de-chlorinated ATZ, and the mass-to-charge ratio changes from $m/z = 216$ to $m/z = 198$ and $m/z = 156$. Therefore, the degradation pathways of ATZ by O_3/H_2O_2 mainly include hydroxylation, de-isopropylation, de-ethylation, and de-chlorination.

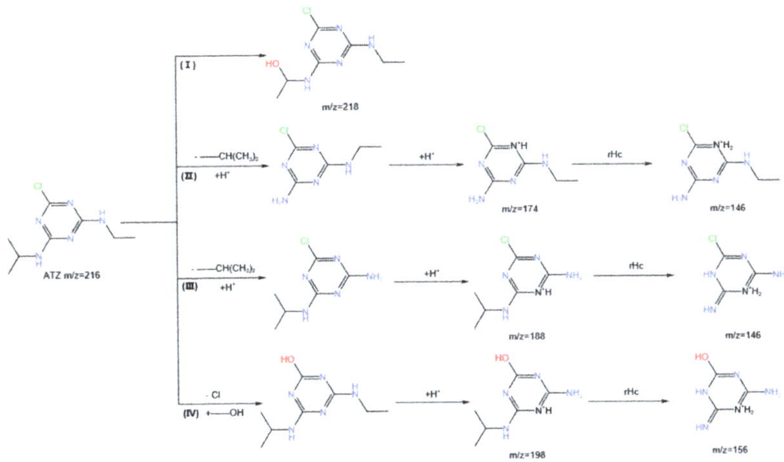

Figure 7. Schematic diagram of ATZ degradation pathway and degradation mechanism (**I**): hydroxylation; (**II**): de-isopropylation; (**III**): de-ethylation; (**IV**): de-chlorination).

K.H Chan et al. [55] indicated that ten intermediate products were found in the degradation of ATZ in Fenton's system of 0.1 mmol/L [Fe^{2+}] and [H_2O_2]. During this process, the main oxidant is HO•, and the presence of HO• triggers the degradation of ATZ through alkyl oxidation, dealkylation, and de-chlorination, and produces the corresponding intermediates. Yuchan Liu et al. [56] proposed that the main pathways for ATZ reduction in UV/PS systems include de-chlorination, demethylation, and de-ethylation to form various intermediates. The intermediates undergo dichlorination, hydroxylation and de-isopropylation to form secondary intermediates. Zeng-Hui Diao et al. [57] used LCQTOF-MS technology to analyze ATZ intermediates in the ATZ degradation process using B-nZVI/FeS_2/H_2O_2. Their results showed that in the degradation process, on the one hand, the degradation of ATZ is caused by the decomposition of the alkyl side chain of ATZ by HO•, which undergoes dealkylation to form an intermediate product. On the other hand, the corresponding intermediates are formed by dealkylation. The intermediate products are then dealkylated and hydroxylated to form secondary intermediates. At the same time, the authors also indicated that with the increase or decrease in the reaction time, the molecular weight of the intermediate product gradually decreased, indicating that the ATZ molecule can be effectively decomposed by HO• [57]. Due to the different degradation methods and operating conditions, the degradation pathways of ATZ are also different, but they all reflect that HO• plays a very important role in the process of degrading ATZ. The degradation pathway is mainly through hydroxylation, de-isopropylation, dealkylation and de-chlorination.

The whole reaction process can be simplified as:

$$\text{ATZ} + \text{HO}\bullet \rightarrow \text{Intermediates} \rightarrow CO_2 + H_20 \tag{13}$$

3.5. *Kinetic Analysis of ATZ Degradation by O_3/H_2O_2*

3.5.1. Kinetic Analysis of ATZ Degradation by O_3/H_2O_2 under Different Temperatures

When the concentrations of ATZ, O_3, and H_2O_2 were 5 mol/L, 20 mol/L, and 20 mol/L, respectively, in phosphate buffer solution at pH7, the quasi-first-order reaction kinetics of ATZ degradation by O_3/H_2O_2 at different temperatures were matched with $Ln(C/C_0)(y)$ as the Y-axis and $t(x)$ as the X-axis. At various temperatures, the quasi-second-order reaction kinetics of ATZ degradation by O_3/H_2O_2 were matched using $1/(1-\beta)(y)$ as the Y-axis and $t(x)$ as the X-axis. The parameters for the dynamic fitting equation are presented in Tables 2 and 3, and the dynamic matching curves are displayed in Figures 8 and 9.

Table 2. The kinetics equations and parameters of quasi-first-order reactions of O_3/H_2O_2 degradation of ATZ at various temperatures.

T(°C)	Fitted Equation	Reaction Order	K_1 (1/min)	R^2
10	Y = −0.11198x − 0.37774	First-order reaction	0.11198	0.91718
15	Y = −0.11688x − 0.63552	First-order reaction	0.11688	0.84758
20	Y = −0.06995x − 0.65722	First-order reaction	0.06995	0.65698
25	Y = −0.10641x − 1.42622	First-order reaction	0.10641	0.48264

Table 3. The kinetics equations and parameters of quasi-second-order reactions of O_3/H_2O_2 degradation of ATZ at different temperatures.

T(°C)	Fitted Equation	Reaction Order	K_1 (1/min)	R^2
10	Y = 0.48111x + 1.00352	Second-order reaction	0.48111	0.99749
15	Y = 0.6592x + 1.38581	Second-order reaction	0.65920	0.99585
20	Y = 0.23151x + 1.08379	Second-order reaction	0.23151	0.87706
25	Y = 0.9533x + 0.83282	Second-order reaction	0.95330	0.82373

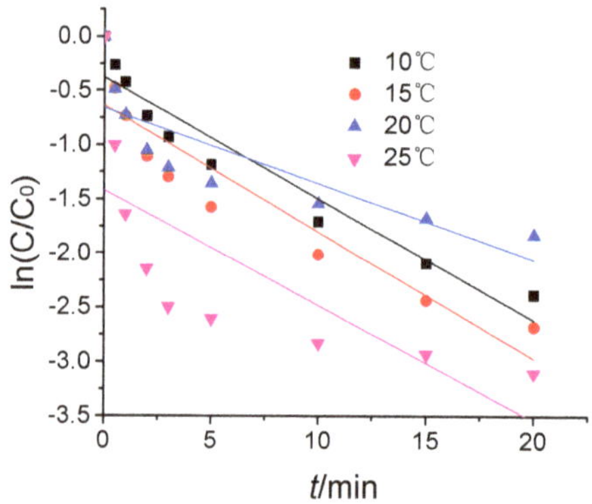

Figure 8. The kinetics of quasi−first−order reactions of O_3/H_2O_2 degradation of ATZ at various temperatures.

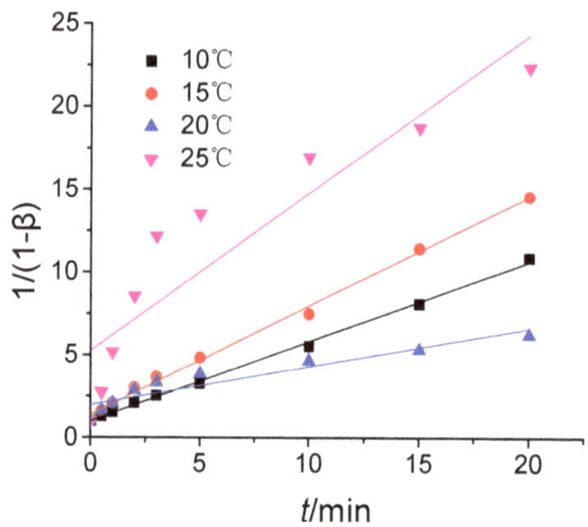

Figure 9. The kinetics of quasi−second−order reactions for ATZ degradation at various temperatures.

From Figures 8 and 9 and Tables 2 and 3, the values of R^2 (linear correlation) fitted by quasi-second-order reaction kinetics were all larger than those fitted by quasi-first-order reaction kinetics, indicating that the reaction kinetics of ATZ degradation by O_3/H_2O_2 at different temperatures are more consistent with quasi-second-order reaction kinetics [58]. The quasi-second-order reaction rate constant increased from 0.48111 to 0.95333 when the reaction temperature rose from 10 °C to 25 °C, with the lowest reaction rate at 20 °C and reaction rate constant at 0.23151. The highest reaction rate increased by 4.12 times compared with the lowest reaction rate. Since the R^2 fitted for the secondary reaction at 25 °C was only 0.82373, the linear correlation was poor, and the actual reaction rate cannot be accurately reflected.

3.5.2. Kinetic Analysis of ATZ Degradation by O_3/H_2O_2 under Different H_2O_2 Concentrations

In phosphate buffer solution at pH7, when the concentrations of ATZ, O_3, and temperature were 5 mol/L, 20 mol/L, and 20 °C, respectively, the quasi-first-order reaction kinetics of ATZ degradation by O_3/H_2O_2 at varied H_2O_2 concentrations were matched using $Ln(C/C_0)$ (y) as the Y-axis and $t(x)$ as the X-axis. The quasi-second-order reaction kinetics of ATZ degradation by O_3/H_2O_2 were matched under different H_2O_2 concentrations adopting $1/(1-\beta)$ (y) as the Y-axis and $t(x)$ as the X-axis. The factors for the dynamic fitting equation are presented in Tables 4 and 5, and the dynamics matching curves are displayed in Figures 10 and 11.

Table 4. The kinetics equations and parameters of quasi-first-order reactions of O_3/H_2O_2 degradation of ATZ at different H_2O_2 densities.

$C(H_2O_2)$ (μmol/L)	Fitted Equation	Reaction Order	K_1 (1/min)	R^2
5	Y = −0.08137x − 0.38598	First-order reaction	0.08137	0.84815
10	Y = −0.11383x − 0.630929	First-order reaction	0.11383	0.85080
15	Y = −0.07185x − 0.70582	First-order reaction	0.07185	0.68089
20	Y = −0.06995x − 0.65722	First-order reaction	0.06995	0.65698

Table 5. The kinetics equations and parameters of quasi-second-order reactions of O_3/H_2O_2 degradation of ATZ at different H_2O_2 densities.

$C(H_2O_2)$ (μmol/L)	Fitted Equation	Reaction Order	K_1 (1/min)	R^2
5	Y = 0.24548x + 1.36493	Second-order reaction	0.24548	0.97377
10	Y = 0.62450x + 1.388	Second-order reaction	0.62450	0.99343
15	Y = 0.25765x + 2.05714	Second-order reaction	0.25765	0.90920
20	Y = 0.23151x + 1.99153	Second-order reaction	0.23151	0.87706

From Figures 10 and 11 and Tables 4 and 5, the values of R^2 (linear correlation) fitted by quasi-second-order reaction kinetics were all larger than those fitted by quasi-first-order reaction kinetics, indicating that the reaction kinetics of ATZ degradation by O_3/H_2O_2 at various H_2O_2 concentrations were more compatible with the quasi-second-order reaction kinetics. The highest and lowest secondary reaction constant was 0.6245 and 0.23151 when the H_2O_2 concentration was 10 and 20 μmol/L, respectively, with the reaction rate increased by 2.70 times.

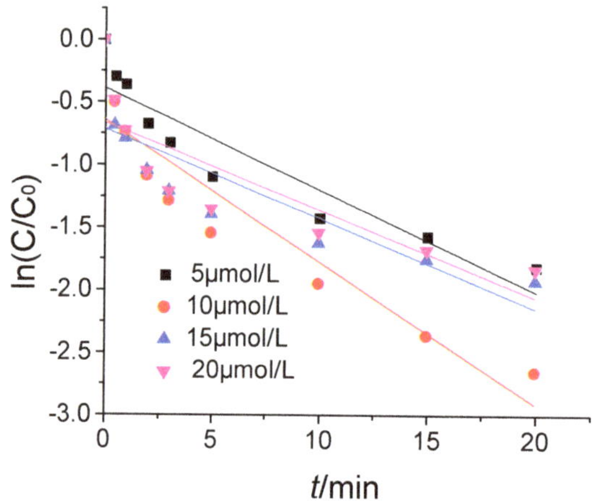

Figure 10. The kinetics of quasi−first−order reactions of O_3/H_2O_2 degradation of ATZ at different H_2O_2 densities.

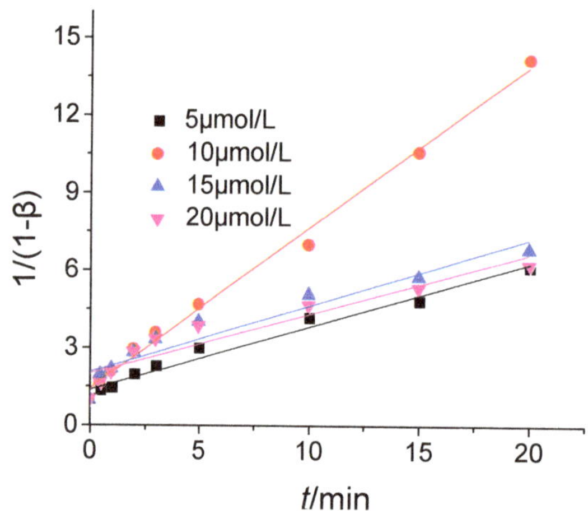

Figure 11. The kinetics of quasi−second−order reactions for ATZ degradation at different H_2O_2 densities.

3.5.3. Kinetic Analysis of ATZ Degradation by O_3/H_2O_2 at Different pH Values

In phosphate buffer solution at 20 °C, the quasi-first-order reaction kinetics of ATZ degradation by O_3/H_2O_2 under different pH values were matched utilizing *Ln(C/C*$_0$*) (y)* as the *Y*-axis and *t(x)* as the *X*-axis when the concentration of ATZ was 5, O_3 was 20μmol/L and H_2O_2 was 20 μmol/L. The quasi-second-order reaction kinetics of ATZ elimination by O_3/H_2O_2 were matched under different pH values adopting $1/(1-\beta)$ *(y)* as the *Y*-axis and *t(x)* as the *X*-axis. The dynamic matching splines are demonstrated in Figures 12 and 13, and the parameters for the dynamic fitting equation are shown in Tables 6 and 7.

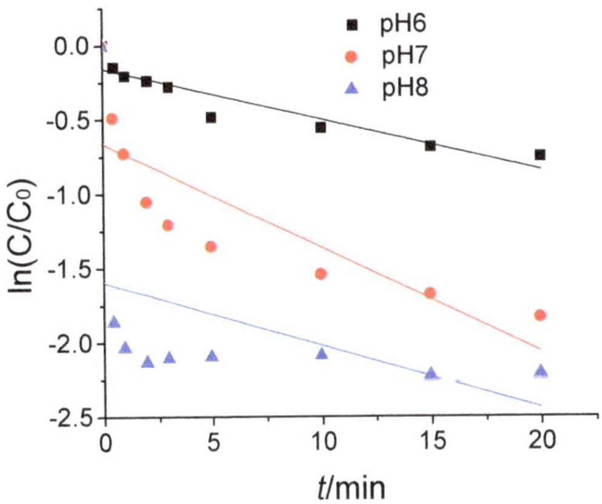

Figure 12. The kinetics of quasi−first−order reactions of O_3/H_2O_2 degradation of ATZ at different pH values.

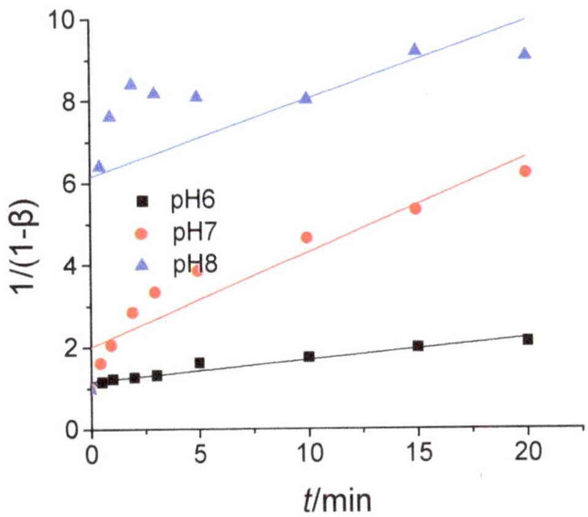

Figure 13. The kinetics of quasi−second−order reactions of O_3/H_2O_2 degradation of ATZ at different pH values.

Table 6. The kinetics equations and parameters of quasi-first-order reactions of O_3/H_2O_2 of degradation ATZ at different pH values.

pH	Fitted Equation	Reaction Order	K_1 (1/min)	R^2
6	Y = −0.03397x − 0.15995	First-order reaction	0.03397	0.86677
7	Y = −0.06995x − 0.65722	First-order reaction	0.06995	0.65698
8	Y = −0.04215x − 1.59326	First-order reaction	0.04215	0.06647

Table 7. The kinetics equations and parameters of quasi-second-order reactions of O_3/H_2O_2 degradation of ATZ at different pH values.

pH	Fitted Equation	Reaction Order	K_1 (1/min)	R^2
6	Y = 0.05299x + 1.16329	Second-order reaction	0.05299	0.92574
7	Y = 0.23151x + 1.99153	Second-order reaction	0.23151	0.87706
8	Y = 0.19067x + 6.14035	Second-order reaction	0.19067	0.19382

The fitting linear correlation of pseudo-first-order reaction kinetics or pseudo-second-order reaction kinetics at pH 8 was not very ideal, and the real reaction rate in the reaction system cannot be accurately determined. Moreover, when the pH value was 8, the degradation reaction of ATZ was very rapid, and the concentration of ATZ in the system tended to be stable after 2 min, which can be seen from Figures 12 and 13. The pseudo-second-order reaction rate constant was 6.62253 and the R^2 value was 0.76401 by fitting the period before 2 min. The R^2 (linear correlation) values of the pseudo-second-order reaction kinetics fit were all larger than the R^2 values of the pseudo-first-order reaction kinetics, and we believe that the reactions were more in line with the pseudo-second-order reaction kinetics model.

4. Conclusions

The addition of a small amounts of H_2O_2 in the O_3/H_2O_2 system can significantly accelerate O_3 chain decomposition, resulting in the generation of a large amount of HO•. The higher the temperature and pH value, the better the degradation effect of O_3/H_2O_2 on ATZ. When the concentration of H_2O_2 went from low to high, the efficiency of O_3/H_2O_2 in degrading the ATZ first increased and then decreased. The optimal reaction ratio of H_2O_2 to O_3 was 1:2. The pH value had a considerable impact on the degradation of ATZ by O_3/H_2O_2, and the degradation efficiency under alkaline conditions was better than that under acidic conditions. Under acidic conditions, the proportion of direct oxidative degradation by O_3 and HO• inside the phase of ATZ decomposition was close to 1:1. Under alkaline and neutral conditions, HO• oxidative degradation dominated the ATZ degradation process. The kinetics of ATZ decomposition by O_3/H_2O_2 were more compatible with quasi-second-order reaction kinetics under varied temperatures, pH values, and H_2O_2 concentrations, according to the kinetic study.

Atrazine is not easy to degrade, and it can exist stably in aqueous environments for a long time, which not only affects the survival of animals and plants, but also threatens human health. Most studies do not clarify the types and hazards of ATZ degradation intermediates. Therefore, paying attention to the degradation efficiency of atrazine while taking into account the toxicity of intermediate degradation products is of great significance for optimizing ATZ degradation technology, which is also one of the future development directions. In addition, current research mainly focuses on the water environment with only ATZ, while ignoring the possible effects of the coexistence of various substances in natural water bodies; the effects of various natural conditions should therefore be fully considered and the cross-combination of various technologies applied to water. The degradation of atrazine can further optimize and improve the degradation efficiency, reduce the operating cost, and thus be more suitable for practical applications.

Author Contributions: Conceptualization, Y.L. (Yixin Lu) and J.C.; methodology, C.T.; software, Y.L. (Yujie Liu); validation, Y.L. (Yixin Lu) and J.C.; data curation, J.C.; writing—preparing the first draft, Y.L. (Yixin Lu); composing—reviewing and revising, C.T. and Y.L. (Yujie Liu); visualization, Y.L. (Yixin Lu); supervision, J.C.; funding acquisition, J.C. All authors have read and agreed to the published version of the manuscript.

Funding: This research was funded by the Sichuan Science and Technology Program, grant numbers: 2022YFG0307, 2022YFG0134, 22YYJC3490; the Open fund of State Environmental Protection Key Laboratory of Synergetic and Joint Remediation for Soil & Water Pollution, grant number: GHBK-2021-004; the School Level Project of Chengdu Technological University, grant numbers: 2021ZR020, QM2021003, QM2021034, QM2021064, QM2021080; and the National Innovation Training Program for College Students, grant numbers: S202011116015, S20211116031.

Institutional Review Board Statement: Not applicable.

Informed Consent Statement: Not applicable.

Data Availability Statement: The article contains all the data.

Acknowledgments: This research is grateful to the Sichuan Provincial Science and Technology Program Project (No.2022YFG0307, No.2022YFG0134, No.22YYJC3490), the National Key Laboratory of

Environmental Protection and Soil Pollution Collaborative Joint Restoration Open Fund (No.GHBK-2021-004), Chengdu University of Technology Self-help for advanced projects (No.2021ZR020, No.QM2021003, No.QM2021034, No.QM202106, No.QM2021080) and National College Student Innovation Training Program (No.S202011116015, No.S20211116031).

Conflicts of Interest: The authors declare no conflict of interest.

References

1. Britt, A.; Bernini, M.; McSweeney, B.; Dalapati, S.; Duchin, S.; Cavanna, K.; Santos, N.; Donovan, G.; O'Byrne, K.; Noyes, S.; et al. The effects of atrazine on the microbiome of the eastern oyster: Crassostrea virginica. *Sci. Rep.* **2020**, *10*, 11088. [CrossRef] [PubMed]
2. Recker, R.A.; Mitchell, P.D.; Stoltenberg, D.E.; Lauer, J.G.; Davis, V.M. Late-Season Weed Escape Survey Reveals Discontinued Atrazine Use Associated with Greater Abundance of Broadleaf Weeds. *Weed Technol.* **2015**, *29*, 451–463. [CrossRef]
3. Ye, X.; Lu, Y.; Zhang, H. Use and harm of the herbicide atrazine. *Environ. Sci. Manag.* **2006**, *8*, 95–97.
4. Khatoon, H.; Rai, J.P.N. Optimization studies on biodegradation of atrazine by Bacillus badius ABP6 strain using response surface methodology. *Biotechnol. Rep.* **2020**, *26*, e00459. [CrossRef]
5. Sun, X.; Liu, F.; Shan, R.; Fan, Y. Spatiotemporal distributions of Cu, Zn, metribuzin, atrazine, and their transformation products in the surface water of a small plain stream in eastern China. *Environ. Monit. Assess.* **2019**, *191*, 433. [CrossRef]
6. Eugenia Taverna, M.; Alberto Busatto, C.; Raquel Lescano, M.; Nicolau, V.V.; Susana Zalazar, C.; Raul Meira, G.; Alejandra Estenoz, D. Microparticles based on ionic and organosolv lignins for the controlled release of atrazine. *J. Hazard. Mater.* **2018**, *359*, 139–147. [CrossRef] [PubMed]
7. Meng, S.; Hu, G.; Zhu, J.; Wu, W.; Chen, J. Research progress on atrazine residues in aquatic environment and their toxicological effects. *Environ. Pollut. Prev.* **2009**, *31*, 64–68+83.
8. Buser, H.R. Atrazine and other s-triazine herbicides in lakes and in Switzerland. *Environ. Sci. Technol.* **1990**, *24*, 1049–1058. [CrossRef]
9. He, H.; Liu, Y.; You, S.; Liu, J.; Xiao, H.; Tu, Z. A Review on Recent Treatment Technology for Herbicide Atrazine in Contaminated Environment. *Int. J. Environ. Res. Public Health* **2019**, *16*, 5129. [CrossRef]
10. Xu, X.; Li, C.; Sun, J.; Wang, H.; Wang, D.; Song, H.; Wang, Z. Residue Characteristics and Ecological Risk Assessment of Twenty-nine Pesticides in Surface Water of Major River-Basin in China. *Asian J. Ecotoxicol.* **2016**, *11*, 347–354.
11. Sun, D.; Gao, G. Effects of atrazine and its degrading bacteria on the germination rate of rice seeds. *Heilongjiang Agric. Sci.* **2017**, *4*, 48–50.
12. Fu, L.; Ni, J.; Ruan, Y.; Dong, R.; Shi, H. Effects of Atrazine on Embryonic Development and Histological Structure of Liver and Kidney in Red-eared Turtle (*Trachemys scripta elegans*). *Fish. Sci.* **2017**, *36*, 104–108.
13. Belloni, V.; Dessi-Fulgheri, F.; Zaccaroni, M.; Di Consiglio, E.; De Angelis, G.; Testai, E.; Santochirico, M.; Alleva, E.; Santucci, D. Early exposure to low doses of atrazine affects behavior in juvenile and adult CD1 mice. *Toxicology* **2011**, *279*, 19–26. [CrossRef]
14. Stradtman, S.C.; Freeman, J.L. Mechanisms of Neurotoxicity Associated with Exposure to the Herbicide Atrazine. *Toxics* **2021**, *9*, 207. [CrossRef] [PubMed]
15. Sagarkar, S.; Gandhi, D.; Devi, S.S.; Sakharkar, A.; Kapley, A. Atrazine exposure causes mitochondrial toxicity in liver and muscle cell lines. *Indian J. Pharmacol.* **2016**, *48*, 200–207. [CrossRef] [PubMed]
16. Marcus, S.R.; Fiumera, A.C. Atrazine exposure affects longevity, development time and body size in Drosophila melanogaster. *J. Insect Physiol.* **2016**, *91–92*, 18–25. [CrossRef] [PubMed]
17. Barchanska, H.; Sajdak, M.; Szczypka, K.; Swientek, A.; Tworek, M.; Kurek, M. Atrazine, triketone herbicides, and their degradation products in sediment, soil and surface water samples in Poland. *Environ. Sci. Pollut. Res.* **2017**, *24*, 644–658. [CrossRef] [PubMed]
18. Sun, J.T.; Pan, L.L.; Zhan, Y.; Tsang, D.C.W.; Zhu, L.Z.; Li, X.D. Atrazine contamination in agricultural soils from the Yangtze River Delta of China and associated health risks. *Environ. Geochem. Health* **2017**, *39*, 369–378. [CrossRef]
19. Ding, Z.; Zhang, J.; He, Y.; Lu, Y. Experimental study on the degradation of atrazine by ozone/peroxymonosulfate in phosphate buffer. *Water Treat. Technol.* **2018**, *44*, 43–47+52. [CrossRef]
20. Wardenier, N.; Liu, Z.; Nikiforov, A.; Van Hulle, S.W.H.; Leys, C. Micropollutant elimination by O_3, UV and plasma-based AOPs: An evaluation of treatment and energy costs. *Chemosphere* **2019**, *234*, 715–724. [CrossRef]
21. Hou, B.; Shen, J.; Li, T.; Xu, P. Ferric silicate catalyzed ozone removal of atrazine and nitrobenzene in water. *J. Nat. Sci. Heilongjiang Univ.* **2015**, *32*, 223–228. [CrossRef]
22. Jing, L.; Chen, B.; Wen, D.Y.; Zheng, J.S.; Zhang, B.Y. Pilot-scale treatment of atrazine production wastewater by UV/O_3/ultrasound: Factor effects and system optimization. *J. Environ. Manag.* **2017**, *203*, 182–190. [CrossRef] [PubMed]
23. Song, W.; Li, J.; Fu, C.X.; Wang, Z.Y.; Zhou, Y.X.; Zhang, X.L.; Yang, J.X.; Wang, K.; Liu, Y.H.; Song, Q. Establishment of sulfate radical advanced oxidation process based on Fe^{2+}/O_{-2}/dithionite for organic contaminants degradation. *Chem. Eng. J.* **2021**, *410*, 128204. [CrossRef]
24. Khandarkhaeva, M.; Batoeva, A.; Aseev, D.; Sizykh, M.; Tsydenova, O. Oxidation of atrazine in aqueous media by solar- enhanced Fenton-like process involving persulfate and ferrous ion. *Ecotoxicol. Environ. Saf.* **2017**, *137*, 35–41. [CrossRef] [PubMed]

25. Zheng, Z.Z.; Zhang, K.F.; Toe, C.Y.; Amal, R.; Zhang, X.W.; McCarthy, D.T.; Deletic, A. Stormwater herbicides removal with a solar-driven advanced oxidation process: A feasibility investigation. *Water Res.* **2021**, *190*, 116783. [CrossRef] [PubMed]
26. Wen, D.Y.; Chen, B.; Liu, B. An ultrasound/O_3 and UV/O_3 process for atrazine manufacturing wastewater treatment: A multiple scale experimental study. *Water Sci. Technol.* **2022**, *85*, 229–243. [CrossRef]
27. Cruz-Alcalde, A.; Esplugas, S.; Sans, C. Continuous versus single H_2O_2 addition in peroxone process: Performance improvement and modelling in wastewater effluents. *J. Hazard. Mater.* **2020**, *387*, 121993. [CrossRef]
28. Kordkandi, S.A.; Motlagh, A.M. Optimization of peroxone reaction rate using metaheuristic approach in the dearomatization and discoloration process. *Environ. Prog. Sustain. Energy* **2018**, *37*, 695–702. [CrossRef]
29. Xu, X.; Zhu, T.; Chen, J. Detection method of ozone concentration in water. *J. Hohai Univ. Chang.* **2007**, *1*, 48–52.
30. Ho, Y.S.; McKay, G. Pseudo-second order model for sorption processes. *Process. Biochem.* **1999**, *34*, 451–465. [CrossRef]
31. Ho, Y.-S. Review of second-order models for adsorption systems. *J. Hazard. Mater.* **2006**, *136*, 681–689. [CrossRef] [PubMed]
32. Zhao, L.; Ma, J.; Sun, Z.Z.; Liu, H.L. Influencing mechanism of temperature on the degradation of nitrobenzene in aqueous solution by ceramic honeycomb catalytic ozonation. *J. Hazard. Mater.* **2009**, *167*, 1119–1125. [CrossRef] [PubMed]
33. Cooper, C.; Burch, R. An investigation of catalytic ozonation for the oxidation of halocarbons in drinking water preparation. *Water Res.* **1999**, *33*, 3695–3700. [CrossRef]
34. Qiang, Z.; Liu, C.; Dong, B.; Zhang, Y. Degradation mechanism of alachlor during direct ozonation and O_3/H_2O_2 advanced oxidation process. *Chemosphere* **2010**, *78*, 517–526. [CrossRef]
35. Cuerda-Correa, E.M.; Alexandre-Franco, M.F.; Fernandez-Gonzalez, C. Advanced Oxidation Processes for the Removal of Antibiotics from Water. An Overview. *Water* **2020**, *12*, 102. [CrossRef]
36. Balci, B.; Oturan, N.; Cherrier, R.; Oturan, M.A. Degradation of atrazine in aqueous medium by electrocatalytically generated hydroxyl radicals. A kinetic and mechanistic study. *Water Res.* **2009**, *43*, 1924–1934. [CrossRef]
37. Kim, H.C.; Park, S.H.; Noh, J.H.; Choi, J.; Lee, S.; Maeng, S.K. Comparison of pre-oxidation between O_3 and O_3/H_2O_2 for subsequent managed aquifer recharge using laboratory-scale columns. *J. Hazard. Mater.* **2019**, *377*, 290–298. [CrossRef]
38. Chen, H.; Wang, J.L. Degradation and mineralization of ofloxacin by ozonation and peroxone (O_3/H_2O_2) process. *Chemosphere* **2021**, *269*. [CrossRef]
39. Gounden, A.N.; Singh, S.; Jonnalagadda, S.B. Simultaneous removal of 2,4,6-tribromophenol from water and bromate ion minimization by ozonation. *J. Hazard. Mater.* **2018**, *357*, 415–423. [CrossRef]
40. Piras, F.; Santoro, O.; Pastore, T.; Pio, I.; De Dominicis, E.; Gritti, E.; Caricato, R.; Lionetto, M.G.; Mele, G.; Santoro, D. Controlling micropollutants in tertiary municipal wastewater by O_3/H_2O_2, granular biofiltration and UV254/H_2O_2 for potable reuse applications. *Chemosphere* **2020**, *239*, 124635. [CrossRef]
41. Biard, P.F.; Dang, T.T.; Bocanegra, J.; Couvert, A. Intensification of the O_3/H_2O_2 advanced oxidation process using a continuous tubular reactor filled with static mixers: Proof of concept. *Chem. Eng. J.* **2018**, *344*, 574–582. [CrossRef]
42. Muhammad, A.; Shafeeq, A.; Butt, M.A.; Rizvi, Z.H.; Chughtai, M.A.; Rehman, S. Decolorization and removal of cod and bod from raw and biotreated textile dye bath effluent through advanced oxidation processes (AOPS). *Braz. J. Chem. Eng.* **2008**, *25*, 453–459. [CrossRef]
43. Gu, Z.P.; Chen, W.M.; Wang, F.; Li, Q.B. Transformation and degradation of recalcitrant organic matter in membrane bioreactor leachate effluent by the O_3/H_2O_2 process. *Environ. Sci. -Water Res. Technol.* **2019**, *5*, 1748–1757. [CrossRef]
44. Qi, S.; Mao, Y.; Lv, M.; Sun, L.; Wang, X.; Yang, H.; Xie, Y.F. Pathway fraction of bromate formation during O_3 and O_3/H_2O_2 processes in drinking water treatment. *Chemosphere* **2016**, *144*, 2436–2442. [CrossRef]
45. Yang, J.X.; Li, J.; Dong, W.Y.; Ma, J.; Cao, J.; Li, T.T.; Li, J.Y.; Gu, J.; Liu, P.X. Study on enhanced degradation of atrazine by ozonation in the presence of hydroxylamine. *J. Hazard. Mater.* **2016**, *316*, 110–121. [CrossRef]
46. Bourgin, M.; Borowska, E.; Helbing, J.; Hollender, J.; Kaiser, H.P.; Kienle, C.; McArdell, C.S.; Simon, E.; von Gunten, U. Effect of operational and water quality parameters on conventional ozonation and the advanced oxidation process O_3/H_2O_2: Kinetics of micropollutant abatement, transformation product and bromate formation in a surface water. *Water Res.* **2017**, *122*, 234–245. [CrossRef]
47. Venta, M.B.; Castro, C.H.; Garcia, L.A.F.; Marzo, A.L.; Lorenzo, E.V.; Alvarez, C.A. Effect of O_3/H_2O_2 molar concentration ratio at different pH values on cyclophosphamide degradation. *J. Water Supply Res. Technol.-Aqua* **2005**, *54*, 403–410. [CrossRef]
48. Feng, H.R.; Liu, M.; Zeng, W.; Chen, Y. Optimization of the O_3/H_2O_2 process with response surface methodology for pretreatment of mother liquor of gas field wastewater. *Front. Environ. Sci. Eng.* **2021**, *15*, 78. [CrossRef]
49. Ochir, D.; Lee, Y.; Shin, J.; Kim, S.; Kwak, J.; Chon, K. Oxidative Treatments of Pesticides in Rainwater Runoff by HOCl, O_3, and O_3/H_2O_2: Effects of pH, Humic Acids and Inorganic Matters. *Separations* **2021**, *8*, 101. [CrossRef]
50. Farzaneh, H.; Loganathan, K.; Saththasivam, J.; McKay, G. Selectivity and competition in the chemical oxidation processes for a binary pharmaceutical system in treated sewage effluent. *Sci. Total Environ.* **2021**, *765*, 142704. [CrossRef]
51. Asaithambi, P.; Sajjadi, B.; Aziz, A.R.A. Integrated ozone-photo-Fenton process for the removal of pollutant from industrial wastewater. *Chin. J. Chem. Eng.* **2017**, *25*, 516–522. [CrossRef]
52. Pillai, K.C.; Kwon, T.O.; Moon, I.S. Degradation of wastewater from terephthalic acid manufacturing process by ozonation catalyzed with Fe^{2+}, H_2O_2 and UV light: Direct versus indirect ozonation reactions. *Appl. Catal. B-Environ.* **2009**, *91*, 319–328. [CrossRef]

53. Li, S.; Liang, Y.; Zhang, R.; Li, P.; Ye, F. Study on Influencing Factors of O_3/H_2O_2 Degradation of Atrazine. *Chin. J. Environ. Eng.* **2008**, *2*, 358–361.
54. Wu, Z.; Chen, H. Research and Application Progress of O_3/H_2O_2 Process in Water Treatment. *Water Treat. Technol.* **2013**, *39*, 1–5. [CrossRef]
55. Chan, K.H.; Chu, W. Model applications and mechanism study on the degradation of atrazine by Fenton's system. *J. Hazard. Mater.* **2005**, *118*, 227–237. [CrossRef]
56. Liu, Y.C.; Ji, X.G.; Yang, J.J.; Tang, W.; Zhu, Y.L.; Wang, Y.; Zhang, Y.X.; Zhang, Y.; Duan, J.M.; Li, W. Degradation of the typical herbicide atrazine by UV/persulfate: Kinetics and mechanisms. *Environ. Sci. Pollut. Res.* **2015**, *22*, 7766–7775. [CrossRef]
57. Diao, Z.H.; Chu, W. FeS_2 assisted degradation of atrazine by bentonite-supported nZVI coupling with hydrogen peroxide process in water: Performance and mechanism. *Sci. Total Environ.* **2021**, *754*, 142155. [CrossRef]
58. Shaban, M.; Abukhadra, M.R.; Ibrahim, S.S.; Shahien, M.G. Photocatalytic degradation and photo-Fenton oxidation of Congo red dye pollutants in water using natural chromite-response surface optimization. *Appl. Water Sci.* **2017**, *7*, 4743–4756. [CrossRef]

MDPI AG
Grosspeteranlage 5
4052 Basel
Switzerland
Tel.: +41 61 683 77 34

Water Editorial Office
E-mail: water@mdpi.com
www.mdpi.com/journal/water

www.ingramcontent.com/pod-product-compliance
Lightning Source LLC
Chambersburg PA
CBHW061614210926
51850CB00009B/2316